Fundamentals of Biomechanics

Third Edition

Nihat Özkaya · Margareta Nordin
David Goldsheyder · Dawn Leger

Fundamentals of Biomechanics

Equilibrium, Motion, and Deformation

Third Edition

 Springer

Nihat Özkaya
Deceased (1956–1998)

David Goldsheyder
New York University Medical Center
New York, NY, USA

Margareta Nordin
New York University Medical Center
New York, NY, USA

Dawn Leger
New York University Medical Center
New York, NY, USA

ISBN 978-1-4899-9378-6   ISBN 978-1-4614-1150-5 (eBook)
DOI 10.1007/978-1-4614-1150-5
Springer New York Dordrecht Heidelberg London

Printed on acid-free paper

Springer is part of Springer Science+Business Media (www.springer.com)

*The third edition of this book is dedicated
to Dr. Nihat Özkaya, to our students,
and to Manya Goldsheyder.*

Foreword

Biomechanics is a discipline utilized by different groups of professionals. It is a required basic science for orthopedic surgeons, osteopaths, physiatrists, rheumatologists, physical and occupational therapists, chiropractors, and athletic trainers. These medical and paramedical specialists usually do not have a strong mathematics and physics background. Biomechanics must be presented to these professionals in a rather nonmathematical way so that they may learn the concepts of mechanics without a rigorous mathematical approach.

On the other hand, many engineers work in fields in which biomechanics plays a significant role. Human factors engineering, ergonomics, biomechanics research, and prosthetic research and development all require that the engineers working in the field have a strong knowledge of biomechanics. They are equipped to learn biomechanics through a rigorous mathematical approach. Classical textbooks in the engineering fields do not approach the biological side of biomechanics.

Fundamentals of Biomechanics (Third Edition) approaches biomechanics through a rigorous mathematical standpoint while emphasizing the biological side. This book will be very useful for engineers studying biomechanics and for medical specialists enrolled in courses who desire a more intensive study of biomechanics and are equipped through previous study of mathematics to develop a deeper comprehension of engineering as it applies to the human body.

Significant progress has been made in the field of biomechanics during the last few decades. Solid knowledge and understanding of biomechanical concepts, principles, assessment methods, and tools are essential components of the study of ergonomics as they help ergonomists in their efforts to reduce the risk and prevent musculoskeletal disorders in the workplace.

This work was prepared in a combined clinical setting at the NYU Hospital for Joint Diseases Orthopedic Institute and teaching setting within the Program of Ergonomics and Biomechanics at New York University. The authors of this volume have the unique experience of teaching biomechanics in a clinical setting to professionals from diverse backgrounds. This work reflects their many years of classroom teaching rehabilitation treatment, and practical and research experience.

New York, NY, USA   Victor Frankel, MD, PhD, KNO

Preface

Even the most "fundamental" of textbooks must undergo a critical revision from time to time, despite the fact that few new discoveries are made in the field. However, using the book as a teaching tool for many years illuminates possible areas for clarification and expansion. With the able assistance of David Goldsheyder, a professional engineer, ergonomist, and instructor at the Graduate Program of Ergonomics and Biomechanics, New York University, who teaches the Biomechanics class each year, we have generated this new edition with help of the graduate students and their comments.

The third edition of this textbook has been prepared with a great deal of input from students, and therefore includes more examples and practice questions intended to deepen their understanding of the concepts covered by the book. Additionally, the order of the chapters has been changed to reflect the natural course of teaching the material. The first section of the book (Chaps. 1–5 and Appendices A and B) introduces the basic concepts of mechanics and provides the mathematical tools necessary to explain these concepts, outlines the procedures for analyzing systems in "equilibrium," and applies this procedure to relatively simple mechanical systems and to the human musculoskeletal system. The second section (Chaps. 6–11 and Appendix C) is devoted to the analysis of "moving" systems and (Chaps. 12–15) provides the techniques for analyzing the "deformation" characteristics of materials with applications to orthopedic biomechanics.

This textbook focuses on the applications of engineering concepts, and attempts to present them, with examples, in the most simple and clearest manner possible, with the understanding that the book will be used by individuals from a wide range of disciplines and preparation. The relationship of engineering concepts to human physiology is the foundation of biomechanics, and one that we sought to illuminate with each example and illustration. We believe that the knowledge of the biomechanical aspects and structural behavior of the human musculoskeletal system is an essential prerequisite for any experimental, theoretical, or analytical approach to analyze its physiological function in the body.

By preparing this text and updating its illustration of these concepts, we hope to contribute to an improved dialogue between those professionals who are primarily interested in the biological and physiological aspects of the human body and those who are interested in the structural behavior of the human body through an engineering approach.

New York, NY, USA

Dawn Leger
Margareta Nordin
David Goldsheyder

Acknowledgements

Special thanks to the many former and current students of the Graduate Program of Ergonomics and Biomechanics at New York University, who provided useful critique to the previous edition of the textbook and highlighted the need for additional topics, practice questions, and example problems. As the majority of students in the program are coming from nonengineering disciplines, with various levels of preparation and some with relatively long breaks in their educational paths, they frequently experience problems in understanding and grasping the basic concepts and principles of the subject matter. Therefore, the primary focus of our work is on the students, their academic performance, and their mastery in applying mechanical principles to the human body. We are grateful to all those who contributed in many different ways to this edition of the textbook and helped make this effort a labor of love.

Special thanks to Angela Maria Lis, PT, Ph.D. for providing a thorough review of the functional anatomy of the major joints of the human musculoskeletal system, and to Jangwhon Yoon, PT, Ph.D. for his help with the dynamics section of the book. Our gratitude is also extended to Dr. Victor H. Frankel for his continuous support and encouragement and to all the staff at the Occupational and Industrial Orthopaedic Center of the Hospital for Joint Diseases Orthopaedic Institute, NYU Langone Medical Center and New York University.

Our gratitude goes to Dr. Nihat Özkaya (who passed away in 1998). His devotion to this work was unparalleled and he hoped to keep the knowledge available and updated to all students and interested readers. It is because of the popularity of the first edition that this book has been translated into Greek and Japanese, and we hope that other translations may follow.

Contents

1.1 Mechanics

Mechanics is a branch of physics that is concerned with the motion and deformation of bodies that are acted on by mechanical disturbances called forces. Mechanics is the oldest of all physical sciences, dating back to the times of Archimedes (287–212 BC). Galileo (1564–1642) and Newton (1642–1727) were the most prominent contributors to this field. Galileo made the first fundamental analyses and experiments in dynamics, and Newton formulated the laws of motion and gravity.

Engineering mechanics or *applied mechanics* is the science of applying the principles of mechanics. Applied mechanics is concerned with both the analysis and design of mechanical systems. The broad field of applied mechanics can be divided into three main parts, as illustrated in Table 1.1.

In general, a material can be categorized as either a solid or fluid. Solid materials can be rigid or deformable. A *rigid body* is one that cannot be deformed. In reality, every object or material does undergo deformation to some extent when acted upon by external forces. In some cases, the amount of deformation is so small that it does not affect the desired analysis. In such cases, it is preferable to consider the body as rigid and carry out the analysis with relatively simple computations.

Statics is the study of forces on rigid bodies at rest or moving with a constant velocity. *Dynamics* deals with bodies in motion. *Kinematics* is a branch of dynamics that deals with the geometry and time-dependent aspects of motion without considering the forces causing the motion. *Kinetics* is based on kinematics, and it includes the effects of forces and masses in the analysis.

Table 1.1 Classification of applied mechanics

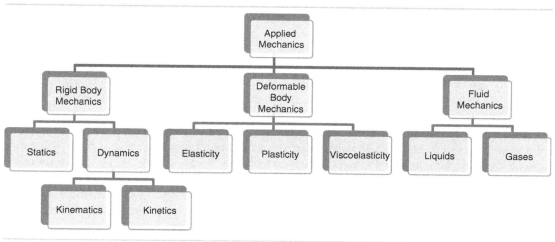

N. Özkaya et al., *Fundamentals of Biomechanics: Equilibrium, Motion, and Deformation,*
DOI 10.1007/978-1-4614-1150-5_1, © Springer Science+Business Media, LLC 2012

Statics and dynamics are devoted primarily to the study of the external effects of forces on rigid bodies, bodies for which the deformation (change in shape) can be neglected. On the other hand, the *mechanics of deformable bodies* deals with the relations between externally applied loads and their internal effects on bodies. This field of applied mechanics does not assume that the bodies of interest are rigid, but considers the true nature of their material properties. The mechanics of deformable bodies has strong ties with the field of material science which deals with the atomic and molecular structure of materials. The principles of deformable body mechanics have important applications in the design of structures and machine elements. In general, analyses in deformable body mechanics are more complex as compared to the analyses required in rigid body mechanics.

The mechanics of deformable bodies is the field that is concerned with the deformability of objects. *Deformable body mechanics* is subdivided into the mechanics of elastic, plastic, and viscoelastic materials, respectively. An *elastic* body is defined as one in which all deformations are recoverable upon removal of external forces. This feature of some materials can easily be visualized by observing a spring or a rubber band. If you gently stretch (deform) a spring and then release it (remove the applied force), it will resume its original (undeformed) size and shape. A *plastic* body, on the other hand, undergoes permanent (unrecoverable) deformations. One can observe this behavior again by using a spring. Apply a large force on a spring so as to stretch the spring extensively, and then release it. The spring will bounce back, but there may be an increase in its length. This increase illustrates the extent of plastic deformation in the spring. Note that depending on the extent and duration of applied forces, a material may exhibit elastic or elastoplastic behavior as in the case of the spring.

To explain viscoelasticity, we must first define what is known as a *fluid*. In general, materials are classified as either solid or fluid. When an external force is applied to a solid body, the body will deform to a certain extent. The continuous application of the same force will not necessarily deform the solid body continuously. On the other hand, a continuously applied force on a fluid body will cause a continuous deformation (flow). *Viscosity* is a fluid property which is a quantitative measure of resistance to flow. In nature, there are some materials that have both fluid and solid properties. The term *viscoelastic* is used to refer to the mechanical properties of such materials. Many biological materials exhibit viscoelastic properties.

The third part of applied mechanics is *fluid mechanics*. This includes the *mechanics of liquids* and the *mechanics of gases*.

Note that the distinctions between the various areas of applied mechanics are not sharp. For example, viscoelasticity simultaneously utilizes the principles of fluid and solid mechanics.

1.2 Biomechanics

In general, biomechanics is concerned with the application of classical mechanics to various biological problems. *Biomechanics* combines the field of engineering mechanics with the fields of biology and physiology. Basically, biomechanics is concerned with the human body. In biomechanics, the principles of mechanics are applied to the conception, design, development, and analysis of equipment and systems in biology and medicine. In essence, biomechanics is a multidisciplinary science concerned with the application of mechanical principles to the human body in motion and at rest.

Although biomechanics is a relatively young and dynamic field, its history can be traced back to the fifteenth century, when Leonardo da Vinci (1452–1519) noted the significance of mechanics in his biological studies. As a result of contributions of researchers in the fields of biology, medicine, basic sciences, and engineering, the interdisciplinary field of biomechanics has been growing steadily in the last five decades.

The development of the field of biomechanics has improved our understanding of many things, including normal and pathological situations, mechanics of neuromuscular control, mechanics of blood flow in the microcirculation, mechanics of air flow in the lung, and mechanics of growth and form. It has contributed to the development of medical diagnostic and treatment procedures. It has provided the means for designing and manufacturing medical equipment, devices, and instruments, assistive technology devices for people with disabilities, and artificial replacements and implants. It has suggested the means for improving human performance in the workplace and in athletic competition.

Different aspects of biomechanics utilize different concepts and methods of applied mechanics. For example, the principles of statics are applied to determine the magnitude and nature of forces involved in various joints and muscles of the musculoskeletal system. The principles of dynamics are utilized for motion description and have many applications in sports mechanics. The principles of the mechanics of deformable bodies provide the necessary tools for developing the field and constitutive equations for biological materials and systems, which in turn are used to evaluate their functional behavior under different conditions. The principles of fluid mechanics are used to investigate the blood flow in the human circulatory system and air flow in the lung.

It is the aim of this textbook to expose the reader to the principles and applications of biomechanics. For this purpose, the basic tools and principles will first be introduced. Next, systematic and comprehensive applications of these principles will be carried out with many solved example problems. Attention will be focused on the applications of statics, dynamics, and mechanics of deformable bodies (i.e., solid mechanics). A limited study of fluid mechanics and its applications in biomechanics will be provided as well.

1.3 Basic Concepts

Engineering mechanics is based on Newtonian mechanics in which the basic concepts are length, time, and mass. These are absolute concepts because they are independent of each other. *Length* is a concept for describing size quantitatively. *Time* is a concept for ordering the flow of events. *Mass* is the property of all matter and is the quantitative measure of inertia. *Inertia* is the resistance to the change in motion of matter. Inertia can also be defined as the ability of a body to maintain its state of rest or uniform motion.

Other important concepts in mechanics are not absolute but derived from the basic concepts. These include force, moment or torque, velocity, acceleration, work, energy, power, impulse, momentum, stress, and strain. *Force* can be defined in many ways, such as mechanical disturbance or load. Force is the action of one body on another. It is the force applied on a body which causes the body to move, deform, or both. *Moment* or *torque* is the quantitative measure of the rotational, bending, or twisting action of a force applied on a body. *Velocity* is defined as the time rate of change of position. The time rate of increase of velocity, on the other hand, is termed *acceleration*. Detailed descriptions of these and other relevant concepts will be provided in subsequent chapters.

1.4 Newton's Laws

The entire field of mechanics rests on a few basic laws. Among these, the laws of mechanics introduced by Sir Isaac Newton form the basis for analyses in statics and dynamics.

Newton's first law states that a body that is originally at rest will remain at rest, or a body in motion will move in a straight line with constant velocity, if the net force acting upon the body is zero. In analyzing this law, we must pay extra attention to a number of key words. The term "rest" implies no motion. For example, a book lying on a desk is said to be at rest. To be able to explain the concept of "net force" fully, we need to introduce vector algebra (see Chap. 2). The net force simply refers to the combined effect of all forces acting on a body. If the net force acting on a body is zero, it does not necessarily mean that there are no forces acting on the body. For example, there may be two equal and opposite forces applied on a body so that the combined effect of the two forces on the body is zero, assuming that the body is rigid. Note that if a body is either at rest or moving in a straight line with a constant velocity, then the body is said to be in *equilibrium*. Therefore, the first law states that if the net force acting on a body is zero, then the body is in equilibrium.

Newton's second law states that a body with a net force acting on it will accelerate in the direction of that force, and that the magnitude of the acceleration will be directly proportional to the magnitude of the net force and inversely proportional to the mass of the body. The important terms in the statement of the second law are "magnitude" and "direction," and they will be explained in detail in Chap. 2, within the context of vector algebra.

Newton's third law states that to every action there is always an equal reaction, and that the forces of action and reaction between interacting bodies are equal in magnitude, opposite in direction, and have the same line of action. This law can be simplified by saying that if you push a body, the body will push you back. This law has important applications in constructing *free-body diagrams* of components constituting large systems. The free-body diagram of a component of a structure is one in which the surrounding parts of the structure are replaced by equivalent forces. It is an effective aid to study the forces involved in the structure.

Newton's laws will be explained in detail in subsequent chapters, and they will be utilized extensively throughout this text.

1.5 Dimensional Analysis

The term "dimension" has several uses in mechanics. It is used to describe space, as, for example, while referring to one-dimensional, two-dimensional, or three-dimensional situations. Dimension is also used to denote the nature of quantities. Every measurable quantity has a *dimension* and a *unit* associated with it. Dimension is a general description of a quantity, whereas unit is associated with a system of units (see Sect. 1.6). Whether a distance is measured in meters or feet, it is a distance. We say that its dimension is "length." Whether a flow of events is measured in seconds, minutes, hours, or even days, it is a point of time when a specific event began and then ended. So we say its dimension is "time."

There are two sets of dimensions. *Primary*, or *basic*, *dimensions* are those associated with the basic concepts of mechanics. In this text, we shall use capital letters L, T, and M to specify the primary dimensions length, time, and mass, respectively. We shall use square brackets to denote the dimensions of physical quantities. The basic dimensions are:

$$[\text{LENGTH}] = \text{L},$$
$$[\text{TIME}] = \text{T},$$
$$[\text{MASS}] = \text{M}.$$

Secondary dimensions are associated with dependent concepts that are derived from basic concepts. For example, the area of a rectangle can be calculated by multiplying its width and length, both of which have the dimension of length. Therefore, the dimension of area is:

$$[\text{AREA}] = [\text{LENGTH}][\text{LENGTH}] = \text{LL} = \text{L}^2.$$

By definition, velocity is the time rate of change of relative position. Change of the relative position is measured in terms of length units. Therefore, the dimension of velocity is:

$$[\text{VELOCITY}] = \frac{[\text{POSITION}]}{[\text{TIME}]} = \frac{\text{L}}{\text{T}}.$$

The secondary dimensional quantities are established as a consequence of certain natural laws. If we know the definition of a physical quantity, we can easily determine the dimension of that quantity in terms of basic dimensions. If the dimension of a physical quantity is known, then the units of that quantity in different systems of units can easily be determined as well. Furthermore, the validity of an equation relating a number of physical quantities can be verified by analyzing the dimensions of terms forming the equation or formula. In this regard, the *law of dimensional homogeneity* imposes restrictions on the formulation of such relations. To explain this law, consider the following arbitrary equation:

$$Z = aX + bY + c.$$

For this equation to be dimensionally homogeneous, every grouping in the equation must have the same dimensional representation. In other words, if Z refers to a quantity whose dimension is length, then products aX and bY, and quantity c must all have the dimension of length. The numerical equality between both sides of the equation must also be maintained for all systems of units.

1.6   Systems of Units

There have been a number of different systems of units adopted in different parts of the world. For example, there is the British gravitational or foot-pound-second system, the Gaussian (metric absolute) or centimeter–gram–second (CGS) system, and the metric gravitational or meter–kilogram–second (MKS) system. The lack of a universal standard in units of measure often causes confusion.

In 1960, an International Conference on Weights and Measures was held to bring an order to the confusion surrounding the units of measure. Based on the metric system, this conference adopted a system called *Le Système International d'Unités* in French, which is abbreviated as SI. In English, it is known as the International System of Units. Today, nearly the entire world is either using this modernized metric system or committed to its adoption. In the International System of Units, the units of length, time, and mass are meter (m), second (s), and kilogram (kg), respectively.

The units of measure of these fundamental concepts in three different systems of units are listed in Table 1.2. Throughout this text, we shall use the International System of Units. Other units will be defined for informational purposes.

Table 1.2 Units of fundamental quantities of mechanics

System	Length	Mass	Time
SI	Meter (m)	Kilogram (kg)	Second (s)
CGS	Centimeter (cm)	Gram (gm)	Second (s)
British	Foot (ft)	Slug (slug)	Second (s)

Once the units of measure for the primary concepts are agreed upon, the units of measure for the derived concepts can easily be determined provided that the dimensional relationship between the basic and derived quantities is known. All that is required is replacing the dimensional representation of length, mass, and time with their appropriate units. For example, the dimension of force is ML/T^2. Therefore, according to the International System of Units, force has the unit of kg-m/s^2, which is also known as Newton (N). Similarly, the unit of force is lb-ft/s^2 in the British system of units, and is g-cm/s^2 or dyne (dyn) in the metric absolute or CGS system. Table 1.3 lists the dimensional representations of some of the derived quantities and their units in the International System of Units.

Table 1.3 Dimensions and units of selected quantities in SI

Quantity	Dimension	SI unit	Special name
Area	L^2	m^2	
Volume	L^3	m^3	
Velocity	L/T	m/s	
Acceleration	L/T^2	m/s^2	
Force	$M\,L/T^2$	kg-m/s^2	Newton (N)
Pressure and stress	$M/L\,T^2$	N/m^2	Pascal (Pa)
Moment (torque)	$M\,L^2/T^2$	N-m	
Work and energy	$M\,L^2/T^2$	N-m	Joule (J)
Power	$M\,L^2/T^3$	J/s	Watt (W)

Note that "kilogram" is the unit of mass in SI. For example, consider a 60 kg object. The weight of the same object in SI is $(60 \text{ kg}) \times (9.8 \text{ m/s}^2) = 588$ N, the factor 9.8 m/s^2 being the magnitude of the gravitational acceleration.

In addition to the primary and secondary units that are associated with the basic and derived concepts in mechanics, there are *supplementary units* such as plane angle and temperature. The common measure of an angle is degree (°). Three hundred and sixty degrees is equal to one revolution (rev) or 2π radians (rad), where $\pi = 3.1416$. The SI unit of temperature is Kelvin (°K). However, degree Celsius (°C) is more commonly used. The British unit of temperature is degree Fahrenheit (°F).

It should be noted that in most cases, a number has a meaning only if the correct unit is displayed with it. In performing calculations, the ideal method is to show the correct units with each number throughout the solution of equations. This approach helps in detecting conceptual errors and eliminates the need for determining the unit of the calculated quantity separately. Another important aspect of using units is consistency. One must not use the units of one system for some quantities and the units of another system for other quantities while carrying out calculations.

1.7 Conversion of Units

The International System of Units is a revised version of the metric system which is based on the decimal system. Table 1.4 lists the SI multiplication factors and corresponding prefixes.

Table 1.4 SI multiplication factors and prefixes

Multiplication factor	SI prefix	SI symbol
$1,000,000,000 = 10^9$	Giga	G
$1,000,000 = 10^6$	Mega	M
$1,000 = 10^3$	Kilo	k
$100 = 10^2$	Hector	h
$10 = 10^1$	Deka	da
$0.1 = 10^{-1}$	Deci	d
$0.01 = 10^{-2}$	Centi	c
$0.001 = 10^{-3}$	Milli	m
$0.000001 = 10^{-6}$	Micro	μ
$0.000000001 = 10^{-9}$	Nano	n
$0.000000000001 = 10^{-12}$	Pico	p

Table 1.5 lists factors needed to convert quantities expressed in British and metric systems to corresponding units in SI.

Table 1.5 Conversion of units

Length	1 centimeter (cm) = 0.01 meter (m) = 0.3937 inch (in.) 1 in. = 2.54 cm = 0.0254 m 1 foot (ft) = 30.48 cm = 0.3048 m 1 m = 3.28 ft = 39.37 in. 1 yard (yd) = 0.9144 m = 3 ft 1 mile = 1,609 m = 1.609 kilometer (km) = 5,280 ft 1 km = 0.6214 mile
Time	1 minute (min) = 60 seconds (s) 1 hour (h) = 60 min = 3,600 s 1 day = 24 h = 1,440 min = 86,400 s
Area	1 cm^2 = 0.155 in.^{2} 1 in.^{2} = 6.452 cm^2 1 m^2 = 10.763 ft^2 1 ft^2 = 0.0929 m^2
Mass	1 pound mass (lbm) = 0.4536 kilogram (kg) 1 kg = 2.2 lbm = 0.0685 slug 1 slug = 14.59 kg = 32.098 lbm
Force	1 kilogram force (kgf) = 9.807 Newton (N) 1 pound force (lbf) = 4.448 N 1 N = 0.2248 lbf 1 dyne (dyn) = 10^{-5} N 1 N = 10^5 dyn
Pressure and stress	1 N/m^2 = 1 Pascal (Pa) = 0.000145 lbf/in.^{2} (psi) 1 psi = 6,895 Pa 1 lbf/ft^2 (psf) = 592,966 Pa 1 dyn/cm^2 = 0.1 Pa
Moment (torque)	1 N-m = 10^7 dyn cm = 0.7376 lbf ft 1 dyn/cm = 10^{-7} N-m 1 lbf ft = 1.356 N-m
Work and energy	1 N-m = 1 Joule (J) = 10^7 erg 1 J = 0.7376 lbf ft 1 lbf ft = 1.356 J
Power	1 kg m^2/s^3 = 1 J/s = 1 Watt (W) 1 horsepower (hp) = 550 lbf ft/s = 746 W 1 lbf ft/s = 1.356 W 1 W = 0.737 lbf ft/s
Plane angle	1 degree (°) = π/180 radian (rad) 1 revolution (rev) = 360° 1 rev = 2π rad = 6.283 rad 1 rad = 57.297° 1° = 0.0175 rad
Temperature	°C = °K − 273.2 °C = (5/9)(°F − 32) °F = (9/5) (°C + 32)

1.8 Mathematics

The applications of biomechanics require some knowledge of mathematics. These include simple geometry, properties of the right-triangle, basic algebra, differentiation, and integration. The appendices that follow the last chapter contain a summary of the mathematical tools and techniques needed to carry out the calculations in this book. The reader may find it useful to examine them now and review them later when those concepts are needed. In subsequent chapters throughout the text, the mathematics required will be reviewed and the corresponding appendix will be indicated.

During the formulation of problems, we shall use Greek letters as well as letters of the Latin alphabet. Greek letters will be used, for example, to refer to angles. The Greek alphabet is provided in Table 1.6 for quick reference.

Table 1.6 Greek alphabet

Alpha	A	α	Iota	I	ι	Rho	P	ρ
Beta	B	β	Kappa	K	κ	Sigma	Σ	σ
Gamma	Γ	γ	Lambda	Λ	λ	Tau	T	τ
Delta	Δ	δ	Mu	M	μ	Upsilon	Y	υ
Epsilon	E	ε	Nu	N	ν	Phi	Φ	φ
Zeta	Z	ζ	Xi	Ξ	ξ	Chi	X	χ
Eta	H	η	Omicron	O	o	Psi	Ψ	ψ
Theta	Θ	θ	Pi	Π	π	Omega	Ω	ω

1.9   Scalars and Vectors

In mechanics, two kinds of quantities are distinguished. A *scalar* quantity, such as mass, temperature, work, and energy, has magnitude only. A *vector* quantity, such as force, velocity, and acceleration, has both a magnitude and a direction. Unlike scalars, vector quantities add according to the rules of vector algebra. Vector algebra will be covered in detail in Chap. 2.

1.10   Modeling and Approximations

One needs to make certain assumptions to simplify complex systems and problems so as to achieve analytical solutions. The complete model is the one that includes the effects of all parts constituting a system. However, the more detailed the model, the more difficult the formulation and solution of the problem. It is not always possible and in some cases it may not be necessary to include every detail in the analysis. For example, during most human activities, there is more than one muscle group activated at a time. If the task is to analyze the forces involved in the joints and muscles during a particular human activity, the best approach is to predict which muscle group is the most active and set up a model that neglects all other muscle groups. As we shall see in the following chapters, bone is a deformable body. If the forces involved are relatively small, then the bone can be treated as a rigid body. This approach may help to reduce the complexity of the problem under consideration.

In general, it is always best to begin with a simple basic model that represents the system. Gradually, the model can be expanded on the basis of experience gained and the results obtained from simpler models. The guiding principle is to make simplifications that are consistent with the required accuracy of the results. In this way, the researcher can set up a model that is simple enough to analyze and exhibit satisfactorily the phenomena under consideration. The more we learn, the more detailed our analysis can become.

1.11   Generalized Procedure

The general method of solving problems in biomechanics may be outlined as follows:

1. Select the system of interest.
2. Postulate the characteristics of the system.
3. Simplify the system by making proper approximations. Explicitly state important assumptions.
4. Form an analogy between the human body parts and basic mechanical elements.
5. Construct a mechanical model of the system.
6. Apply principles of mechanics to formulate the problem.
7. Solve the problem for the unknowns.
8. Compare the results with the behavior of the actual system. This may involve tests and experiments.
9. If satisfactory agreement is not achieved, steps 3 through 7 must be repeated by considering different assumptions and a new model of the system.

1.12   Scope of the Text

Courses in biomechanics are taught within a wide variety of academic programs to students with quite different backgrounds and different levels of preparation coming from various disciplines of engineering as well as other academic disciplines. This text is prepared to provide a teaching and learning tool primarily to health care professionals who are seeking a graduate degree in biomechanics but have limited backgrounds in calculus, physics, and engineering mechanics. This text can also be a useful reference for undergraduate biomedical, biomechanical, or bioengineering programs.

This text is divided into three parts. The first part (Chaps. 1 through 5, and Appendices A and B) will introduce the basic concepts of mechanics including force and moment vectors, provide the mathematical tools (geometry, algebra, and vector algebra) so that complete definitions of these concepts can be given, explain the procedure for analyzing the systems at "static equilibrium," and apply this procedure to analyze simple mechanical systems and the forces involved at various muscles and joints of the human musculoskeletal system. It should be noted here that the topics covered in the first part of this text are prerequisites for both second and third parts.

The second part of the text (Chaps. 6 through 11) is devoted to "dynamic" analyses. The concepts introduced in the second part are position, velocity and acceleration vectors, work, energy, power, impulse, and momentum. Also provided in the second part are the techniques for kinetic and kinematic analyses of systems undergoing translational and rotational motions. These techniques are applied for human motion analyses of various sports activities.

The third part of the text (Chaps. 12 through 15) provides the techniques for analyzing the "deformation" characteristics of materials under different load conditions. For this purpose, the concepts of stress and strain are defined. Classifications of materials based on their stress–strain diagrams are given. The concepts of elasticity, plasticity, and viscoelasticity are also introduced and explained. Topics such as torsion, bending, fatigue, endurance, and factors affecting the strength of materials are provided. The emphasis is placed upon applications to orthopedic biomechanics.

1.13 Notation

While preparing this text, special attention was given to the consistent use of notation. Important terms are italicized where they are defined or described (such as, *force* is defined as load or mechanical disturbance). Symbols for quantities are also italicized (e.g., *m* for mass). Units are not italicized (e.g., kg for kilogram). Underlined letters are used to refer to vector quantities (e.g., force vector $\underline{F}$). Sections and subsections marked with a star (*) are considered optional. In other words, the reader can omit a section or subsection marked with a star without losing the continuity of the topics covered in the text.

Suggested Reading

We believe that this text is a self-sufficient teaching and learning tool. While preparing it, we utilized the information provided in a number of sources, some of which are listed below. Note, however, that it is not our intention to promote these publications or to suggest that these are the only texts available on the subject matter. The field of biomechanics has been growing very rapidly. There are many other sources of readily available information, including scientific journals presenting peer-reviewed research articles in biomechanics.

Ackland, T., Elliot, B., & Bloomfield, J. 2009. Applied Anatomy and Biomechanics on Sport (2nd ed.). Champaign, IL: Human Kinetics.

Black, J. 1988. Orthopaedic Biomaterials in Research and Practice. New York: Churchill Livingstone.

Brinckmann, P., Frobin, W., & Leivseth, G. 2002. Musculoskeletal Biomechanics. Stuttgart-New York: Thieme.

Burnstein, A.H., & Wright, T.M. 1994. Fundamentals of Orthopaedic Biomechanics. Philadelphia, PA: Williams & Wilkins.

Chaffin, D.B., Andersson, G.B.J., & Martin, B.J. 1999. Occupational Biomechanics (3rd ed.). New York: John Wiley & Sons.

Griffiths, I.W. 2005. Principles of Biomechanics & Motion Analysis. Philadelphia, PA: Lippincott Williams & Wilkins.

Hamill, J. 2008. Biomechanical Basis of Human Movement. Philadelphia, PA: Lippincott Williams & Wilkins.

Hamill, J., & Knutzen, K.M. 2003. Biomechanical Basis of Human Movement (2nd ed.). Philadelphia, PA: Lippincott Williams & Wilkins.

Hay, J.G. 1985. The Biomechanics of Sports Techniques (3rd ed.). Englewood Cliffs, NJ: Prentice Hall.

Hay, J.G., & Reid, J.G. 1988. Anatomy, Mechanics and Human Motion (2nd ed.). Englewood Cliffs, NJ: Prentice Hall.

Kelly, D.L. 1971. Kinesiology: Fundamentals of Motion Description. Englewood Cliffs, NJ: Prentice Hall.

Knudson, D. 2007. Fundamentals of Biomechanics (2nd ed.). New York: Springer.

Knudson, D., & Morrison, C. 2002. Qualitative Analysis of Human Movement. Champaign, IL: Human Kinetics.

Marras, W.S. 2008. The Working Back: A System View. Hoboken, NJ: John Wiley & Sons, Inc.

McGinnis, P. 2005. Biomechanics of Sport and Exercise (2nd ed.). Champaign, IL: Human Kinetics.

Mow, V.C., & Huiskes, R. 2004. Basic Orthopaedic Biomechanics and Mechano-Biology (3rd ed.). Philadelphia, PA: Lippincott Williams & Wilkins.

Nahum, A.M., & Melvin, J. (Eds.). 1985. The Biomechanics of Trauma. Norwalk, CT: Appleton-Century-Crofts.

Nordin, M., & Frankel, V.H. (Eds.). 2012. Basic Biomechanics of the Musculoskeletal System (4th ed.). Philadelphia, PA: Lippincott Williams & Wilkins.

Thompson, D.M., Hamill, J., & Knutzen, K.M. 2006. Biomechanical Basis of Human Movement (2nd ed.). Baltimore, MD: Lippincott Williams & Wilkins.

Williams, M., & Lissner, H.R. 1977. Biomechanics of Human Motion (2nd ed.). (B. LeVeau, Ed.). Philadelphia, PA: Saunders.

Winter, D.A. 2004. Biomechanics and Motor Control of Human Movement (3rd ed.). New York: John Wiley & Sons.

Zatsiorsky, V.M. 1998. Kinematics of Human Motion. Champaign, IL: Human Kinetics.

Zatsiorsky, V.M. 2002. Kinetics of Human Motion. Champaign, IL: Human Kinetics.

The following sources contain advanced topics and up-to-date information in the fields of biomechanics and bioengineering.

Bronzino, J.D. (Ed.) 1995. The Biomedical Engineering Handbook. Boca Raton, FL: CRC Press.

Fung, Y.C. 1981. Biomechanics: Mechanical Properties of Living Tissues. New York: Springer-Verlag.

Fung, Y.C. 1997. Biomechanics: Circulation. (2nd ed.). New York: Springer.

Fung, Y.C. 1998. Biomechanics: Motion, Flow, Stress, and Growth (2nd printing). New York: Springer.

Mow, V.C., Ratcliff, A., & Woo, S.L.-Y. (Eds.). 1990. Biomechanics of Diarthrodial Joints. New York: Springer-Verlag.

Schmid-Schonbein, G.W., Woo, S.L.-Y., & Zweifach, B.W. (Eds.). 1985. Frontiers in Biomechanics. New York: Springer-Verlag.

Skalak, R., & Chien, S. (Eds.). 1987. Handbook of Bioengineering. New York: McGraw-Hill.

Whiting, W.C., & Zernicke, R.F. 1998. Biomechanics of Musculoskel-
etal Injury. Champaign, IL: Human Kinetics.
Winters, J.M., & Woo, S.L.-Y. (Eds.). 1990. Multiple Muscle Systems.
New York: Springer-Verlag.

In addition to the biomechanics related books listed above, there are many physics and engineering mechanics books dedicated to undergraduate engineering education. These books may be useful for reviewing the basic concepts of mechanics. Listed below are examples of such titles.

Bedford, A.M., & Fowler, W. 2008. Engineering Mechanics: Statics &
Dynamics. Englewood Cliffs, NJ: Prentice Hall.
Hibbeler, R.C. 2009. Engineering Mechanics: Combined Statics &
Dynamics (12th ed.). Englewood Cliffs, NJ: Prentice Hall.
Sandor, B.I. 1987. Engineering Mechanics: Statics (2nd ed.).
Englewood Cliffs, NJ: Prentice Hall.
Sandor, B.I. 1987. Engineering Mechanics: Dynamics (2nd ed.).
Englewood Cliffs, NJ: Prentice Hall.
Servay, R.A. 1997. Physics for Scientists and Engineers. Philadelphia,
PA: Saunders College Publishing.
Shames, I.H. 1996. Engineering Mechanics: Statics (4th ed.).
Englewood Cliffs, NJ: Prentice Hall.
Shames, I.H. 1996. Engineering Mechanics: Dynamics (4th ed.).
Englewood Cliffs, NJ: Prentice Hall.

The third part of this text will introduce the principles of mechanics of deformable bodies and use these principles to explain certain aspects of biomechanics. The following books can be reviewed to gain more detailed information on these principles.

Crandall, S.H., Dahl, N.C., & Lardner, T.J. 1978. An Introduction to the
Mechanics of Solids (2nd ed.). New York: McGraw-Hill.
Popov, E.P. 1978. Mechanics of Materials (2nd ed.). Englewood Cliffs,
NJ: Prentice Hall.
Popov, E.P. 1999. Engineering Mechanics of Solids (2nd ed.).
Englewood Cliffs, NJ: Prentice Hall.
Pytel, A., & Singer, F.L. 1987. Strength of Materials (4th ed.).
New York: Harper & Row.

Biomechanics Societies

American Society of Biomechanics: http://www.asbweb.org/.
Australian and New Zealand Society of Biomechanics: http://www.anzsb.asn.au/.
Canadian Society of Biomechanics: http://www.health.uottawa.ca/biomech/csb/.
German Society of Biomechanics: http://www.biomechanics.de/dgbiomech/dgbiomech.html.
International Society of Biomechanics: http://isbweb.org/.
The European Society of Biomechanics: http://www.esbiomech.org/.

Biomechanics Journals

Applied Bionics and Biomechanics: http://www.editorialmanager.com/abbi/mainpage.html.
Clinical Biomechanics: http://www.elsevier.com/wps/find/journaldescription.cws_home/30397/description#description.
Journal of Applied Biomechanics: http://hk.humankinetics.com/JAB/journalAbout.cfm.
Journal of Biomechanics: http://www.elsevier.com/wps/find/journaldescription.cws_home/321/description#description.

2.1   Definition of Force

Force may be defined as mechanical disturbance or load. When you pull or push an object, you apply a force to it. You also exert a force when you throw or kick a ball. In all of these cases, the force is associated with the result of muscular activity. Forces acting on an object can deform it, change its state of motion, or both. Although forces cause motion, it does not necessarily follow that force is always associated with motion. For example, a person sitting on a chair applies his/her weight on the chair, and yet the chair remains stationary. There are relatively few basic laws that govern the relationship between force and motion. These laws will be discussed in detail in later chapters.

2.2   Properties of Force as a Vector Quantity

Forces are vector quantities and the principles of vector algebra (see Appendix B) must be applied to analyze problems involving forces. To describe a force fully, its magnitude and direction must be specified. As illustrated in Figure 2.1, a force vector can be illustrated graphically with an arrow such that the orientation of the arrow indicates the line of action of the force vector, the arrowhead identifies the direction and sense along which the force is acting, and the base of the arrow represents the point of application of the force vector. If there is a need for showing more than one force vector in a single drawing, then the length of each arrow must be proportional to the magnitude of the force vector it is representing.

Like other vector quantities, forces may be added by utilizing graphical and trigonometric methods. For example, consider the partial knee illustrated in Fig. 2.2. Forces applied by the quadriceps $\underline{F}_Q$ and patellar tendon $\underline{F}_P$ on the patella are shown. The resultant force $\underline{F}_R$ on the patella due to the forces applied by the quadriceps and patellar tendon can be determined by considering the vector sum of these forces:

$$\underline{F}_R = \underline{F}_Q + \underline{F}_P. \tag{2.1}$$

If the magnitude of the resultant force needs to be calculated, then the Pythagorean theorem can be utilized:

$$F_R = \sqrt{F_Q{}^2 + F_P{}^2}. \tag{2.2}$$

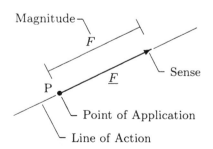

Fig. 2.1 Graphical representation of the force vector

Fig. 2.2 Resultant force

N. Özkaya et al., *Fundamentals of Biomechanics: Equilibrium, Motion, and Deformation*, DOI 10.1007/978-1-4614-1150-5_2, © Springer Science+Business Media, LLC 2012

2.3　Dimension and Units of Force

By definition, force is equal to mass times acceleration, acceleration is the time rate of change of velocity, and velocity is the time rate of change of relative position. The change in position is measured in terms of length units. Therefore, velocity has a dimension of length divided by time, acceleration has a dimension of velocity divided by time, and force has a dimension of mass times acceleration:

$$[\text{VELOCITY}] = \frac{[\text{POSITION}]}{[\text{TIME}]} = \frac{L}{T},$$

$$[\text{ACCELERATION}] = \frac{[\text{VELOCITY}]}{[\text{TIME}]} = \frac{L/T}{T} = \frac{L}{T^2},$$

$$[\text{FORCE}] = [\text{MASS}][\text{ACCELERATION}] = \frac{ML}{T^2}.$$

Units of force in different unit systems are provided in Table 2.1.

Table 2.1　Units of force (1 N = 10^5 dyn, 1 N = 0.225 lb)

System	Units of force	Special name
SI	kg-m/s^2	Newton (N)
CGS	g-cm/s^2	dyne (dyn)
British	Slug foot/s^2	Pound (lb)

2.4　Force Systems

Any two or more forces acting on a single body form a *force system*. Forces constituting a force system may be classified in various ways. Forces may be classified according to their effect on the bodies upon which they are applied or according to their orientation as compared to one another.

2.5　External and Internal Forces

A force may be broadly classified as *external* or *internal*. Almost all commonly known forces are external forces. For example, when you push a cart, hammer a nail, sit on a chair, kick a football, or shoot a basketball, you apply an external force on the cart, nail, chair, football, or basketball. Internal forces, on the other hand, are the ones that hold a body together when the body is under the effect of externally applied forces. For example, a piece of string does not necessarily break when it is pulled from both ends. When a rubber band is stretched, the band elongates to a certain extent. What holds any material together under externally applied forces is the internal forces generated within that material. If we consider the human body as a whole, then the forces generated by muscle contractions are also internal forces. The significance and details of internal forces will be studied by introducing the concept of "stress" in later chapters.

2.6　Normal and Tangential Forces

In mechanics, the word "normal" implies perpendicular. If a force acting on a surface is applied in a direction perpendicular to that surface, then the force is called a *normal force*. For example, a book resting on a flat horizontal desk applies a normal force on the desk, the magnitude of which is equal to the weight of the book (Fig. 2.3).

Fig. 2.3　Forces normal to the surfaces in contact

A *tangential force* is that applied on a surface in the direction parallel to the surface. A good example of tangential forces is the frictional force. As illustrated in Fig. 2.4, pushing or pulling a block will cause a frictional force to occur between the bottom surface of the block and the floor. The line of action of the frictional force is always tangential to the surfaces in contact.

Fig. 2.4　Frictional forces are tangential forces

2.7 Tensile and Compressive Forces

A *tensile force* applied on a body will tend to stretch or elongate the body, whereas a *compressive force* will tend to shrink the body in the direction of the applied force (Fig. 2.5). For example, a tensile force applied on a rubber band will stretch the band. Poking into an inflated balloon will produce a compressive force on the balloon. It must be noted that there are certain materials upon which only tensile forces can be applied. For example, a rope, a cable, or a string cannot withstand compressive forces. The shapes of these materials will be completely distorted under compressive forces. Similarly, muscles contract to produce tensile forces that pull together the bones to which they are attached to. Muscles can neither produce compressive forces nor exert a push.

Fig. 2.5 (**a**) Tensile and (**b**) compressive forces

2.8 Coplanar Forces

A system of forces is said to be *coplanar* if all the forces are acting on a two-dimensional (plane) surface. Forces forming a coplanar system have at most two nonzero components. Therefore, with respect to the Cartesian (rectangular) coordinate frame, it is sufficient to analyze coplanar force systems by considering the x and y components of the forces involved.

2.9 Collinear Forces

A system of forces is *collinear* if all the forces have a common line of action. For example, the forces applied on a rope in a rope-pulling contest form a collinear force system (Fig. 2.6).

Fig. 2.6 Collinear forces

2.10 Concurrent Forces

A system of forces is *concurrent* if the lines of action of the forces have a common point of intersection. Examples of concurrent force systems can be seen in various traction devices, as illustrated in Fig. 2.7. Due to the weight in the weight pan, the cables stretch and forces are applied on the pulleys and the leg. The force applied on the leg holds the leg in place.

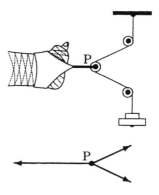

Fig. 2.7 Concurrent forces

2.11 Parallel Force

A set of forces form a *parallel* force system if the lines of action of the forces are parallel to each other. An example of parallel force systems is illustrated in Fig. 2.8 by a human arm flexed at a right angle and holding an object. The forces acting on the forearm are the weight of the object W_1, the weight of the arm itself W_2, the tension in the biceps muscle F_M, and the joint reaction force at the elbow F_j. These forces are parallel to each other, thus forming a system of parallel forces.

Fig. 2.8 Parallel forces

2.12 Gravitational Force or Weight

The force exerted by Earth on an object is called the *gravitational force* or *weight* of the object. The magnitude of weight of an object is equal to the mass of the object times the magnitude of gravitational acceleration, $w = m \times g$ where w is the weight of the object, m is the mass of the object, and g is the gravitational acceleration. The magnitude of gravitational acceleration for different unit systems is listed in Table 2.2. These values are valid only on the surface of Earth. The magnitude of the gravitational acceleration can vary slightly with altitude.

Table 2.2 Gravitational acceleration on Earth

System	Gravitational acceleration
SI	9.81 m/s^2
CGS	981 cm/s^2
British	32.2 ft/s^2

For our applications, we shall assume g to be a constant.

The terms mass and weight are often confused with one another. Mass is a property of a body. Weight is the force of gravity acting on the mass of the body. A body has the same mass on Earth and on the Moon. However, the weight of a body is about six times as much on Earth as on the Moon, because the magnitude of the gravitational acceleration on the Moon is about one-sixth of what it is on Earth. Therefore, a 10 kg mass on Earth weighs about 98 N on the Earth, while it weighs about 17 N on the Moon.

Like force, acceleration is a vector quantity. The direction of gravitational acceleration and gravitational force vectors is always toward the center of Earth or always vertically downward. The force of gravity acts on an object at all times. If we drop an object from a height, it is the force of gravity that will pull the object downward.

When an object is at rest on the ground, the gravitational force does not disappear. An object at rest or in static equilibrium simply means that the net force acting on the object is zero (Fig. 2.9).

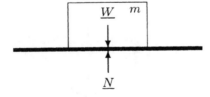

Fig. 2.9 The net force on an object at rest in zero

2.13 Distributed Force Systems and Pressure

Consider a pile of sand lying on a flat horizontal surface, as illustrated in Fig. 2.10a. The sand exerts force or load on the surface, which is distributed over the area under the sand. The load is not uniformly distributed over this area. The marginal regions under the pile are loaded less as compared

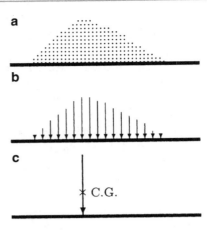

Fig. 2.10 A pile of sand (**a**), distributed load on the ground (**b**), and an equivalent force (**c**)

to the central regions (Fig. 2.10b). For practical purposes, the distributed load applied by the sand may be represented by a single force, called the *equivalent force* or *concentrated load*. The magnitude of the equivalent force would be equal to the total weight of the sand (Fig. 2.10c). The line of action of this force would pass through a point, called the *center of gravity*. For some applications, we can assume that the entire weight of the pile is concentrated at the center of gravity of the load. For uniformly distributed loads, such as the load applied by the rectangular block on the horizontal surface shown in Fig. 2.11, the center of gravity coincides with the geometric center of the load. For nonuniformly distributed loads, the center of gravity can be determined by experimentation (see Chap. 4).

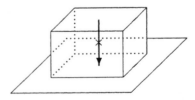

Fig. 2.11 Rectangular block

Center of gravity is associated with the gravitational force of Earth. There is another concept called *center of mass*, which is independent of gravitational effects. For a large object or a structure, such as the Empire State building in New York City, the center of gravity may be different than the center of mass because the magnitude of gravitational acceleration varies with altitude. For relatively small objects and for our applications, the difference between the two can be ignored.

Another important concept associated with distributed force systems is *pressure*, which is a measure of the intensity of distributed loads. By definition, *average pressure* is equal to total applied force divided by the area of the surface over which the force is applied in a direction perpendicular to the surface. It is also known as *load intensity*. For example, if the bottom surface area of the rectangular block in Fig. 2.11 is A and the total weight of the block is W, then the magnitude p of the pressure exerted by the block on the horizontal surface can be calculated by:

$$p = \frac{W}{A}. \qquad (2.3)$$

It follows that the dimension of pressure has the dimension of force (ML/T^2) divided by the dimension of area (L^2):

$$[\text{PRESSURE}] = \frac{[\text{FORCE}]}{[\text{AREA}]} = \frac{(M \cdot L)/T^2}{L^2} = \frac{M}{LT^2}.$$

Units of pressure in different unit systems are listed in Table 2.3.

Table 2.3 Limits of pressure

System	Units of pressure	Special name
SI	kg/m s^2 or N/m^2	Pascal (Pa)
CGS	g/cm s^2 or dyn/cm^2	
British	lb/ft^2 or lb/in.^{2}	psf or psi

The principles behind the concept of pressure have many applications. Note that the larger the area over which a force is applied, the lower the magnitude of pressure. If we observe two people standing on soft snow, one wearing a pair of boots and the other wearing skis, we can easily notice that the person wearing boots stands deeper in the snow than the skier. This is simply because the weight of the person wearing boots is distributed over a smaller area on the snow, and therefore applies a larger force per unit area of snow (Fig. 2.12). It is obvious that the sensation and pain induced by a sharp object is much more severe than that produced by a force that is applied by a dull object.

A prosthesis that fits the amputated limb, or a set of dentures that fits the gum and the bony structure properly, would feel and function better than an improperly fitted implant or replacement device. The idea is to distribute the forces involved as uniformly as possible over a large area.

2.14 Frictional Forces

Frictional forces occur between two surfaces in contact when one surface slides or tends to slide over the other. When a body is in motion on a rough surface or when an object moves in a fluid (a viscous medium such as water), there is resistance to motion because of the interaction of the body with its surroundings. In some applications friction may be desirable, while in others it may have to be reduced to a minimum. For example, it would be impossible to start walking in the absence of frictional forces. Automobile, bicycle, and wheelchair brakes utilize the principles of friction. On the other hand, friction can cause heat to be generated between the surfaces in contact. Excess heat can cause early, unexpected failure of machine parts. Friction may also cause wear.

There are several factors that influence frictional forces. Friction depends on the nature of the two sliding surfaces. For example, if all other conditions are the same, the friction between two metal surfaces would be different than the friction between two wood surfaces in contact. Friction is larger for materials that strongly interact. Friction depends on the surface quality and surface finish. A good surface finish can reduce frictional effects. The frictional force does not depend on the total surface area of contact.

Consider the block resting on the floor, as shown in Fig. 2.13. The block is applying its weight $\underline{W}$ on the floor. In return, the floor is applying a normal force $\underline{N}$ on the block, such that the magnitudes of the two forces are equal ($N = W$). Now consider that a horizontal force $\underline{F}$ is applied on the block to move it toward the right. This will cause a frictional force $\underline{f}$ to develop between the block and the floor. As long as the block remains stationary (in static equilibrium), the magnitude f of the frictional force would be equal to the magnitude F of the applied force. This frictional force is called the *static friction*.

Fig. 2.12 Intensity of force (pressure) applied on the snow by a pair of boots is higher than that applied by a pair of skis

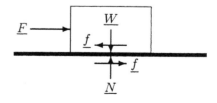

Fig. 2.13 Friction occurs on surfaces when one surface slides or tends to slide over the other

If the magnitude of the applied force is increased, the block will eventually slip or begin sliding over the floor. When the block is on the verge of sliding (the instant just before sliding occurs), the magnitude of the static friction is maximum (f_{max}). When the magnitude of the applied force exceeds f_{max}, the block moves toward the right. When the block is in motion, the resistance to motion at the surfaces of contact is called the *kinetic* or *dynamic friction, f_k*. In general, the magnitude of the force of kinetic friction is lower than the maximum static friction ($f_k < f_{max}$) and the magnitude of the applied force ($f_k < F$). The difference between the magnitudes of the applied force and kinetic friction causes the block to accelerate toward the right.

It has been experimentally determined that the magnitudes of both static and kinetic friction are directly proportional to the normal force ($\underline{N}$ in Fig. 2.13) acting on the surfaces in contact. The constant of proportionality is commonly referred to with μ (mu) and is called the *coefficient of friction*, which depends on such factors as the material properties, the quality of the surface finish, and the conditions of the surfaces in contact. The coefficient of friction also varies depending on whether the bodies in contact are stationary or sliding over each other. To be able to distinguish the frictional forces involved at static and dynamic conditions, two different friction coefficients are defined. The *coefficient of static friction* (μ_s) is associated with static friction, and the *coefficient of kinetic friction* (μ_k) is associated with kinetic or dynamic friction. The magnitude of the static frictional force is such that $f_s = \mu_s N = f_{max}$ when the block is on the verge of sliding, and $f_s < \mu_s N$ when the magnitude of the applied force is less than the maximum frictional force, in which case the magnitude of the force of static friction is equal in magnitude to the applied force ($f_s = F$). The formula relating the kinetic friction and the normal force is:

$$F_k = \mu_k N. \tag{2.4}$$

The variations of frictional force with respect to the force applied in a direction parallel (tangential) to the surfaces in contact are shown in Fig. 2.14. For any given pair of materials, the coefficient of kinetic friction is usually lower than the coefficient of static friction. The coefficient of kinetic friction is approximately constant at moderate sliding speeds. At higher speeds, μ_k may decrease because of the heat generated by friction. Sample coefficients of friction are listed in Table 2.4. Note that the figures provided in Table 2.4 are some average ranges and do not distinguish between static and kinetic friction coefficients.

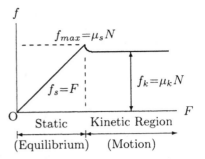

Fig. 2.14 The variation of frictional force as a function of applied force

Table 2.4 Coefficients of friction

Friction coefficient	Surfaces in contact
Wood on wood	0.25–0.50
Metal on metal	0.30–0.80
Plastic on plastic	0.10–0.30
Metal on plastic	0.10–0.20
Rubber on concrete	0.60–0.70
Rubber on tile	0.20–0.40
Rubber on wood	0.70–0.75
Bone on metal	0.10–0.20
Cartilage on cartilage	0.001–0.002

Frictional forces always act in a direction tangent to the surfaces in contact. If one of the two bodies in contact is moving, then the frictional force acting on that body has a direction opposite to the direction of motion. For example, under the action of applied force, the block in Fig. 2.13 tends to move toward the right. The direction of the frictional force on the block is toward the left, trying to stop the motion of the block. The frictional forces always occur in pairs because there are always two surfaces in contact for friction to occur. Therefore, in Fig. 2.13, a frictional force is also acting on the floor. The magnitude of the frictional force on the floor is equal to that of the frictional force acting on the block. However, the direction of the frictional force on the floor is toward the right.

The effects of friction and wear may be reduced by introducing additional materials between the sliding surfaces. These materials may be solids or fluids and are called *lubricants*. Lubricants placed between the moving parts reduce frictional effects and wear by reducing direct contact between the moving parts. In the case of the human body, the diarthrodial joints (such as the elbow, hip, and knee joints) are lubricated by the *synovial fluid*. The synovial fluid is a viscous material that reduces frictional effects, reduces wear and tear of articulating surfaces by limiting direct contact between them, and nourishes the articular cartilage lining the joint surfaces.

Although diarthrodial joints are subjected to very large loading conditions, the cartilage surfaces undergo little wear under normal, daily conditions. It is important to note that introducing a fluid as a lubricant between two solid surfaces undergoing relative motion changes the discussion of how to assess the frictional effects. For example, frictional force with a viscous medium present is not only a function of the normal forces (pressure) involved, but also depends on the relative velocity of the moving parts. A number of lubrication modes have been defined to account for frictional effects at diarthrodial joints under different loading and motion conditions. These modes include hydrodynamic, boundary, elastohydrodynamic, squeeze-film, weeping, and boosted lubrication.

2.15 Exercise Problems

Problem 2.1 As illustrated in Fig. 2.15, consider two workers who are trying to move a block. Assume that both workers are applying equal magnitude forces of 200 N. One of the workers is pushing the block toward the north and the other worker is pushing it toward the east. Determine the magnitude and direction of the net force applied by the workers on the block.

Answer: 283 N, northeast.

Fig. 2.15 Problem 2.1

Problem 2.2 As illustrated in Fig. 2.16, consider two workers who are trying to move a block. Assume that both workers are applying equal magnitude forces of 200 N. One of the workers

Fig. 2.16 Problem 2.2

is pushing the block toward the northeast, while the other is pulling it in the same direction. Determine the magnitude and direction of the net force applied by the workers on the block.

Answer: 400 N, northeast.

Problem 2.3 Consider the two forces, $\underline{F}_1$ and $\underline{F}_2$, shown in Fig. 2.17. Assume that these forces are applied on an object in the *xy*-plane. The first force has a magnitude $F_1 = 15$ N and is applied in a direction that makes an angle $\alpha = 30°$ with the positive *x* axis, and the second force has a magnitude $F_2 = 10$ N and is applied in a direction that makes an angle $\beta = 45°$ with the negative *x* axis.

(a) Calculate the scalar components of $\underline{F}_1$ and $\underline{F}_2$ along the *x* and *y* directions.
(b) Express $\underline{F}_1$ and $\underline{F}_2$ in terms of their components.
(c) Determine an expression for the resultant force vector, $\underline{F}_R$.
(d) Calculate the magnitude of the resultant force vector.
(e) Calculate angle θ that $\underline{F}_R$ makes with the positive *y* axis.

Answers:
(a) $F_{1x} = 13.0$ N, $F_{1y} = 7.5$ N, $F_{2x} = 7.1$ N, $F_{2y} = 7.1$ N.
(b) $\underline{F}_1 = 13.0_i + 7.5_j$ and $\underline{F}_2 = -7.1_i + 7.1_j$.
(c) $\underline{F}_R = 5.9_i + 14.6_j$.
(d) $F_R = 15.7$ N.
(e) $\theta = 22°$.

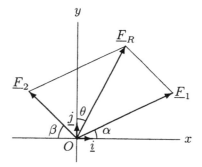

Fig. 2.17 Problem 2.3

Problem 2.4 As illustrated in Fig. 2.18, consider a 2 kg, 20 cm × 30 cm book resting on a table. Calculate the average pressure applied by the book on the table top.

Answer: 327 Pa.

Fig. 2.18 Problem 2.4

Problem 2.5 As shown in Fig. 2.19, consider a 50 kg cylindrical barrel resting on a wooden pallet. The pressure applied on the pallet by the barrel is 260 Pa. What is the radius r of the barrel?

Answer: $r = 0.77$ m.

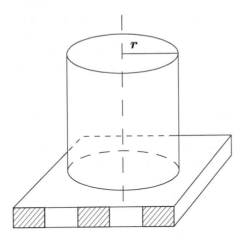

Fig. 2.19 Problem 2.5

Problem 2.6 As illustrated in Fig. 2.20, a stack of three identical boxes of 10 kg each are placed on top of the table. The bottom area of the box is 40 cm × 50 cm. Calculate the average pressure applied by the boxes on the table.

Answer: $P = 1,470$ Pa.

Fig. 2.20 Problem 2.6

Problem 2.7 As illustrated in Fig. 2.21, consider a block that weighs 400 N and is resting on a horizontal surface. Assume that the coefficient of static friction between the block and the horizontal surface is 0.3. What is the minimum horizontal force required to move the block toward the right?

Answer: Slightly greater than 120 N.

Problem 2.8 As shown in Fig. 2.21, consider a block moving over the floor to the right as the result of externally applied force $F = 280$ N. Assume that the coefficient of kinetic friction between the block and the floor is 0.35. What is the mass (m) of the block?

Fig. 2.21 Problems 2.7, 2.8, 2.9

Answer: $m = 28.6$ kg.

Problem 2.9 As shown in Fig. 2.21, consider a block moving over the floor to the right as the result of horizontal force $F = 62.5$ N applied on the block. The coefficient of friction between the block and the floor is 0.25. What is the weight of the block?

Answer: $W = 250$ N.

Problem 2.10 As illustrated in Fig. 2.22, consider a person pushing a 50 kg file cabinet over a tile-covered floor by applying 74 N horizontal force. What is the coefficient of friction between the file cabinet and the floor?

Answer: $\mu = 0.15$.

Fig. 2.22 Problem 2.10

Problem 2.11 As shown in Fig. 2.23, consider a block that weighs W. Due to the effect of gravity, the block is sliding down a slope that makes an angle θ with the horizontal. The coefficient of kinetic friction between the block and the slope is μ_k.

Show that the magnitude of the frictional force generated between the block and the slope is $f = \mu_k W \cos \theta$.

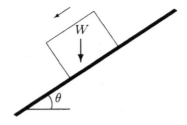

Fig. 2.23 Problem 2.11

Problem 2.12 As shown in Fig. 2.24, a person is trying to push a box weighing 500 N up an inclined surface by applying a force parallel to the incline. If the coefficient of friction between the box and the incline is 0.4, and the incline makes an angle $\theta = 25°$ with the horizontal, determine the magnitude of the frictional force(s) acting on the box.

Answer: $f = 181.3$ N.

Fig. 2.24 Problem 2.12

Problem 2.13 Figure 2.25 shows a simple experimental method to determine the coefficient of static friction between surfaces in contact. This method is applied by placing a block on a horizontal plate, tilting the plate slowly until the block starts sliding over the plate, and recording the angle the plate makes with the horizontal at the instant when the sliding occurs. This critical angle (θ_c) is called the *angle of repose*. At the instant just before the sliding occurs, the block is in static equilibrium.

Through force equilibrium, show that the coefficient of static friction just before motion starts is $\mu = \tan\theta_c$.

Fig. 2.25 Problem 2.13

3.1 Definitions of Moment and Torque Vectors

A force applied to an object can translate, rotate, and/or deform the object. The effect of a force on the object to which it is applied depends on how the force is applied and how the object is supported. For example, when pulled, an open door will swing about the edge along which it is hinged to the door frame (Fig. 3.1). What causes the door to swing is the torque generated by the applied force about an axis that passes through the hinges of the door. If one stands on the free-end of a diving board, the board will bend (Fig. 3.2). What bends the board is the moment of the body weight about the fixed end of the board.

Fig. 3.1 Rotational effect

In general, *torque* is associated with the rotational and twisting actions of applied forces, while *moment* is related to their bending effect. However, the mathematical definition of moment and torque is the same. Therefore, it is sufficient to use moment to discuss the common properties of moment and torque vectors.

3.2 Magnitude of Moment

The magnitude of the moment of a force about a point is equal to the magnitude of the force times the length of the shortest distance between the point and the line of action of the force, which is known as the *lever* or *moment arm*. Consider a person on an exercise apparatus who is holding a handle that is attached to a cable (Fig. 3.3). The cable is wrapped around a pulley and attached to a weight pan. The weight in the weight pan stretches the cable and produces a tensile force $\underline{F}$ in the cable. This force is transmitted to the person's hand through the handle. Assume that the magnitude of the moment of force $\underline{F}$ about point O at the elbow joint is to be determined. To determine the shortest distance between O and the line of action of the force, extend the line of action of $\underline{F}$ and drop a line from O that cuts the line of action of $\underline{F}$ at right angles. If the point of intersection of the two lines is R, then the distance d between O and R is the

Fig. 3.2 Bending effect

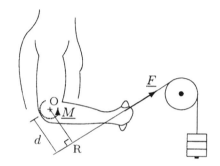

Fig. 3.3 Moment of a force about a point

N. Özkaya et al., *Fundamentals of Biomechanics: Equilibrium, Motion, and Deformation*,
DOI 10.1007/978-1-4614-1150-5_3, © Springer Science+Business Media, LLC 2012

lever arm, and the magnitude of the moment $\underline{M}$ of force $\underline{F}$ about point O is:

$$M = dF. \qquad (3.1)$$

3.3   Direction of Moment

The moment of a force about a point acts in a direction perpendicular to the plane upon which the point and the force lie. For example, in Fig. 3.4, point O and the line of action of force $\underline{F}$ lie on plane A. The line of action of moment $\underline{M}$ of force $\underline{F}$ about point O is perpendicular to plane A. The direction and sense of the moment vector along its line of action can be determined using the *right-hand rule*. As illustrated in Fig. 3.5, when the fingers of the right hand curl in the direction that the applied force tends to rotate the body about point O, the right-hand thumb points in the direction of the moment vector. More specifically, extend the right hand with the thumb at a right angle to the rest of the fingers, position the finger tips in the direction of the applied force, and position the hand so that the point about which the moment is to be determined faces toward the palm. The tip of the thumb points in the direction of the moment vector.

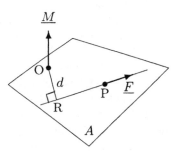

Fig. 3.4   Direction of the moment vector

Fig. 3.5   The right-hand rule

The line of action and direction of the moment vector can also be explained using a wrench and a right-treaded bolt (Fig. 3.6). When a force is applied on the handle of the wrench, a torque is generated that rotates the wrench. The line of action of this torque coincides with the centerline of the bolt. Due to the torque, the bolt will either advance into or retract from the board depending on how the force is applied. As in Fig. 3.6, if the force causes a clockwise rotation, then the direction of torque is "into" the board and the bolt will advance into the board. If the force causes a counterclockwise rotation, then the direction of torque is "out of" the board and the bolt will retract from the board.

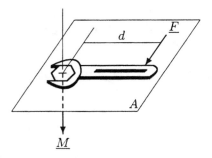

Fig. 3.6   Wrench and bolt

In Fig. 3.7, if point O and force $\underline{F}$ lie on the surface of the page, then the line of action of moment $\underline{M}$ is perpendicular to the page. If you pin the otherwise unencumbered page at point O, $\underline{F}$ will rotate the page in the counterclockwise direction. Using the right-hand rule, that corresponds to the direction away from the page. To refer to the direction of the moments of coplanar force systems, it may be sufficient to say that a particular moment is either clockwise (cw) or counterclockwise (ccw).

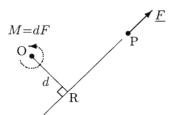

Fig. 3.7   The magnitude and direction of moment of $\underline{F}$ about O

3.4   Dimension and Units of Moment

By definition, moment is equal to the product of applied force and the length of the moment arm. Therefore, the dimension of moment is equal to the dimension of force (ML/T^2) times the dimension of length (L):

$$[\text{MOMENT}] = [\text{FORCE}][\text{MOMENT ARM}] = \frac{ML}{T^2}L = \frac{ML^2}{T^2}.$$

The units of moment in different systems are listed in Table 3.1.

Table 3.1 Units of moment and torque[a]

System	Units of moment and torque
SI	Newton meter (N-m)
CGS	Dyne centimeter (dyn-cm)
British	Pound foot (lb-ft)

[a]1 lb-ft = 1.3573 N-m; 1 N-m = 1 × 10^7 dyn-cm; 1 N-m = 0.738 lb-ft

3.5 Some Fine Points About the Moment Vector

- The moment of a force is invariant under the operation of sliding the force vector along its line of action, which is illustrated in Fig. 3.8. For all cases illustrated, the moment of force F applied at points P_1, P_2, and P_3 about point O is:

$$M = dF \quad (\text{ccw})$$

where length d is always the shortest distance between point O and the line of action of $\underline{F}$. Again for all three cases shown in Fig. 3.8, the forces generate a counterclockwise moment.

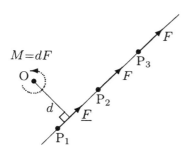

Fig. 3.8 Moment is invariant under the operation of sliding the force along its line of action

- Let $\underline{F}_1$ and $\underline{F}_2$ (Fig. 3.9) be two forces with equal magnitude ($\underline{F}_1 = \underline{F}_2 = F$) and the same line of action, but acting in opposite directions. The moment $\underline{M}_1$ of force $\underline{F}_1$ and the moment $\underline{M}_2$ of force $\underline{F}_2$ about point O have an equal magnitude ($\underline{M}_1 = \underline{M}_2 = M = dF$) but opposite directions ($\underline{M}_1 = -\underline{M}_2$).
- The magnitude of the moment of an applied force increases with an increase in the length of the moment arm. That is, the greater the distance of the point about which the moment is to be calculated from the line of action of the force vector, the higher the magnitude of the corresponding moment vector.

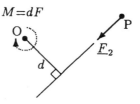

Fig. 3.9 Opposite moments

- The moment of a force about a point that lies on the line of action of the force is zero, because the length of the moment arm is zero (Fig. 3.10).

Fig. 3.10 The moment of $\underline{F}$ about point O is zero

- A force applied to a body may tend to rotate or bend the body in one direction with respect to one point and in the opposite direction with respect to another point in the same plane.
- The principles of resolution of forces into their components along appropriate directions can be utilized to simplify the calculation of moments. For example, in Fig. 3.11, the applied force $\underline{F}$ is resolved into its components $\underline{F}_x$ and $\underline{F}_y$ along the x and y directions, such that

$$\underline{F}_x = F \cos \theta$$
$$\underline{F}_y = F \sin \theta.$$

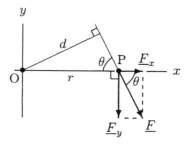

Fig. 3.11 Component of $\underline{F}$

Since point O lies on the line of action of $\underline{F}_x$, the moment arm of $\underline{F}_x$ relative to point O is zero. Therefore, the moment of $\underline{F}_x$ about point O is zero. On the other hand, r is the length of the moment arm for force $\underline{F}$ relative to point O. Therefore, the moment of force $\underline{F}_y$ about point O is

$$M = rF_y = rF \sin \theta \quad \text{(cw)}.$$

Note that this is also the moment dF generated by the resultant force vector $\underline{F}$ about point O, because $d = r \sin \theta$.

3.6   The Net or Resultant Moment

When there is more than one force applied on a body, the net or resultant moment can be calculated by considering the vector sum of the moments of all forces. For example, consider the coplanar three-force system shown in Fig. 3.12. Let d_1, d_2, and d_3 be the moment arms of $\underline{F}_1$, $\underline{F}_2$, and $\underline{F}_3$ relative to point O. These forces produce moments $\underline{M}_1$, $\underline{M}_2$, and $\underline{M}_3$ about point O, which can be calculated as follows:

$$M_1 = d_1F_1 \quad \text{(cw)}$$
$$M_2 = d_2F_2 \quad \text{(cw)}$$
$$M_3 = d_3F_3 \quad \text{(ccw)}.$$

The net moment $\underline{M}_{\text{net}}$ generated on the body due to forces $\underline{F}_1$, $\underline{F}_2$, and $\underline{F}_3$ about point O is equal to the vector sum of the moments of all forces about the same point:

$$M_{\text{net}} = M_1 + M_2 + M_3. \tag{3.2}$$

A practical way of determining the magnitude and direction of the net moment for coplanar force systems is discussed next. Note that for the case illustrated in Fig. 3.12, the individual moments are either clockwise or counterclockwise. Therefore, the resultant moment must be either clockwise or counterclockwise. Choose or guess the direction of the resultant moment. For example, if we assume that the resultant moment vector is clockwise, then

the clockwise moments $\underline{M}_1$ and $\underline{M}_2$ are positive and the counterclockwise moment $\underline{M}_3$ is negative. The magnitude of the net moment can now be determined by simply adding the magnitudes of the positive moments and subtracting the negatives:

$$M_{\text{net}} = M_1 + M_2 - M_3. \tag{3.3}$$

Depending on the numerical values of M_1, M_2, and M_3, this equation will give a positive, negative, or zero value for the net moment. If the computed value is positive, then it is actually the magnitude of the net moment and the chosen direction for the net moment was correct (in this case, clockwise). If the value calculated is negative, then the chosen direction for the net moment was wrong, but can readily be corrected. If the chosen direction was clockwise, a negative value will indicate that the correct direction for the net moment is counterclockwise. (Note that magnitudes of vector quantities are scalar quantities that are always positive.) Once the correct direction for the net moment is indicated, the negative sign in front of the value calculated can be eliminated. The third possibility is that the value calculated from equation (3.3) may be zero. If the net moment is equal to zero, then the body is said to be in *rotational equilibrium*. This case is discussed in detail in the following chapter.

Example 3.1 Figure 3.13 illustrates a person preparing to dive into a pool. The horizontal diving board has a uniform thickness, mounted to the ground at point O, has a mass of 120 kg, and $l = 4$ m in length. The person has a mass of 90 kg and stands at point B which is the free-end of the board. Point A indicates the location of the center of gravity of the board. Point A is equidistant from points O and B.

Determine the moments generated about point O by the weights of the person and the board. Calculate the net moment about point O.

Solution
Let m_1 and m_2 be the masses and W_1 and W_2 be the weights of the person and diving board, respectively. W_1 and W_2 can be calculated if the masses of the person and the board are given:

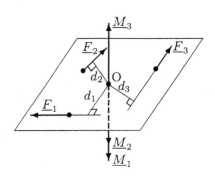

Fig. 3.12 The net moment

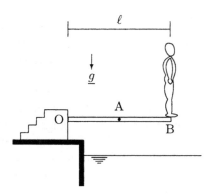

Fig. 3.13 A person is preparing to dive

$$W_1 = m_1 g = (90 \text{ kg})(9.8 \text{ m/s}^2) = 882 \text{ N}$$
$$W_2 = m_2 g = (120 \text{ kg})(9.8 \text{ m/s}^2) = 1,176 \text{ N}.$$

The person is standing at point B. Therefore, the weight $\underline{W}_1$ of the person is applied on the board at point B. The mass of the diving board produces a force system distributed over the entire length of the board. The resultant of this distributed force system is equal to the weight $\underline{W}_2$ of the board. For practical purposes and since the board has a uniform thickness, we can assume that the weight of the board is a concentrated force acting at A which is the center of gravity of the board.

As shown in Fig. 3.14, weights $\underline{W}_1$ and $\underline{W}_2$ act vertically downward or in the direction of gravitational acceleration. The diving board is horizontal. Therefore, the distance between points O and B(l) is the length of the moment arm for $\underline{W}_1$ and the distance between points O and A($l/2$) is the length of the moment arm for $\underline{W}_2$. Therefore, moments $\underline{M}_1$ and $\underline{M}_2$ due to $\underline{W}_1$ and $\underline{W}_2$ about point O are

$$M_1 = lW_1 = (4 \text{ m})(882 \text{ N}) = 3,528 \text{ N-m} \quad (\text{cw})$$
$$M_2 = \frac{l}{2}W_2 = (2 \text{ m})(1,176 \text{ N}) = 2,352 \text{ N-m} \quad (\text{cw}).$$

Fig. 3.15 Example 3.2

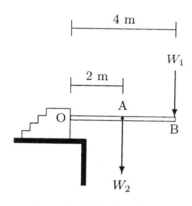

Fig. 3.14 Forces acting on the diving board

Since both moments have a clockwise direction, the net moment must have a clockwise direction as well. The magnitude of the net moment about point O is

$$M_{\text{net}} = M_1 + M_2 = 880 \text{ N-m} \quad (\text{cw}).$$

Example 3.2 As illustrated in Fig. 3.15a, consider an athlete wearing a weight boot, and from a sitting position, doing lower leg flexion/extension exercises to strengthen quadriceps muscles. The weight of the athlete's lower leg is $W_1 = 50 \text{ N}$ and the weight of the boot is $W_2 = 100 \text{ N}$. As measured from the knee joint at point O, the center of gravity (point A) of the lower leg is located at a distance $a = 20 \text{ cm}$ and the center of gravity (point B) of the weight boot is located at a distance $b = 50 \text{ cm}$.

Determine the net moment generated about the knee joint when the lower leg is extended horizontally (position 1),

and when the lower leg makes an angle of 30° (position 2), 60° (position 3), and 90° (position 4) with the horizontal (Fig. 3.15b).

Solution

At position 1, the lower leg is extended horizontally and the long axis of the leg is perpendicular to the lines of action of $\underline{W}_1$ and $\underline{W}_2$. Therefore, a and b are the lengths of the moment arms for $\underline{W}_1$ and $\underline{W}_2$, respectively. Both $\underline{W}_1$ and $\underline{W}_2$ apply clockwise moments about the knee joint. The net moment $\underline{M}_O$ about the knee joint when the lower leg is at position 1 is

$$M_O = aW_1 + bW_2$$
$$= (0.20)(50) + (0.50)(100)$$
$$= 60 \text{ N-m} \quad (\text{cw}).$$

Figure 3.16 illustrates the external forces acting on the lower leg and their moment arms (d_1 and d_2) when the lower

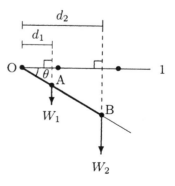

Fig. 3.16 Forces and moment arms when the lower leg makes an angle θ with the horizontal

leg makes an angle θ with the horizontal. From the geometry of the problem,

Therefore, the net moment about point O is

(a)

$$d_1 = a \cos \theta$$
$$d_2 = b \cos \theta.$$

$$\begin{aligned} M_O &= d_1 W_1 + d_2 W_2 \\ &= a \cos \theta \, W_1 + b \cos \theta \, W_2 \\ &= (aW_1 + bW_2) \cos \theta. \end{aligned}$$

The term in the parentheses has already been calculated as 60 N-m. Therefore, we can write

$$M_O = 60 \cos \theta.$$

For position 1: $\theta = 0°$, $M_O = 60$ N-m (cw).
For position 2: $\theta = 30°$, $M_O = 52$ N-m (cw).
For position 3: $\theta = 60°$, $M_O = 30$ N-m (cw).
For position 4: $\theta = 90°$, $M_O = 0$ (cw).

In Fig. 3.17, the moment generated about the knee joint is plotted as a function of angle θ.

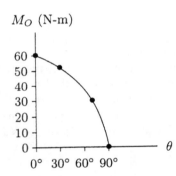

Fig. 3.17 Variation of moment with angle θ

Example 3.3 Figure 3.18a illustrates an athlete doing shoulder muscle strengthening exercises by lowering and raising a barbell with straight arms. The position of the arms when they make an angle θ with the vertical is simplified in Fig. 3.18b. Point O represents the shoulder joint, A is the center of gravity of one arm, and B is a point of intersection of the centerline of the barbell and the extension of line OA. The distance between points O and A is $a = 24$ cm and the distance between points O and B is $b = 60$ cm. Each arm weighs $W_1 = 50$ N and the total weight of the barbell is $W_2 = 300$ N.

Determine the net moment due to $\underline{W}_1$ and $\underline{W}_2$ about the shoulder joint as a function of θ, which is the angle the arm makes with the vertical. Calculate the moments for $\theta = 0°$, $15°$, $30°$, $45°$, and $60°$.

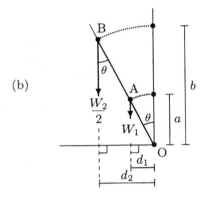

Fig. 3.18 An exercise to strengthen the shoulder muscles, and a simple model of the arm

Solution

To calculate the moments generated about the shoulder joint by $\underline{W}_1$ and $\underline{W}_2$, we need to determine the moment arms d_1 and d_2 of forces $\underline{W}_1$ and $\underline{W}_2$ relative to point O. From the geometry of the problem (Fig. 3.18b), the lengths of the moment arms are

$$d_1 = a \sin \theta$$
$$d_2 = b \sin \theta.$$

Since the athlete is using both arms, the total weight of the barbell is assumed to be shared equally by each arm. Also note that relative to the shoulder joint, both the weight of the arm and the weight of the barbell are trying to rotate the arm in the counterclockwise direction. Moments $\underline{M}_1$ and $\underline{M}_2$ due to $\underline{W}_1$ and $\underline{W}_2$ about point O are

$$M_1 = d_1 W_1 = aW_1 \sin \theta = (0.24)(50) \sin \theta = 12 \sin \theta$$
$$M_2 = d_2 \frac{W_2}{2} = b\frac{W_2}{2} \sin \theta = (0.60)\left(\frac{300}{2}\right) \sin \theta = 90 \sin \theta.$$

Since both moments are counterclockwise, the net moment must be counterclockwise as well. Therefore, the net moment $\underline{M}_O$ generated about the shoulder joint is

$$M_O = M_1 + M_2 = 12 \sin \theta + 90 \sin \theta = 102 \sin \theta \, \text{N-m}\ \ (\text{ccw}).$$

To determine the magnitude of the moment about point O, for $\theta = 0°$, $15°$, $30°$, $45°$, and $60°$, all we need to do is evaluate the sines and carry out the multiplications. The results are provided in Table 3.2.

Table 3.2 Moment about the shoulder joint (Example 3.3)

θ (°)	$\sin \theta$	M_O (N-m)
0	0.000	0.0
15	0.259	26.4
30	0.500	51.0
45	0.707	72.1
60	0.866	88.3

Example 3.4 Consider the total hip joint prosthesis shown in Fig. 3.19. The geometric parameters of the prosthesis are such that $l_1 = 50$ mm, $l_2 = 50$ mm, $\theta_1 = 45°$, $\theta_2 = 90°$. Assume that, when standing symmetrically on both feet, a joint reaction force of $F = 400$ N is acting at the femoral head due to the body weight of the patient. For the sake of illustration, consider three different lines of action for the applied force, which are shown in Fig. 3.20.

Fig. 3.19 Total hip joint prosthesis

Determine the moments generated about points B and C on the prosthesis for all cases shown.

Solution
For each case shown in Fig. 3.20, the line of action of the joint reaction force is different, and therefore the lengths of the moment arms are different.

From the geometry of the problem in Fig. 3.20a, we can see that the moment arm of force F about points B and C are the same:

$$d_1 = l_1 \cos \theta_1 = (50)(\cos 45°) = 35 \text{ mm}.$$

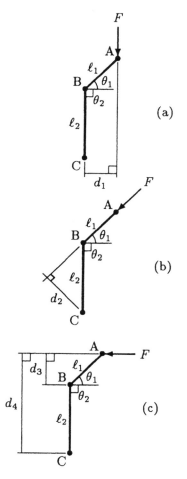

Fig. 3.20 Example 3.4

Therefore, the moments generated about points B and C are

$$M_B = M_C = d_1 F = (0.035)(400) = 14 \text{ N-m} \quad (\text{cw}).$$

For the case shown in Fig. 3.20b, point B lies on the line of action of the joint reaction force. Therefore, the length of the moment arm for point B is zero, and

$$M_B = 0.$$

For the same case, the length of the moment arm and the moment about point C are

$$d_2 = l_2 \cos \theta_1 = (100)(\cos 45°) = 71 \text{ mm}$$
$$M_C = d_2 F = (0.071)(400) = 28 \text{ N-m} \quad (\text{ccw}).$$

For the case shown in Fig. 3.20c, the moment arms relative to points B and C are

$$d_3 = l_1 \sin \theta_1 = (50)(\sin 45°) = 35 \text{ mm}$$
$$d_4 = d_3 + l_2 = (35) + (100) = 135 \text{ mm}.$$

Therefore, the moments generated about points B and C are

$$M_B = d_3 F = (0.035)(400) = 14 \text{ N-m} \quad (ccw)$$
$$M_C = d_4 F = (0.135)(400) = 54 \text{ N-m} \quad (ccw).$$

3.7 The Couple and Couple-Moment

A special arrangement of forces that is of importance is called *couple*, which is formed by two parallel forces with equal magnitude and opposite directions. On a rigid body, the couple has a pure rotational effect. The rotational effect of a couple is quantified with *couple-moment*.

Consider the forces shown in Fig. 3.21, which are applied at points A and B. Note that the net moment about point A is $M = dF$(cw), which is due to the force applied at point B. The net moment about point B is also $M = dF$(cw), which is due to the force applied at point A. Consider point C. The distance between points C and B is b, and therefore, the distance between points C and A is $d - b$. The net moment about point C is equal to the sum of the clockwise moments of forces applied at points A and B with moment arms $d - b$ and b. Therefore,

$$M = (d - b)F + bF = dF \quad (cw).$$

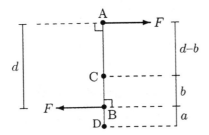

Fig. 3.21 A couple

Consider point D. The distance between points B and D is a. The net moment about point D is equal to the difference between the moments of force F applied at points A and B, respectively,

$$M_D = (a + d)F - aF = dF + aF - aF = dF.$$

It can be concluded without further proof that the couple has the same moment about every point in space. If F is the magnitude of the forces forming the couple and d is the perpendicular distance between the lines of actions of the forces, then the magnitude of the couple-moment is:

$$M = dF. \quad (3.4)$$

The direction of the couple-moment can be determined by the right-hand rule.

3.8 Translation of Forces

The overall effect of a pair of forces applied on a rigid body is zero if the forces have an equal magnitude and the same line of action, but are acting in opposite directions. Keeping this in mind, consider the force with magnitude F applied at point P_1 in Fig. 3.22a. As illustrated in Fig. 3.22b, this force may be translated to point P_2 by placing a pair of forces at P_2 with equal magnitude (F), having the same line of action, but acting in opposite directions. Note that the original force at point P_1 and the force at point P_2 that is acting in a direction opposite to that of the original force form a couple. This couple produces a counterclockwise moment with magnitude $M = dF$, where d is the shortest distance between the lines of action of forces at P_1 and P_2. Therefore, as illustrated in Fig. 3.22c, the couple can be replaced by the couple-moment.

Fig. 3.22 Translation of a force from point P_1 to P_2

Provided that the original force was applied to a rigid body, the one-force system in Fig. 3.22a, the three-force system in Fig. 3.22b, and the one-force and one-couple-moment system in Fig. 3.22c are mechanically equivalent.

3.9 Moment as a Vector Product

We have been applying the scalar method of determining the moment of a force about a point. The scalar method is satisfactory to analyze relatively simple coplanar force systems and systems in which the perpendicular distance between the point and the line of action of the applied force are easy to determine. The analysis of more complex problems can be simplified by utilizing additional mathematical tools.

The concept of the *vector (cross) product* of two vectors is introduced in Appendix B and is reviewed here. Consider vectors $\underline{A}$ and $\underline{B}$, shown in Fig. 3.23. The cross product of $\underline{A}$ and $\underline{B}$ is equal to a third vector, $\underline{C}$:

$$\underline{C} = \underline{A} \times \underline{B} \tag{3.5}$$

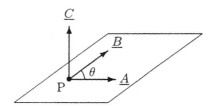

Fig. 3.23 $\underline{C}$ is the vector product of $\underline{A}$ and $\underline{B}$

The product vector $\underline{C}$ has the following properties:

- The magnitude of $\underline{C}$ is equal to the product of the magnitude of $\underline{A}$, the magnitude of $\underline{B}$, and $\sin\theta$, where θ is the smaller angle between $\underline{A}$ and $\underline{B}$.

$$C = AB \sin\theta \tag{3.6}$$

- The line of action of $\underline{C}$ is perpendicular to the plane formed by vectors $\underline{A}$ and $\underline{B}$.
- The direction and sense of $\underline{C}$ obeys the right-hand rule.

The principle of vector or cross product can be applied to determine the moments of forces. The moment of a force about a point is defined as the vector product of the position and force vectors. The *position vector* of a point P with respect to another point O is defined by an arrow drawn from point O to point P. To help understand the definition of moment as a vector product, consider Fig. 3.24. Force $\underline{F}$ acts in the xy-plane and has a point of application at point P. Force $\underline{F}$ can be expressed in terms of its components F_x and F_y along the x and y directions:

$$\underline{F} = F_x\underline{i} + F_y\underline{j}. \tag{3.7}$$

The position vector of point P with respect to point O is represented by vector $\underline{r}$, which can be written in terms of its components:

$$\underline{r} = r_x\,\underline{i} + r_y\,\underline{j} \tag{3.8}$$

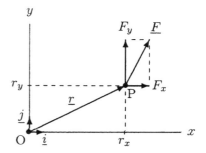

Fig. 3.24 The moment about O is $\underline{M} = \underline{r} \times \underline{F}$

The components r_x and r_y of the position vector are simply the x and y coordinates of point P as measured from point O.

The moment of force $\underline{F}$ about point O is equal to the vector product of the position vector $\underline{r}$ and force vector $\underline{F}$:

$$\underline{M} = \underline{r} \times \underline{F}. \tag{3.9}$$

Using (3.7) and (3.8), (3.9) can alternatively be written as:

$$
\begin{aligned}
\underline{M} &= (r_x\underline{i} + r_y\underline{j}) \times (F_x\underline{i} + F_y\underline{j}) \\
&= r_xF_x(\underline{i} \times \underline{i}) + r_xF_y(\underline{i} \times \underline{j}) \\
&\quad + r_yF_x(\underline{j} \times \underline{i}) + r_yF_y(\underline{j} \times \underline{j}).
\end{aligned}
\tag{3.10}
$$

Recall that $\underline{i} \times \underline{j} = \underline{j} \times \underline{j} = 0$ since the angle that a unit vector makes with itself is zero, and $\sin 0° = 0$. $\underline{i} \times \underline{j} = \underline{k}$ because the angle between the positive x axis and the positive y axis is $90°$ ($\sin 90° = 1$). On the other hand, $\underline{j} \times \underline{i} = -\underline{k}$. For the last two cases, the product is either in the positive z (counterclockwise or out of the page) or negative z (clockwise or into the page) direction. z and unit vector $\underline{k}$ designate the direction perpendicular to the xy-plane (Fig. 3.25). Now, (3.10) can be simplified as:

$$\underline{M} = (r_xF_y - r_yF_x)\underline{k}. \tag{3.11}$$

To show that the definition of moment as the vector product of the position and force vectors is consistent with the scalar method of finding the moment, consider the simple

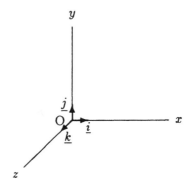

Fig. 3.25 The right-hand rule applies to Cartesian coordinate directions as well

case illustrated in Fig. 3.26. The force vector $\underline{F}$ is acting in the positive y direction and its line of action is d distance away from point O. Applying the scalar method, the moment about point O is

$$M = dF \quad \text{(ccw)}. \tag{3.12}$$

The force vector is acting in the positive y direction. Therefore,

$$\underline{F} = F\underline{j}. \tag{3.13}$$

If b is the y coordinate of the point of application of the force, then the position vector of point P is

$$\underline{r} = d\underline{i} + b\underline{j}. \tag{3.14}$$

Therefore, the moment of $\underline{F}$ about point O is

$$
\begin{aligned}
\underline{M} &= \underline{r} \times \underline{F} \\
&= (d\underline{i} + b\underline{j}) \times (F\underline{j}) \\
&= dF(\underline{i} \times \underline{j}) + bF(\underline{j} \times \underline{j}) \\
&= dF\underline{k}.
\end{aligned}
\tag{3.15}
$$

Equations (3.12) and (3.15) carry exactly the same information, in two different ways. Furthermore, the y coordinate (b) of point P does not appear in the solution. This is consistent with the definition of the moment, which is the vector product of the position vector of any point on the line of action of the force and the force itself. In Fig. 3.26, if C is the point of intersection of the line of action of force $\underline{F}$ and the x axis, then the position vector r' of point C with respect to point O is

$$\underline{r}' = d\underline{i}. \tag{3.16}$$

Therefore, the moment of $\underline{F}$ about point O can alternatively be determined as:

$$
\begin{aligned}
\underline{M} &= \underline{r}'F \\
&= (d\underline{i}) \times (F\underline{j}) \\
&= dF\underline{k}.
\end{aligned}
\tag{3.17}
$$

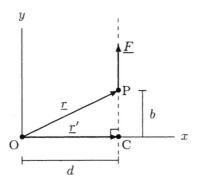

Fig. 3.26 $M = dF$ (ccw)

For any two-dimensional problem composed of a system of coplanar forces in the xy-plane, the resultant moment vector has only one component. The resultant moment vector has a direction perpendicular to the xy-plane, acting in the positive or negative z direction.

The method we have outlined to study coplanar force systems using the concept of the vector (cross) product can easily be expanded to analyze three-dimensional situations. In general, the force vector $\underline{F}$ and the position vector $\underline{r}$ of a point on the line of action of $\underline{F}$ about a point O would have up to three components. With respect to the Cartesian coordinate frame,

$$\underline{F} = F_x\underline{i} + F_y\underline{j} + F_z\underline{k}, \tag{3.18}$$

$$\underline{r} = r_x\underline{i} + r_y\underline{j} + r_z\underline{k}. \tag{3.19}$$

The moment of $\underline{F}$ about point O can be determined as:

$$
\begin{aligned}
\underline{M} &= \underline{r} \times \underline{F} \\
&= (r_x\underline{i} + r_y\underline{j} + r_z\underline{k}) \times (F_x\underline{i} + F_y\underline{j} + F_z\underline{k}) \\
&= (r_yF_z - r_zF_y)\underline{i} + (r_zF_x - r_xF_z)\underline{j} \\
&\quad + (r_xF_y - r_yF_x)\underline{k}.
\end{aligned}
\tag{3.20}
$$

The moment vector can be expressed in terms of its components along the x, y, and z directions:

$$\underline{M} = M_x\underline{i} + M_y\underline{j} + M_z\underline{k}. \tag{3.21}$$

By comparing Eqs. (3.20) and (3.21), we can conclude that

$$
\begin{aligned}
M_x &= r_yF_z - r_zF_y \\
M_y &= r_zF_x - r_xF_z \\
M_z &= r_xF_y - r_yF_x.
\end{aligned}
\tag{3.22}
$$

The following example provides an application of the analysis outlined in the last three sections of this chapter.

Example 3.5 Figure 3.27a illustrates a person using an exercise machine. The "L"-shaped beam shown in Fig. 3.27b represents the left arm of the person. Points A and B correspond to the shoulder and elbow joints, respectively. Relative to the person, the upper arm (AB) is extended toward the left (x direction) and the lower arm (BC) is extended forward (z direction). At this instant, the person is holding a handle that is connected by a cable to a suspending weight. The weight applies an upward (in the y direction) force with magnitude F on the arm at point C. The lengths of the upper arm and lower arm are $a = 25$ cm and $b = 30$ cm, respectively, and the magnitude of the applied force is $F = 200$ N.

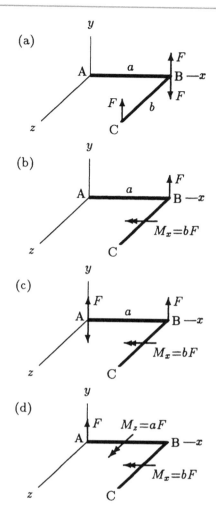

Fig. 3.27 Example 3.5

Explain how force F can be translated to the shoulder joint at point A, and determine the magnitudes and directions of moments developed at the lower and upper arms by $\underline{F}$.

Solution 1: Scalar Method

The scalar method of finding the moments generated on the lower arm (BC) and upper arm (AB) is illustrated in Fig. 3.28, and it utilizes the concepts of couple and couple-moment.

The first step is placing a pair of forces at point B with equal magnitude (F) and opposite directions, both being parallel to the original force at point C (Fig. 3.28a). The upward force at point C and the downward force at point B form a couple. Therefore, they can be replaced by a couple-moment (shown by a double-headed arrow in Fig. 3.28b). The magnitude of the couple-moment is bF. Applying the right-hand rule, we can see that the couple-moment acts in the negative x direction. If M_x refers to the magnitude of this couple-moment, then

$$M_x = bF \quad (-x \text{ direction}).$$

The next step is to place another pair of forces at point A where the shoulder joint is located (Fig. 3.28c). This time, the upward force at point B and the downward force at point A form a couple, and again, they can be replaced by a couple-moment (Fig. 3.28d). The magnitude of this couple-moment is aF, and it has a direction perpendicular to the xy-plane, or, it is acting in the positive z direction. Referring to the magnitude of this moment as M_z, then

$$M_z = aF \quad (+z \text{ direction}).$$

Fig. 3.28 Scalar method

To evaluate the magnitudes of the couple-moments, the numerical values of a, b, and F must be substituted in the above equations:

$$M_x = (0.30)(200) = 60 \text{ N-m}$$
$$M_z = (0.25)(200) = 50 \text{ N-m}.$$

The effect of the force applied at point C is such that at the elbow joint, the person feels an upward force with magnitude F and a moment with magnitude M_x that is trying to *rotate* the lower arm in the yz-plane. At the shoulder joint, the feeling is such that there is an upward force of F, a torque with magnitude M_x that is trying to twist the upper arm in the yz-plane, and a moment M_z that is trying to rotate or bend the upper arm in the xy-plane. If the person is able to hold the arm in this position, then he/she is producing sufficient muscle forces to counter-balance these applied forces and moments.

Solution 2: Vector Product Method

The definition of moment as the vector product of the position and force vectors is more straightforward to apply. The position vector of point C (where the force is applied) with respect to point A (where the shoulder joint is located) and the force vector shown in Fig. 3.29 can be expressed as follows:

$$\underline{r} = a\underline{i} + b\underline{k}$$
$$\underline{F} = F\underline{j}.$$

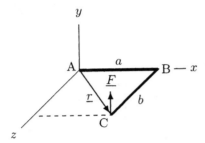

Fig. 3.29 Vector product method (Example 3.5)

The cross product of $\underline{r}$ and $\underline{F}$ will yield the moment of force $\underline{F}$ about point A:

$$\begin{aligned}
\underline{M} &= \underline{r} \times \underline{F} \\
&= (a\underline{i} + b\underline{k}) \times (F\underline{j}) \\
&= aF(\underline{i} \times \underline{j}) + bF(\underline{k} \times \underline{j}) \\
&= aF\underline{k} - bF\underline{i} \\
&= (0.25)(200)\underline{k} - (0.30)(200)\underline{i}. \\
&= 50k - 60i
\end{aligned}$$

The negative sign in front of $60\underline{i}$ indicates that the x component of the moment vector is acting in the negative x direction. Furthermore, there is no component associated with the unit vector $\underline{j}$, which implies that the y component of the moment vector is zero. Therefore, the moment about point A has the following components:

$$\begin{aligned}
M_x &= 60 \text{ N-m} \quad (-x \text{ direction}) \\
M_y &= 0 \\
M_z &= 50 \text{ N-m} \quad (+z \text{ direction}).
\end{aligned}$$

These results are consistent with those obtained using the scalar method.

3.10 Exercise Problems

Problem 3.1 As illustrated in Fig. 3.13, consider a diver of 88 kg mass standing on the free end of a horizontal diving board at point B, preparing for a jump. The diving board has a uniform thickness and a mass of 56 kg, and it is mounted to the ground at point O. Point A indicates the center of gravity of the diving board, and it is equidistant from points O and B. Determine the length of the diving board (l) if the net moment generated about point O by the weight of the diver and the diving board is $M_O = 3,979$ N-m.

Answer: $l - 3.5$ m.

Problem 3.2 As illustrated in Fig. 3.30, consider two divers preparing to dive into a pool in a sequence. The horizontal diving board of uniform thickness is mounted to the ground at point O, has a mass of 170 kg, and $l = 4$ m in length. The first diver has a mass of 86 kg and he stands at point B, which is the free end of the diving board. The second diver has a mass of 82 kg and he stands at point A, which is the center of gravity of the board. Furthermore, point A is equidistant from points O and B.

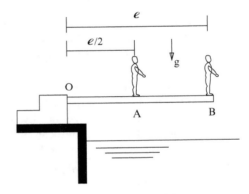

Fig. 3.30 Divers preparing for sequential jump into the pool

Determine the moment generated about point O by the weights of the divers and the board. Calculate the net moment about point O.

Answer: $M_1 = 3,371.2$ N-m; $M_2 = 1,607.2$ N-m; $M_b = 2,548$ N-m; $M_{net} = 7,526.4$ N-m.

Problem 3.3 Consider a person using an exercise apparatus who is holding a handle that is attached to a cable (Fig. 3.31). The cable is wrapped around a pulley and attached to a weight pan. The weight in the weight pan stretches the cable and produces a tensile force $\underline{F}$ in the cable. This force is transmitted to the person's hand through the handle.

Fig. 3.31 Problem 3.3

The force makes an angle θ with the horizontal and applied to the hand at point B. Point A represents the center of gravity of the person's lower arm and O is a point along the center of rotation of the elbow joint. Assume that points O, A, and B and force F all lie on a plane surface.

If the horizontal distance between point O and A is $a = 15$ cm, distance between point O and B is $b = 35$ cm, total weight of the lower arm is $W = 20$ N, magnitude of the applied force is $F = 50$ N, and angle $\theta = 30°$, determine the net moment generated about O by F and W.

Answer: $M_O = 5.75$ N-m   (ccw).

Problem 3.4 Figure 3.32 illustrates a simplified version of a hamstring strength training system for rehabilitation and athlete training protocols. From a seated position, a patient or athlete flexes the lower leg against a set resistance provided through a cylindrical pad that is attached to a load. For the position illustrated, the lower leg makes an angle θ with the horizontal. Point O represents the knee joint, point A is the center of gravity of the lower leg, W is the total weight of the lower leg, F is the magnitude of the force applied by the pad on the lower leg in a direction perpendicular to the

long axis of the lower leg, a is the distance between points O and A, and b is the distance between point O and the line of action of F measured along the long axis of the lower leg.

(a) Determine an expression for the net moment about O due to W and F.
(b) If $a = 20$ cm, $b = 40$ cm, $\theta = 30°$, $W = 60$ N, and $F = 200$ N, calculate the net moment about O.

Answers: (a) $M_O = bF - aW \cos\theta$, (b) $M_O = 69.6$ N-m (ccw).

Problem 3.5 As illustrated in Fig. 3.33, consider an athlete performing flexion/extension exercises of the lower arm to strengthen the biceps muscles. The athlete is holding the weight of $W_1 = 150$ N in his hand, and the weight of his lower arm is $W_2 = 20$ N. As measured from the elbow joint at point O, the center of gravity of the lower arm (point A) is located at a distance $a = 7.5$ cm and the center of gravity of the weight held in the hand is located at a distance $b = 32$ cm.

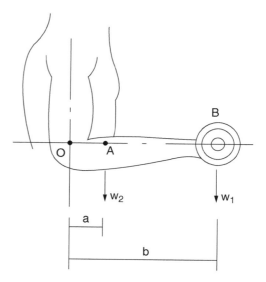

Fig. 3.33 An athlete performing lower arm exercises

Determine the net moment generated about the elbow joint, when the lower arm is extended horizontally and when the long axis of the lower arm makes an angle f = 30° and f = 60°, respectively, with the horizontal.

Answer:
$M_{net}(\theta = 0°) = 49.5$ N-m; $M_{net}(\theta = 30°) = 42.9$ N-m; $M_{net}(\theta = 60°) = 24.8$ N-m.

Problem 3.6 Figure 3.34 illustrates a bench experiment designed to test the strength of materials. In the case illustrated, an intertrochanteric nail that is commonly used to stabilize fractured femoral heads is firmly clamped to the

Fig. 3.32 Problem 3.4

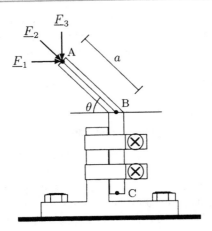

Fig. 3.34 A bench test

bench such that the distal arm (BC) of the nail is aligned vertically. The proximal arm (AB) of the nail has a length a and makes an angle θ with the horizontal.

As illustrated in Fig. 3.34, the intertrochanteric nail is subjected to three experiments by applying forces $\underline{F}_1$ (horizontal, toward the right), $\underline{F}_2$ (aligned with AB, toward A), and $\underline{F}_3$ (vertically downward). Determine expressions for the moment generated at point B by the three forces in terms of force magnitudes and geometric parameters a and θ.

Answers: $M_1 = aF_1 \sin\theta$ (cw), $M_2 = 0$, $M_3 = aF_3 \cos\theta$(ccw).

Problem 3.7 The simple structure shown in Fig. 3.35 is called a *cantilever beam* and is one of the fundamental mechanical elements in engineering. A cantilever beam is fixed at one end and free at the other. In Fig. 3.35, the fixed and free ends of the beam are identified as points A and C, respectively. Point B corresponds to the center of gravity of the beam.

Assume that the beam shown has a weight $W = 100$ N and a length $l = 1$ m. A force with magnitude $F = 150$ N is applied at the free-end of the beam in a direction that makes an angle $\theta = 45°$ with the horizontal.

Determine the magnitude and direction of the net moment developed at the fixed-end of the beam.

Answer: $M_A = 56$ N-m (ccw).

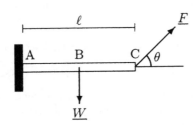

Fig. 3.35 A cantilever beam

Problem 3.8 As illustrated in Fig. 3.36, consider a cantilever beam of 9 kg mass. The beam is fixed to the wall at point A and at point B. The length of the beam is $l = 4$ m. Point C is the center of gravity of the beam. The weight of $W_1 = 50$ N is attached at the free end of the beam such that the distance between point B and the line of gravity of the weight is $l = 35$ cm. A force with magnitude $F = 180$ N is applied at the free end of the beam to keep the beam in place. The line of action of the force makes an angle $\beta = 35°$ with the horizontal.

Determine the magnitude and direction of the net moment generated about point A.

Answer: $M_{net} = 54.1$ N (ccw).

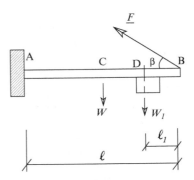

Fig. 3.36 A cantilever beam with weight W$_1$ attached at free end

Problem 3.9 Consider the L-shaped beam illustrated in Fig. 3.37. The beam is mounted to the wall at point A, the arm AB extends in the z direction, and the arm BC extends in the x direction. A force $\underline{F}$ is applied in the z direction at the free-end of the beam.

(a) If the lengths of arms AB and BC are a and b, respectively, and the magnitude of the applied force is F, observe that the position vector of point C relative to point A can be written as $\underline{r} = b\underline{i} + a\underline{k}$ and the force vector can be expressed as $\underline{F} = F\underline{k}$, where $\underline{i}, \underline{j}$, and $\underline{k}$ are unit vectors indicating positive x, y, and z directions, respectively.

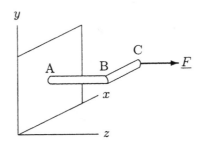

Fig. 3.37 Problem 3.9

(b) Using the cross product of position and force vectors, determine an expression for the moment generated by $\underline{F}$ about point A in terms of a, b, and F.

(c) If $a = b = 30$ cm and $F = 20$ N, calculate the magnitude of the moment about point A due to $\underline{F}$.

Answers: (b) $\underline{M}_A = -bF\underline{j}$, (c) $M_A = 6$ N-m.

Problem 3.10 As shown in Fig. 3.38, consider a worker using a special wrench to tighten bolts. The couple moment generated about the long axis of the wrench is $M_c = 100$ N-m and the distance between the handles of the wrench and the long axis is $r = 25$ cm.

Determine the force exerted by the worker on the handles of the wrench.

Answer: $F = 200$ N.

Fig. 3.38 Worker using a special wrench

Statics: Systems in Equilibrium

4.1 Overview

Statics is an area within the field of applied mechanics, which is concerned with the analysis of rigid bodies in equilibrium. In mechanics, the term *equilibrium* implies that the body of concern is either at rest or moving with a constant velocity. A *rigid body* is one that is assumed to undergo no deformation under the effect of externally applied forces. In reality, there is no rigid material and the concept is an approximation. For some applications, the extent of deformations involved may be so small that the inclusion of the deformation characteristics of the material may not influence the desired analysis. In such cases, the material may be treated as a rigid body.

It should be noted that although the field of statics deals with the analyses of rigid bodies in equilibrium, analyses that deal with the deformation characteristics and strength of materials also start with static analysis. In this chapter, the principles of statics will be introduced and the applications of these principles to relatively simple systems will be provided. In Chap. 5, applications of the same principles to analyze forces involved at and around the major joints of the human body will be demonstrated.

4.2 Newton's Laws of Mechanics

The entire structure of mechanics is based on a few basic laws that were established by Sir Isaac Newton. *Newton's first law*, the law of inertia, states that a body that is originally at rest will remain at rest, or a body moving with a constant velocity in a straight line will maintain its motion unless an external nonzero resultant force acts on the body. Newton's first law must be considered in conjunction with his second law.

Newton's second law, the law of acceleration, states that if the net or the resultant force acting on a body is not zero then the body will accelerate in the direction of the resultant force. Furthermore, the magnitude of the acceleration of the body will be directly proportional to the magnitude of the net force acting on the body and inversely proportional to its mass. Newton's second law can be formulated as:

$$\underline{F} = m\underline{a}. \tag{4.1}$$

This is also known as the *equation of motion*. In Eq. (4.1), $\underline{F}$ is the net or the resultant force (vector sum of all forces) acting on the body, m is the mass of the body, and $\underline{a}$ is its acceleration. Note that both force and acceleration are vector quantities while mass is a scalar quantity. Mass is a quantitative measure of the inertia of a body. *Inertia* is defined as the tendency of a body to maintain its state of rest or uniform motion along a straight line. Inertia can also be defined as the resistance to change in the motion of a body. The more inertia a body has, the more difficult it is to start moving it from rest, to change its motion, or to change its direction of motion.

Equation (4.1) is valid for translational or linear motion analyses such as those along a straight path. As illustrated in Fig. 4.1a, *linear motion* occurs if all parts of a body move the same distance at the same time and in the same direction. A typical example of linear motion is the vertical movement of an elevator in a shaft.

Fig. 4.1 (a) Linear and (b) angular movements

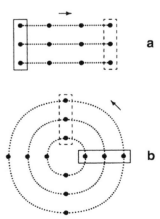

N. Özkaya et al., *Fundamentals of Biomechanics: Equilibrium, Motion, and Deformation*,
DOI 10.1007/978-1-4614-1150-5_4, © Springer Science+Business Media, LLC 2012

Equation (4.1) is only one way of formulating Newton's second law which can alternatively be formulated for rotational or angular motion analysis as:

$$\underline{M} = I\underline{\alpha} \qquad (4.2)$$

In Eq. (4.2), $\underline{M}$ is the net moment or torque acting on the body, I is the *mass moment of inertia* of the body, and α is its *angular acceleration*. As illustrated in Fig. 4.1b, *angular motion* occurs when a body moves in a circular path such that all parts of the body move in the same direction through the same angle at the same time. These concepts are discussed in detail in later chapters that cover dynamic analysis.

Newton's third law, the law of action–reaction, is based on the observation that there are always two sides when it comes to forces. A force applied to an object is always applied by another object. When a worker pushes a cart, a child pulls a wagon, and a hammer hits a nail, force is applied by one body onto another. Newton's third law states that if two bodies are in contact and body 1 is exerting a force on body 2, then body 2 will apply a force on body 1 in such a way that the two forces will have an equal magnitude but in opposite directions and they will act along the same line of action.

For example, consider the case of a hammer pushing a nail (Fig. 4.2). It is the force applied by the hammer on the nail that causes the nail to advance into the wall. However, it is the force of the nail applied back on the hammer that causes the hammer to stop after every impact upon the nail.

Fig. 4.2 Hammering

If you press your hand against the edge of a desk, you can see the shape of your hand change, and also feel the force exerted by the desk on your hand (Fig. 4.3). The harder you press your hand against the desk, the harder the desk will push your hand back.

Fig. 4.3 The harder you push, the harder you will be pushed

Perhaps the best example that can help us understand Newton's third law is a skater applying a force on a wall (Fig. 4.4). By pushing against the wall, the skater can move backward. It is the force exerted by the wall back on the skater that causes the motion of the skater.

Fig. 4.4 An ice skater

Newton's third law can be summarized as "to every action there is an equal and opposite reaction." This law is particularly useful in analyzing complex problems in mechanics and biomechanics in which there are several interacting bodies.

4.3 Conditions for Equilibrium

According to Newton's second law as formulated in Eqs. (4.1) and (4.2), a body undergoing translational or rotational motion will have linear or angular accelerations if the net force or the net moment acting on the body are not zero. If the net force or the net moment are zero, then the acceleration (linear or angular) of the body is zero, and consequently, the velocity (linear or angular) of the body is either constant or zero. When the acceleration is zero, the body is said to be in *equilibrium*. If the velocity is zero as well, then the body is in *static equilibrium* or at *rest*.

Therefore, there are two conditions that need to be satisfied for a body to be in equilibrium. The body is said to be in *translational equilibrium* if the net force (vector sum of all forces) acting on it is zero:

$$\sum \underline{F} = 0. \qquad (4.3)$$

The body is in *rotational equilibrium* if the net moment (vector sum of all moments) acting on it is zero:

$$\sum \underline{M} = 0. \qquad (4.4)$$

Note that there may be a number of forces acting on a body. For example, consider the coplanar force system in Fig. 4.5. Assume that the three forces are acting on an object in the *xy*-plane. These forces can be expressed in terms of their components along the x and y directions:

$$\begin{aligned} \underline{F}_1 &= F_{1x}\underline{i} + F_{1y}\underline{j} \\ \underline{F}_2 &= -F_{2x}\underline{i} + F_{2y}\underline{j} \\ \underline{F}_3 &= -F_{3x}\underline{i} + F_{3y}\underline{j} \end{aligned} \qquad (4.5)$$

Fig. 4.5 A coplanar force system in the xy-plane

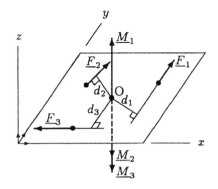

Fig. 4.6 Moments due to coplanar forces

In Eq. (4.5), F_{1x}, F_{2x}, and F_{3x} are the scalar components of $\underline{F}_1$, $\underline{F}_2$, and $\underline{F}_3$ in the x direction, F_{1y}, F_{2y}, and F_{3y} are their scalar components in the y direction, and $\underline{i}$ and $\underline{j}$ are the unit vectors indicating positive x and y directions. These equations can be substituted into Eq. (4.3), and the x and y components of all forces can be grouped together to write:

$$\sum \underline{F} = (F_{1x} - F_{2x} - F_{3x})\underline{i} + (F_{1y} + F_{2y} - F_{3y})\underline{j} = 0.$$

For this equilibrium to hold, each group must individually be equal to zero. That is,

$$\begin{aligned} F_{1x} - F_{2x} - F_{3x} &= 0 \\ F_{1y} + F_{2y} - F_{3y} &= 0 \end{aligned} \tag{4.6}$$

In other words, for a body to be in translational equilibrium, the net force acting in the x and y directions must be equal to zero. For three-dimensional force systems, the net force in the z direction must also be equal to zero. These results may be summarized as follows:

$$\begin{aligned} \sum F_x &= 0 \\ \sum F_y &= 0 \\ \sum F_z &= 0 \end{aligned} \tag{4.7}$$

Caution. The equilibrium conditions given in Eq. (4.7) are in scalar form and must be handled properly. For example, while applying the translational equilibrium condition in the x direction, the forces acting in the positive x direction must be added together and the forces acting in the negative x direction must be subtracted.

As stated earlier, for a body to be in equilibrium, it has to be both in translational and rotational equilibrium. For a body to be in rotational equilibrium, the net moment about any point must be zero. Consider the coplanar system of forces illustrated in Fig. 4.6. Assume that these forces are acting on an object in the xy-plane. Let d_1, d_2, and d_3 be the moment arms of forces F_1, F_2, and F_3 relative to point O. Therefore, the moments due to these forces about point O are

$$\begin{aligned} \underline{M}_1 &= d_1 F_1 \underline{k} \\ \underline{M}_2 &= -d_2 F_2 \underline{k} \\ \underline{M}_3 &= -d_3 F_3 \underline{k} \end{aligned} \tag{4.8}$$

In (4.8), $\underline{k}$ is the unit vector indicating the positive z direction. Moment $\underline{M}_1$ is acting in the positive z direction while $\underline{M}_2$ and $\underline{M}_3$ are acting in the negative z direction. Equation (4.8) can be substituted into the rotational equilibrium condition in Eq. (4.4) to obtain:

$$\sum \underline{M} = \underline{M}_1 + \underline{M}_2 + \underline{M}_3 = (d_1 F_1 - d_2 F_2 - d_3 F_3)\underline{k} = 0.$$

For this equilibrium to hold,

$$d_1 F_1 - d_2 F_2 - d_3 F_3 = 0. \tag{4.9}$$

That is, in this case, the resultant moment in the z direction must be zero. Instead of a coplanar system, if we had a three-dimensional force system, then we could have moments in the x and y directions as well. The rotational equilibrium condition would also require us to have zero net moment in the x and y directions. For three-dimensional systems,

$$\begin{aligned} \sum M_x &= 0 \\ \sum M_y &= 0 \\ \sum M_z &= 0 \end{aligned} \tag{4.10}$$

Caution. The rotational equilibrium conditions given in Eq. (4.10) are also in scalar form and must be handled properly. For example, while applying the rotational equilibrium condition in the z direction, moments acting in the positive z direction must be added together and the moments acting in the negative z direction must be subtracted.

Note that for the coplanar force system shown in Fig. 4.6, the moments can alternatively be expressed as:

$$\begin{aligned} M_1 &= d_1 F_1 \quad &\text{(ccw)} \\ M_2 &= d_2 F_2 \quad &\text{(cw)} \\ M_3 &= d_3 F_3 \quad &\text{(cw)} \end{aligned} \tag{4.11}$$

In other words, with respect to the xy-plane, moments are either clockwise or counterclockwise. Depending on the choice of the positive direction as either clockwise or counterclockwise, some of the moments are positive and others are negative. If we consider the counterclockwise moments to be positive, then the rotational equilibrium about point O would again yield Eq. (4.9):

$$d_1F_1 - d_2F_2 - d_3F_3 = 0.$$

4.4 Free-Body Diagrams

Free-body diagrams are constructed to help identify the forces and moments acting on individual parts of a system and to ensure the correct use of the equations of statics. For this purpose, the parts constituting a system are isolated from their surroundings, and the effects of the surroundings are replaced by proper forces and moments. In a free-body diagram, all known and unknown forces and moments are shown. A force is unknown if its magnitude or direction is not known. For the known forces, we indicate the correct directions. If the direction of a force is not known, then we try to predict it. Our prediction may be correct or not. However, the correct direction of the force will be an outcome of the static analysis.

For example, consider a person trying to push a file cabinet to the right on a horizontal surface (Fig. 4.7). There are three parts constituting this system: the person, the file cabinet, and the horizontal surface representing the floor. The free-body diagrams of these parts are shown in Fig. 4.8. F is the magnitude of the horizontal force applied by the person on the file cabinet to move the file cabinet to

Fig. 4.7 A person pushing a file cabinet

Fig. 4.8 Free-body diagrams

the right. Since forces of action and reaction must have equal magnitudes, F is also the magnitude of the force applied by the file cabinet on the person. Since forces of action and reaction must have opposite directions and the same line of action, the force applied by the file cabinet on the person tends to push the person to the left. W_1 and W_2 are the weights of the file cabinet and the person, respectively, and they are always directed vertically downward. N_1 and N_2 are the magnitudes of the normal forces on the horizontal surface applied by the file cabinet and the person, respectively. They are also the magnitudes of the forces applied by the horizontal surface on the file cabinet and the person. f_1 and f_2 are the magnitudes of the frictional forces f_1 and f_2 between the file cabinet, the person, and the horizontal surface, both acting in the horizontal direction parallel to the surfaces in contact. Since the file cabinet tends to move toward the right, f_1 on the file cabinet acts toward the left. Since the person tends to move toward the left, f_2 on the person acts toward the right and it is the driving force for the person to push the file cabinet.

For many applications, it is sufficient to consider the free-body diagrams of only a few of the individual parts forming a system. For the example illustrated in Figs. 4.7 and 4.8, the free-body diagram of the file cabinet provides sufficient detail to proceed with further analysis.

4.5 Procedure to Analyze Systems in Equilibrium

The general procedure for analyzing the forces and moments acting on rigid bodies in equilibrium is outlined below.

- Draw a simple, neat diagram of the system to be analyzed.
- Draw the free-body diagrams of the parts constituting the system. Show all of the known and unknown forces and moments on the free-body diagrams. Indicate the correct directions of the known forces and moments. If the direction of a force or a moment is not known, try to predict the correct direction, indicate it on the free-body diagram, and do not change it in the middle of the analysis. The correct direction for an unknown force or moment will appear in the solutions. For example, a positive numerical value for an unknown force in the solution will indicate that the correct direction was assumed. A negative numerical value, on the other hand, will indicate that the force has a direction opposite to that assumed. In this case, the force magnitude must be made positive, the proper unit of force must be indicated, and the correct direction must be identified in the solution.
- Adopt a proper coordinate system. If the rectangular coordinate system (x, y, z) will be used, one can orient the coordinate axes differently for different free-body

diagrams within a single system. A good orientation of the coordinate axes may help simplify the analyses. Resolve all forces and moments into their components, and express the forces and moments in terms of their components.

- For each free-body diagram, apply the translational and rotational equilibrium conditions. For three-dimensional problems, the number of equations available are six (three translational and three rotational):

$$\sum F_x = 0, \qquad \sum M_x = 0$$
$$\sum F_y = 0, \qquad \sum M_y = 0$$
$$\sum F_z = 0, \qquad \sum M_z = 0$$

For two-dimensional force systems in the xy-plane, the number of equations available is three (two translational and one rotational, and the other three are automatically satisfied):

$$\sum F_x = 0$$
$$\sum F_y = 0$$
$$\sum M_z = 0$$

- Solve these equations simultaneously for the unknowns. Include the correct directions and the units of forces and moments in the solution.

4.6 Notes Concerning the Equilibrium Equations

- Remember that the equilibrium equations given above are in "scalar" form and they do not account for the directions of forces and moments involved. Therefore, for example, while applying the translational equilibrium condition in the x direction, let a force be positive if it is acting in the positive x direction and let it be a negative force if it is acting in the negative x direction. Apply this rule in every direction and also for moments.
- For coplanar force systems while applying the rotational equilibrium condition, choose a proper point about which all moments are to be calculated. The choice of this point is arbitrary, but a good choice can help simplify the calculations.
- For some problems, you may not need to use all of the equilibrium conditions available. For example, if there are no forces in the x direction, then the translational equilibrium condition in the x direction is satisfied automatically.
- Sometimes, it may be more convenient to apply the rotational equilibrium condition more than once. For example, for a coplanar force system in the xy-plane,

the rotational equilibrium condition in the z direction may be applied twice by considering the moments of forces about two different points. In such cases, the third independent equation that might be used is the translational equilibrium condition either in the x or y direction. The rotational equilibrium condition may also be applied three times in a single problem. However, the three points about which moments are calculated must not all lie along a straight line.

- *Two-force systems*: If there are only two forces acting on a body, then the body will be in equilibrium if and only if the forces are collinear, have an equal magnitude, and act in opposite directions (Fig. 4.9).

Fig. 4.9 A collinear, two-force system

- *Three-force systems*: If there are only three forces acting on a body and if the body is in equilibrium, then the forces must form either a parallel or a concurrent system of forces. Furthermore, for either case, the forces form a coplanar force system. A parallel three-force system is illustrated in Fig. 4.10. If two of the three forces acting on a body are known to be parallel to one another and the body is in equilibrium, then we can readily assume that the third force is parallel to the other two and establish the line of action of the unknown force.

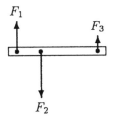

Fig. 4.10 A parallel, three-force system

A concurrent three-force system is illustrated in Fig. 4.11. In this case, the lines of action of the forces have a common point of intersection. If the lines of action of two of three forces acting on a body are known and if they are not parallel and the body is in equilibrium, then the line of action of the third force can be determined by extending the lines of action of the first two forces until they meet and by drawing a straight line that connects the point of intersection and the point of application of the unknown force. The straight line thus obtained will represent the line of action of the unknown force. To analyze a concurrent force system, the forces

Fig. 4.11 A concurrent, three-force system

(a)

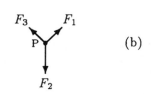

(b)

forming the system can be translated along their lines of action to the point of intersection (Fig. 4.11b). The forces can then be resolved into their components along two perpendicular directions and the translational equilibrium conditions can be employed to determine the unknowns. In such cases, the rotational equilibrium condition is satisfied automatically.

- *Statically determinate and statically indeterminate systems*: For two-dimensional problems in statics, the number of equilibrium equations available is three. Therefore, the maximum number of unknowns (forces and/or moments) that can be determined by applying the equations of equilibrium is limited to three. For two-dimensional problems, if there are three or less unknowns, then the problem is a *statically determinate* one. If the number of unknowns exceeds three, then the problem is *statically indeterminate*. This means that the information we have through statics is not sufficient to solve the problem fully and that we need additional information. Usually, the additional information comes from the material characteristics and properties of the parts constituting the system. The extension of this argument for three-dimensional problems is straightforward. Since there are six equilibrium equations, the maximum number of unknowns that can be determined is limited to six.

4.7   Constraints and Reactions

One way of classifying forces is by saying that they are either active or reactive. Active forces include externally applied loads and gravitational forces. The constraining forces and moments acting on a body are called *reactions*. Usually, reactions are unknowns and must be determined by applying the equations of equilibrium.

Reactive forces and moments are those exerted by the ground, supporting elements such as rollers, wedges, and knife-edges, and connecting members such as cables, pivots, and hinges. Knowing the common characteristics of these supporting elements and connective members, which are summarized in Table 4.1, can help facilitate drawing free-body diagrams and solve particular problems.

4.8   Simply Supported Structures

Mechanical systems are composed of a number of elements connected in various ways. "Beams" constitute the fundamental elements that form the building blocks in mechanics. A beam connected to other structural elements or to the ground by means of rollers, knife-edges, hinges, pin connections, pivots, or cables forms a mechanical system. We shall first analyze relatively simple examples of such cases to demonstrate the use of the equations of equilibrium.

Example 4.1 As illustrated in Fig. 4.12, consider a person standing on a uniform, horizontal beam that is resting on frictionless knife-edge (wedge) and roller supports. Let A and B be two points where the knife-edge and roller supports contact the beam, C be the center of gravity of the beam, and D be a point on the beam directly under the center of gravity of the person. Assume that the length of the beam (the distance between A and B) is $l = 5\,\mathrm{m}$, the distance between points A and D is $d = 3\,\mathrm{m}$, the weight of the beam is $W_1 = 900\,\mathrm{N}$, and the mass of the person is $m = 60\,\mathrm{kg}$.

Calculate the reactions on the beam at points A and B.

Solution
The free-body diagram of the beam is shown in Fig. 4.13. The forces acting on the beam include the weights $\underline{W}_1$ and $\underline{W}_2$ of the beam and the person, respectively. Note that the weight of the beam is assumed to be a concentrated load applied at the center of gravity of the beam which is equidistant from A and B. The magnitude of the weight of the beam is given as $W_1 = 900\,\mathrm{N}$. The mass of the person is given as $m = 60\,\mathrm{kg}$. Therefore, the weight of the person is $W_2 = mg = 60 \times 9.8 = 588\,\mathrm{N}$.

Also acting on the beam are reaction forces $\underline{R}_A$ and $\underline{R}_B$ due to the knife-edge and roller supports, respectively. To a horizontal beam, frictionless knife-edge and roller supports can only apply forces in the vertical direction. Furthermore, the direction of these forces is toward the beam, not away from the beam. Therefore, we know the directions of the reaction forces, but we do not know their magnitudes. Note that the forces acting on the beam form a parallel force system.

There are two unknowns: $\underline{R}_A$ and $\underline{R}_B$. For the solution of the problem, we can either apply the vertical equilibrium condition along with the rotational equilibrium condition

Table 4.1 Commonly encountered supports, connections and connecting elements, and some of their characteristics

TYPE OF SUPPORT OR JOINT	SAMPLE GRAPHICAL REPRESENTATIONS	REPRESENTATIONS FOR USE IN FREE-BODY DIAG.	BIOMECHANICS EXAMPLES	UNKNOWNS
Flexible members (cables, ropes)		T	Muscles, ligaments	Magnitude of the tension in the cable or muscle
Two-force members		F / F	Muscles	Force magnitude
Rollers and other simple supports (no friction)		F	Bone-to-bone contact	Force magnitude
Hinge or pin connection		F_x / F_y	Elbow	Magnitude and direction of force (force components)
Ball-and-socket		F_y / F_x / F_z	Hip	Magnitude and direction of force
Fixed, welded, or built-in		F_y / M_x / F_x / F_z / M_z / M_y	Skull	Magnitude and direction of force and moment

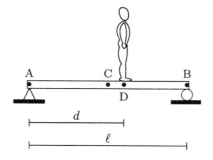

Fig. 4.12 Example 4.1

about point A or B, or consider the rotational equilibrium of the beam about points A and B. Here, we shall use the latter approach, and check the results by considering the equilibrium in the vertical direction. Note that the horizontal equilibrium condition is automatically satisfied because there are no forces acting in that direction.

Consider the rotational equilibrium of the beam about point A. Relative to point A, the length of the moment arm of $\underline{R}_A$ is zero, the moment arm of $\underline{W}_1$ is $l/2$, the moment arm of $\underline{W}_2$ is d, and the moment arm of $\underline{R}_B$ is l. Assuming that clockwise moments are positive, then the moments due to $\underline{W}_1$ and $\underline{W}_2$

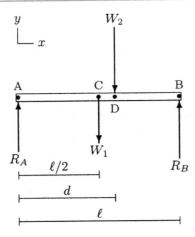

Fig. 4.13 Free-body diagram of the beam

are positive, while the moment due to $\underline{R}_B$ is negative. Therefore, for the rotational equilibrium of the beam about point A:

$$\sum M_A = 0: \quad \frac{l}{2}W_1 + dW_2 - lR_B = 0$$
$$R_B = \frac{1}{l}\left(\frac{l}{2}W_1 + dW_2\right)$$
$$R_B = \frac{1}{5}\left(\frac{5}{2}900 + 3 \times 588\right) = 802.8\,\text{N}(\uparrow)$$

Now, consider the rotational equilibrium of the beam about point B. Relative to point B, the length of the moment arm of $\underline{R}_B$ is zero, the moment arm of $\underline{W}_2$ is $l - d$, the moment arm of $\underline{W}_1$ is $l/2$, and the moment arm of $\underline{R}_A$ is l. Assuming that clockwise moments are positive, then the moments due to $\underline{W}_1$ and $\underline{W}_2$ are negative while the moment due to $\underline{R}_A$ is positive. Therefore, for the rotational equilibrium of the beam about point B:

$$\sum M_B = 0: \quad lR_A - \frac{l}{2}W_1 - (l - d)W_2 = 0$$
$$R_A = \frac{1}{l}\left(\frac{l}{2}W_1 + (l-d)W_2\right)$$
$$R_A = \frac{1}{5}\left(\frac{5}{2}900 + (5-3)588\right) = 685.2\,\text{N}(\uparrow)$$

Remarks
- The arrows in parentheses in the solutions indicate the correct directions of the forces.
- We can check these results by considering the equilibrium of the beam in the vertical direction:

$$\sum F_y = 0: \quad R_A - W_1 - W_2 + R_B \overset{?}{=} 0$$
$$685.2 - 900 - 588 + 802.8 \overset{\checkmark}{=} 0$$

- Because of the type of supports used, the beam analyzed in this example has a limited use. A frictionless knife-edge, point support, fulcrum, or roller can provide a support only in the direction perpendicular (normal) to

the surfaces in contact. Such supports cannot provide reaction forces in the direction tangent to the contact surfaces. Therefore, if there were active forces on the beam applied along the long axis (x) of the beam, these supports could not provide the necessary reaction forces to maintain the horizontal equilibrium of the beam.

Example 4.2 The uniform, horizontal beam shown in Fig. 4.14 is hinged to the ground at point A. A frictionless roller is placed between the beam and the ceiling at point D to constrain the counterclockwise rotation of the beam about the hinge joint. A force that makes an angle $\beta = 60°$ with the horizontal is applied at point B. The magnitude of the applied force is $P = 1,000\,\text{N}$. Point C represents the center of gravity of the beam. The distance between points A and B is $l = 4\,\text{m}$ and the distance between points A and D is $d = 3\,\text{m}$. The beam weighs $W = 800\,\text{N}$.

Calculate the reactions on the beam at points A and D.

Fig. 4.14 Example 4.2

Solution
Figure 4.15 illustrates the free-body diagram of the beam under consideration. The horizontal and vertical directions are identified by the x and y axes, respectively. The hinge joint at point A constrains the translational movement of the beam both in the x and y directions. Therefore, there exists a reaction force $\underline{R}_A$ at point A. We know neither the magnitude

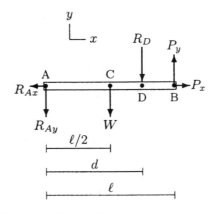

Fig. 4.15 Free-body diagram of the beam

nor the direction of this force (two unknowns). As illustrated in Fig. 4.15, instead of a single resultant force with two unknowns (magnitude and direction), the reaction force at point A can be represented in terms of its components $\underline{R}_{Ax}$ and $\underline{R}_{Ay}$ (still two unknowns).

The frictionless roller at point D functions as a "stop." It prevents the counterclockwise rotation of the beam. Under the effect of the applied force $\underline{P}$, the beam compresses the roller, and as a reaction, the roller applies a force $\underline{R}_D$ back on the beam. This force is applied in a direction perpendicular to the beam, or vertically downward. The magnitude R_D of this force is not known either. The weight of the beam is given as $W = 800\,\text{N}$ and is assumed to be acting at the center of gravity of the beam. P_x and P_y are the scalar components of the applied force along the x and y directions such that:

$$P_x = P \cos \beta = (1,000)(\cos 60) = 500\,\text{N}$$
$$P_y = P \sin \beta = (1,000)(\sin 60) = 866\,\text{N}$$

There are three unknowns $(\underline{R}_{Ax}, \underline{R}_{Ay}, R_D)$ and we need three equations to solve this problem. Consider the equilibrium condition in the x direction:

$$\sum F_x = 0: \quad -R_{Ax} + P_x = 0$$
$$R_{Ax} = P_x = 500\,\text{N}(\leftarrow)$$

Next, consider the rotational equilibrium of the beam about point A. Relative to point A, lengths of moment arms for R_{Ax} and R_{Ay} are zero. The length of the moment arm for P_x is zero as well because its line of action passes through point A. On the other hand, $l/2$ is the length of the moment arm for $\underline{W}$, d is the moment arm for R_D, and l is the moment arm for P_y. Assuming that clockwise moments are positive:

$$\sum M_A = 0: \quad \tfrac{l}{2}W + dR_D - lP_y = 0$$
$$R_D = \tfrac{1}{d}\left(lP_y - \tfrac{l}{2}W\right)$$
$$R_D = \tfrac{1}{3}\left(4 \times 866 - \tfrac{4}{2}800\right) = 621\,\text{N}(\downarrow)$$

Now, consider the translational equilibrium of the beam in the y direction:

$$\sum F_y = 0: \quad -R_{Ay} - W - R_D + P_y = 0$$
$$R_{Ay} = P_y - W - R_D$$
$$R_{Ay} = 866 - 800 - 621 = 555\,\text{N}(\downarrow)$$

R_{Ay} is determined to have a negative value. This is not permitted because R_{Ay} corresponds to a scalar quantity and scalar quantities cannot take negative values. The negative sign implies that while drawing the free-body diagram of the beam, we assumed the wrong direction (downward) for the

vertical component of the reaction force at point A. We can now correct it by writing:

$$R_{Ay} = 555\,\text{N}(\uparrow).$$

The corrected free-body diagram of the beam is shown in Fig. 4.16. Since we already calculated the components of the reaction force at A, we can also determine the magnitude and direction of the resultant reaction force $\underline{R}_A$ at point A. The magnitude of $\underline{R}_A$ is

$$R_A = \sqrt{(R_{Ax})^2 + (R_{Ay})^2} = 747\,\text{N}.$$

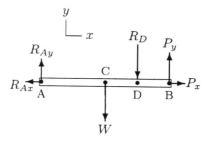

Fig. 4.16 Corrected free-body diagram of the beam

If α is the angle $\underline{R}_A$ makes with the horizontal, then:

$$\alpha = \tan^{-1}\left(\frac{R_{Ay}}{R_{Ax}}\right) = \tan^{-1}\left(\frac{555}{500}\right) = 50°.$$

The modified free-body diagram of the beam showing the force resultants is illustrated in Fig. 4.17.

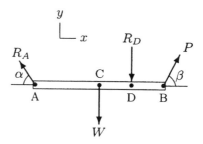

Fig. 4.17 Resultant forces acting on the beam

Example 4.3 The uniform, horizontal beam shown in Fig. 4.18 is hinged to the wall at point A and supported by a cable attached to the beam at point B. At the other end, the cable is attached to the wall such that it makes an angle $\beta = 53°$ with the horizontal. Point C represents the center of gravity of the beam which is equidistant from A and B. A load that weighs $W_2 = 400\,\text{N}$ is placed on the beam such that its center of gravity is directly above point C. If the length of the beam is $l = 4\,\text{m}$ and the weight of the beam is

Fig. 4.18 Example 4.3

$W_1 = 600\,\text{N}$, calculate the tension T in the cable and the reaction force on the beam at point A.

Solution

Figure 4.19 illustrates the free-body diagram of the beam under consideration. The horizontal and vertical directions are identified by the x and y axes, respectively. The hinge joint at point A constrains the translational movement of the beam both in the x and y directions. In Fig. 4.19b, the reaction force $\underline{R}_A$ at A is represented in terms of its components R_{Ax} and R_{Ay} (two unknowns). Because of the weight of the beam and the load, the cable is stretched, or a tensile force $\underline{T}$ is developed in the cable. This force is applied by the cable on the beam in a direction along the length of the cable that makes an angle β with the horizontal. In other words, we know the direction of $\underline{T}$, but we do not know its

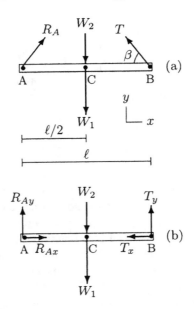

Fig. 4.19 Forces acting on the beam

magnitude (another unknown). In Fig. 4.19b, the force applied by the cable on the beam is expressed in terms of its scalar components T_x and T_y.

We have three unknowns: R_{Ax}, R_{Ay}, and T. First, consider the rotational equilibrium of the beam about point A. Relative to A, there are three moment producing forces: W_1, W_2, and T_y with moment arms $l/2$, $l/2$, and l, respectively. Assuming that clockwise moments are positive:

$$\sum M_A = 0: \quad \tfrac{l}{2}W_1 + \tfrac{l}{2}W_2 - lT_y = 0$$
$$T_y = \tfrac{1}{2}(W_1 + W_2)$$
$$T_y = \tfrac{1}{2}(600 + 400) = 500(\uparrow)$$

T_y is the vertical component of $\underline{T}$ that makes an angle $\beta = 53°$ with the horizontal. Therefore,

$$T = \frac{T_y}{\sin \beta} = \frac{500}{\sin 53°} = 626.1\,\text{N}.$$

The x component of the tensile force applied by the cable on the beam is

$$T_x = T \cos \beta = 626.1 \cos 53° = 376.8\,\text{N}(\leftarrow).$$

Next, consider the translational equilibrium of the beam in the x direction:

$$\sum F_x = 0: \quad R_{Ax} - T_x = 0$$
$$R_{Ax} = T_x = 376.8\,\text{N}(\rightarrow)$$

Finally, consider the translational equilibrium of the beam in the y direction:

$$\sum F_y = 0: \quad R_{Ay} - W_1 - W_2 + T_y = 0$$
$$R_{Ay} = W_1 + W_2 - T_y$$
$$R_{Ay} = 600 + 400 - 500 = 500\,\text{N}(\uparrow)$$

Now that we determined the scalar components of the reaction force at point A, we can also calculate its magnitude:

$$R_A = \sqrt{(R_{Ax})^2 + (R_{Ay})^2} = 626.1\,\text{N}.$$

If α is the angle $\underline{R}_A$ makes with the horizontal, then:

$$\alpha = \tan^{-1}\left(\frac{R_{Ay}}{R_{Ax}}\right) = \tan^{-1}\left(\frac{500}{376.8}\right) = 53°.$$

Remarks

- The weights of the beam and load on the beam have a common line of action. As illustrated in Fig. 4.20a, their effects can be combined and expressed by a single weight

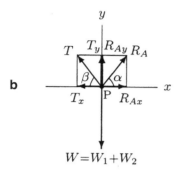

a

b

Fig. 4.20 A concurrent force system

$W = W_1 + W_2 = 1{,}000\,\mathrm{N}$ that acts at the center of gravity of the beam. In addition to $\underline{W}$, we also have $\underline{R}_A$ and $\underline{T}$ applied on the beam. In other words, we have a three-force system. Since these forces do not form a parallel force system, they have to be concurrent. Therefore, the lines of action of the forces must meet at a single point, say P. As illustrated in Fig. 4.20b, if we slide the forces to point P and express them in terms of their components, we can observe equilibrium in the x and y directions such that

In the x direction: $R_{Ax} = T_x$
In the y direction: $R_{Ay} + T_y = W$

- The fact that $\underline{R}_A$ and $\underline{T}$ have an equal magnitude and they both make a 53° angle with the horizontal is due to the symmetry of the problem with respect to a plane perpendicular to the xy-plane that passes through the center of gravity of the beam.

4.9   Cable-Pulley Systems and Traction Devices

Cable-pulley arrangements are commonly used to elevate weights and have applications in the design of traction devices used in patient rehabilitation. For example, consider

the simple arrangement in Fig. 4.21 where a person is trying to lift a load through the use of a cable-pulley system. Assume that the person lifted the load from the floor and is holding it in equilibrium. The cable is wrapped around the pulley which is housed in a case that is attached to the ceiling. Figure 4.22 shows free-body diagrams of the pulley and the load. r is the radius of the pulley and O represents a point along the centerline (axle or shaft) of the pulley. When the person pulls the cable to lift the load, a force is applied on the pulley that is transmitted to the ceiling via the case housing the pulley. As a reaction, the ceiling applies a force back on the pulley through the shaft that connects the pulley and the case housing the pulley. In other words, there exists a reaction force $\underline{R}$ on the pulley. In Fig. 4.22, the reaction force at point O is represented by its scalar components R_x and R_y.

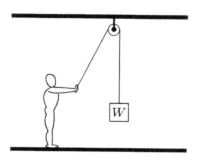

Fig. 4.21 A cable-pulley arrangement

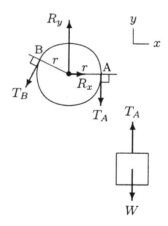

Fig. 4.22 Pulley and load

The cable is wrapped around the pulley between points A and B. If we ignore the frictional effects between the cable and pulley, then the magnitude T of the tension in the cable is constant everywhere in the cable. To prove this point, let us assume that the magnitude of the tensile force generated in the cable is not constant. Let T_A and T_B be the magnitudes of the tensile forces applied by the cable on the pulley at points A and B, respectively. Regardless of the way the cable is wrapped around the pulley, the cable is tangent to the circumference of the pulley at the initial and final contact points. This implies that straight lines drawn from point O toward points A and B will cut the cable at right angles. Therefore, as measured from point O, the moment arms of the forces applied by the cable on the pulley are always equal to the radius of the pulley. Now, consider the rotational equilibrium of the pulley about point O:

$$\sum M_O = 0 : \quad rT_B - rT_A = 0$$
$$T_B = T_A$$

Therefore, the tension T in the cable is the same on the two sides of the pulley. For the case illustrated in Figs. 4.21 and 4.22, the vertical equilibrium of the load requires that the tension in the cable is equal to the weight W of the load to be lifted.

Note that R_x and R_y applied by the ceiling on the pulley do not produce any moment about O. If needed, these forces could be determined by considering the horizontal and vertical equilibrium of the pulley.

Figures 4.23 and 4.24 illustrate examples of simple traction devices. Such devices are designed to maintain parts of the human body in particular positions for healing purposes. For such devices to be effective, they must be designed to transmit forces properly to the body part in terms of force direction and magnitude. Different arrangements of cables and pulleys

can transmit different magnitudes of forces and in different directions. For example, the traction in Fig. 4.23 applies a horizontal force to the leg with magnitude equal to the weight in the weight pan. On the other hand, the traction in Fig. 4.24 applies a horizontal force to the leg with magnitude twice as great as the weight in the weight pan.

Example 4.4 Using three different cable-pulley arrangements shown in Fig. 4.25, a block of weight W is elevated to a certain height. For each system, determine how much force is applied to the person holding the cable.

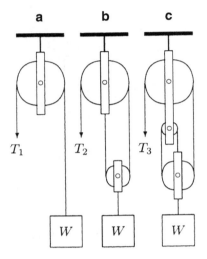

Fig. 4.25 Example 4.4

Solutions The necessary free-body diagrams to analyze each system in Fig. 4.25 are shown in Fig. 4.26. For the analysis of the system in Fig. 4.25a, all we need is the free-body diagram of the block (Fig. 4.26a). For the vertical equilibrium of the block, tension in the cable must be $T_1 = W$. Therefore, a force with magnitude equal to the weight of the block is applied to the person holding the cable.

Fig. 4.23 One-pulley traction

Fig. 4.24 Three-pulley traction

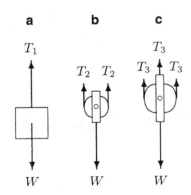

Fig. 4.26 Free-body diagrams

For the analysis of the system in Fig. 4.25b, we need to examine the free-body diagram of the pulley closest to the block (Fig. 4.26b). For the vertical equilibrium of the pulley, the tension in the cable must be $T_2 = W/2$. T_2 is the tension in the cable that is wrapped around the two pulleys and is held by the person. Therefore, a force with magnitude equal to half of the weight of the block is applied to the person holding the cable.

For the analysis of the system in Fig. 4.25c, we again need to examine the free-body diagram of the pulley closest to the block (Fig. 4.26c). For the vertical equilibrium of the pulley, the tension in the cable must be $T_3 = W/3$. T_3 is the tension in the cable that is wrapped around the three pulleys and is held by the person. Therefore, a force with a magnitude equal to one-third of the weight of the block is applied to the person holding the cable.

4.10 Built-in Structures

The beam shown in Fig. 4.27 is built-in or welded to a wall and is called a *cantilever beam*. Cantilever beams can withstand externally applied moments as well as forces. The beam in Fig. 4.27 is welded to the wall at point A and a downward force with magnitude P is applied at the free end at point B. The length of the beam is l. For the sake of simplicity, ignore the weight of the beam. The free-body diagram of the beam is shown in Fig. 4.28. Instead of welding, if the beam were hinged to the wall at point A, then under the effect of the applied force, the beam would undergo a clockwise rotation. The fact that the beam is not rotating indicates that there is rotational equilibrium. This rotational equilibrium is due to a reactive moment generated at the welded end of the beam. The counterclockwise moment with magnitude M in Fig. 4.28 represents this reactive moment at point A. Furthermore, the applied force tends to translate the beam downward. However,

the beam is not translating. This indicates the presence of an upward reaction force at point A that balances the effect of the applied force. Force R_A represents the reactive force at point A. Since the beam is in equilibrium, we can utilize the conditions of equilibrium to determine the reactions. For example, consider the translational equilibrium of the beam in the y direction:

$$\sum F_y = 0: \quad R_A - P = 0$$
$$R_A = P$$

Now, consider the rotational equilibrium of the beam about point A and assume that counterclockwise moments are positive:

$$\sum M_A = 0: \quad M - lP = 0$$
$$M = lP$$

Note that the reactive moment with magnitude M is a "free vector" that acts everywhere along the beam.

Example 4.5 Consider the uniform, horizontal beam shown in Fig. 4.29. The beam is fixed at point A and a force that makes an angle $\beta = 60°$ with the horizontal is applied at point B. The magnitude of the applied force is $P = 100$ N. Point C is the center of gravity of the beam. The beam weighs $W = 50$ N and has a length $l = 2$ m.

Determine the reactions generated at the fixed end of the beam.

Fig. 4.29 Example 4.5

Solution

The free-body diagram of the beam is shown in Fig. 4.30. The horizontal and vertical directions are indicated by the x and y axes, respectively. P_x and P_y are the scalar components of the applied force $\underline{P}$. Since we know the magnitude and direction of $\underline{P}$, we can readily calculate P_x and P_y:

$$P_x = P \cos \beta = 100 \cos 60° = 50.0 \ N \ (+x)$$
$$P_y = P \sin \beta = 100 \sin 60° = 86.6 \ N \ (+y)$$

In Fig. 4.30, the reactive force $\underline{R}_A$ at point A is represented in terms of its scalar components R_{Ax} and R_{Ay}. We know neither the magnitude nor the direction of $\underline{R}_A$ (two unknowns). We also have a reactive moment at point A with magnitude M acting in a direction perpendicular to the xy-plane.

In this problem, we have three unknowns: R_{Ax}, R_{Ay}, and M. To solve the problem, we need to apply three

Fig. 4.27 Cantilever beam

Fig. 4.28 Free-body diagram of the cantilever beam

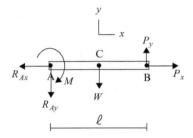

Fig. 4.30 Free-body diagram of the beam

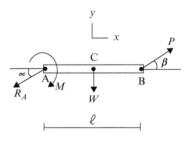

Fig. 4.31 Forces and moment acting on the beam

equilibrium conditions. First, consider the translational equilibrium of the beam in the x direction:

$$\sum F_x = 0: \quad -R_{Ax} + P_x = 0$$
$$R_{Ax} = P_x = 50 \, \text{N} \, (-x)$$

Next, consider the translational equilibrium of the beam in the y direction:

$$\sum F_y = 0: \quad -R_{Ay} - W + P_y = 0$$
$$R_{Ay} = P_y - W$$
$$R_{Ay} = 86.6 - 50 = 36.6 \, \text{N} \, (-y)$$

Once the magnitude and the direction of the components of the reaction force $\underline{R}_A$ are determined, we can then determine the magnitude and the direction of the reaction force at point A:

$$R_A^2 = R_{Ax}^2 + R_{Ay}^2$$
$$R_A = \sqrt{R_{Ax}^2 + R_{Ay}^2}$$
$$R_A = \sqrt{50^2 + 36.6^2} = 62 \, \text{N}$$

Finally, consider the rotational equilibrium of the beam about point A. Assuming that counterclockwise moments are positive:

$$\sum M_A = 0: \quad M + W \times \frac{l}{2} - P_y \times l = 0$$
$$M = P_y \times l - W \times \frac{l}{2}$$
$$M = 86.6 \times 2 - 50 \times \frac{2}{2} = 123.2 \, \text{N-m} \, (\text{cw})$$

As we determined the magnitude of the reaction force and the magnitude and the direction of the reactive moment at point A, we can also determine the direction of the reaction force R_A. If α is an angle that the line of action of the reaction force R_A makes with the horizontal, then:

$$\tan \alpha = \frac{R_{Ay}}{R_{Ax}}$$
$$\alpha = arc \, \tan \left(\frac{R_{Ay}}{R_{Ax}} \right) = 36.2°$$

The final free-body diagram of the beam is shown in Fig. 4.31.

Example 4.6 Consider the L-shaped beam illustrated in Fig. 4.32. The beam is welded to the wall at point A, the arm AB extends in the positive z direction, and the arm BC extends in the negative x direction. A force $\underline{P}$ is applied in the negative y direction at point B. The lengths of arms AB and BC are $a = 20 \, \text{cm}$ and $b = 30 \, \text{cm}$, respectively, and the magnitude of the applied force is $P = 120 \, \text{N}$.

Assuming that the weight of the beam is negligibly small as compared to the magnitude of the applied force, determine the reactions at the fixed end of the beam.

This is a three-dimensional problem and we shall employ two methods to analyze it. The first method utilizes the concepts of couple and couple-moment introduced in the previous chapter, and the second method utilizes the vector properties of forces and moments.

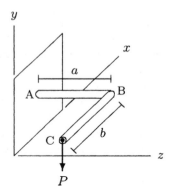

Fig. 4.32 Example 4.6

Solution A: Scalar method
The scalar method of analyzing the problem is described in Fig. 4.33 and it involves the translation of the force applied on the beam from point C to point A. First, to translate $\underline{P}$ from point C to point B, place a pair of forces at point B that are equal in magnitude (P) and acting in opposite directions such that the common line of action of the new forces is parallel to the line of action of the original force at point C (Fig. 4.33a). The downward force at point C and the upward force at point B form a couple. Therefore, they can be replaced by a couple-moment illustrated by a double-headed arrow in Fig. 4.33b. The magnitude of the

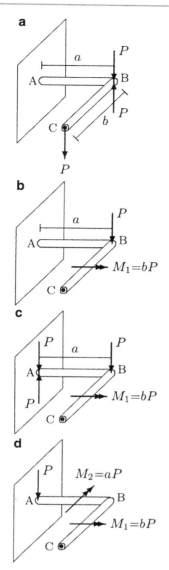

Fig. 4.33 Scalar method

couple-moment is $M_1 = bP$. Applying the right-hand rule, we can see that the couple-moment acts in the positive z direction. Therefore,

$$M_1 = bP = (0.30)(120) = 36 \text{ N-m}(+z).$$

As illustrated in Fig. 4.33c, to translate the force from point B to point A, place another pair of forces at point A with equal magnitude and opposite directions. This time, the downward force at point B and the upward force at point A form a couple, and again, they can be replaced by a couple-moment (Fig. 4.33d). The magnitude of this couple-moment is $M_2 = aP$ and it acts in the positive x direction. Therefore,

$$M_2 = aP = (0.20)(120) = 24 \text{ N-m}(+x).$$

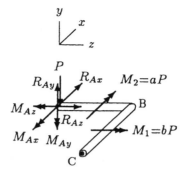

Fig. 4.34 Free-body diagram of the beam

Figure 4.34 shows the free-body diagram of the beam. P is the magnitude of the force applied at point C which is translated to point A, and M_1 and M_2 are the magnitudes of the couple-moments. R_{Ax}, R_{Ay}, and R_{Az} are the scalar components of the reactive force at point A, and M_{Ax}, M_{Ay}, and M_{Az} are the scalar components of the reactive moment at point A. Consider the translational equilibrium of the beam in the x direction:

$$\sum F_x = 0: \qquad R_{Ax} = 0.$$

The translational equilibrium of the beam in the y direction requires that

$$\sum F_y = 0: \qquad R_{Ay} = P = 120 \text{ N}(+y).$$

For the translational equilibrium of the beam in the z direction,

$$\sum F_z = 0: \qquad R_{Az} = 0.$$

Therefore, there is only one nonzero component of the reactive force on the beam at point A and it acts in the positive y direction. Now, consider the rotational equilibrium of the beam in the x direction:

$$\sum M_x = 0: \qquad M_{Ax} = M_2 = 24 \text{ N-m}(-x).$$

For the rotational equilibrium of the beam in the y direction,

$$\sum M_y = 0: \qquad M_{Ay} = 0.$$

Finally, the rotational equilibrium of the beam in the z direction requires that

$$\sum M_z = 0: \qquad M_{Az} = M_1 = 36 \text{ N-m}(-z).$$

Therefore, the reactive moment at point A has two non-zero components in the x and y directions. Now that we determined the components of the reactive force and moment at point A, we can also calculate the magnitudes of the resultant force and moment at point A:

$$R_A = \sqrt{R_{Ax^2} + R_{Ay^2} + R_{Az^2}} = R_{Ay} = 120\,\text{N}$$
$$M_A = \sqrt{M_{Ax^2} + M_{Ay^2} + M_{Az^2}} = 43.3\,\text{N-M}$$

Solution B: Vector method
The second method of analyzing the same problem utilizes the vector properties of the parameters involved. For example, the force applied at point C and the position vector of point C relative to point A can be expressed as (Fig. 4.35):

$$\underline{P} = -P\underline{j} = -120\underline{j}$$
$$\underline{r} = -b\underline{i} + a\underline{k} = -0.30\underline{i} + 0.20\underline{k}$$

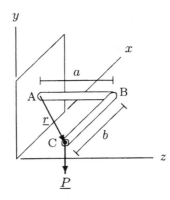

Fig. 4.35 Vector method

Here, $\underline{i}, \underline{j}$, and $\underline{k}$ are unit vectors indicating positive x, y, and z directions, respectively. The free-body diagram of the beam is shown in Fig. 4.36, where the reactive forces and moments are represented by their scalar components such that

$$\underline{R}_A = R_{Ax}\underline{i} + R_{Ay}\underline{j} + R_{Az}\underline{k}$$
$$\underline{M}_A = M_{Ax}\underline{i} + M_{Ay}\underline{j} + M_{Az}\underline{k}$$

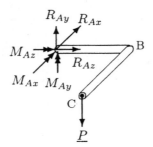

Fig. 4.36 Free-body diagram of the beam

First, consider the translational equilibrium of the beam:

$$\sum \underline{F} = 0: \quad \underline{R}_A + \underline{P} = 0$$
$$(R_{Ax}\underline{i} + R_{Ay}\underline{j} + R_{Az}\underline{k}) + (-120\underline{j}) = 0$$
$$R_{Ax}\underline{i} + (R_{Ay} - 120)\underline{j} + R_{Az}\underline{k} = 0$$

For this equilibrium to hold,

$$R_{Ax} = 0$$
$$R_{Ay} = 120\,\text{N}(+y)$$
$$R_{Az} = 0$$

As discussed in the previous chapter, by definition, moment is the cross (vector) product of the position and force vectors. Therefore, the moment $\underline{M}_C$ relative to point A due to force $\underline{P}$ applied at point C is

$$\underline{M}_C = \underline{r} \times \underline{P}$$
$$= (-0.30\underline{i} + 0.20\underline{k}) \times (-120\underline{j})$$
$$= (-0.30)(-120)(\underline{i} \times \underline{j}) + (0.20)(-120)(\underline{k} \times \underline{j})$$
$$= 36\underline{k} + 24\underline{i}$$

Now consider the rotational equilibrium of the beam about point A

$$\sum \underline{M} = 0: \quad = \underline{M}_A + \underline{M}_C = 0$$
$$= (M_{Ax}\underline{i} + M_{Ay}\underline{j} + M_{Az}\underline{k}) + (36\underline{k} + 24\underline{i}) = 0$$
$$= (M_{Ax} + 24)\underline{i} + M_{Ay}\underline{j} + (M_{Az} + 36)\underline{k} = 0$$

For this equilibrium to hold,

$$M_{Ax} = 24\,\text{N-m}(-x)$$
$$M_{Ay} = 0$$
$$M_{Az} = 36\,\text{N-m}(-z)$$

All of these results are consistent with those obtained using the scalar method of analysis.

Remarks
- We analyzed this problem in two ways. It is clear that the scalar method of analysis requires less rigorous mathematical manipulations. However, it has its limitations. For instance, in this example, the force applied at point C had only one nonzero component. What would happen if we had a force at point C with nonzero components in the x and z directions as well as the y direction? The scalar method could still be applied in a step-by-step manner by considering one component of the applied force at a time, solving the problem for that component alone, repeating this for all three components, and then superimposing the three solutions to obtain the final solution. Obviously, this would be quite time consuming. On the other hand, the

extension of the vector method to analyze such problems is very straightforward. All one needs to do is to redefine the applied force vector as $P = P_x\underline{i} + P_y\underline{j} + P_z\underline{k}$, and carry out exactly the same procedure outlined above.

- In this example, it is stated that the weight of the beam was negligibly small as compared to the force applied at point C. What would happen if the weight of the beam was not negligible? As illustrated in Fig. 4.37, assume that we know the weights W_1 and W_2 of the arms AB and BC of the beam along with their centers of gravity (points D and E). Also, assume that D is located in the middle of arm AB and E is located in the middle of arm BC. We have already discussed the vector representation of $\underline{R}_A$ and $\underline{P}$. The vector representations of $\underline{W}_1$ and $\underline{W}_2$ are

$$\underline{W}_1 = -W_1\underline{j}$$
$$\underline{W}_2 = -W_2\underline{j}$$

The position vectors of points D and E relative to point A are

$$\underline{r}_1 = \frac{a}{2}\underline{k}$$
$$\underline{r}_2 = -\frac{b}{2}\underline{i} + a\underline{k}$$

Therefore, the moments due to $\underline{W}_1$ and $\underline{W}_2$ relative to point A are

$$\underline{M}_D = \underline{r}_1 \times \underline{W}_1 = \frac{aW_1}{2}\underline{i}$$
$$\underline{M}_E = \underline{r}_2 \times \underline{W}_2 = aW_2\underline{i} - \frac{bW_2}{2}\underline{k}$$

The translational equilibrium of the beam would yield

$$R_{Ax} = 0$$
$$R_{Ay} = W_1 + W_2 + P$$
$$R_{Az} = 0$$

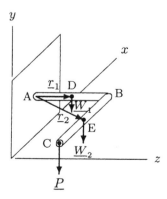

Fig. 4.37 Including the weights of the arms

The rotational equilibrium of the beam would yield

$$M_{Ax} = -\frac{aW_1}{2} - aW_2 + aP$$
$$M_{Ay} = 0$$
$$M_{Az} = \frac{bW_2}{2} - bP$$

- Note the similarities between this and the shoulder example in the previous chapter (Example 3.5).

4.11 Systems Involving Friction

Frictional forces were discussed in detail in Chap. 2. Here, we shall analyze a problem in which frictional forces play an important role.

Example 4.7 Figure 4.38 illustrates a person trying to push a block up an inclined surface by applying a force parallel to the incline. The weight of the block is W, the coefficient of maximum friction between the block and the incline is μ, and the incline makes an angle θ with the horizontal.

Determine the magnitude P of the minimum force the person must apply in order to overcome the frictional and gravitational effects to start moving the block in terms of W, μ, and θ.

Fig. 4.38 Example 4.7

Solution

Note that if the person pushes the block by applying a force closer to the top of the block, the block may tilt (rotate in the clockwise direction) about its bottom right edge. Here, we shall assume that there is no such effect and that the bottom surface of the block remains in full contact with the ground.

The free-body diagram of the block is shown in Fig. 4.39. x and y correspond to the directions parallel and perpendicular to the incline, respectively. P is the magnitude of the force applied by the person on the block in the x direction, f is the frictional force applied by the inclined surface on the block in the negative x direction, N is the normal force applied by the inclined surface on the block in the positive y direction, and W is the weight of the block acting vertically downward. The weight of the block has components in the x and y

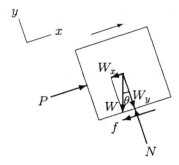

Fig. 4.39 Free-body diagram of the block

directions that can be determined from the geometry of the problem (see Fig. A.3 in Appendix A):

$$W_x = W \sin \theta$$
$$W_y = W \cos \theta$$

Since the intended direction of motion of the block is in the positive x direction, the frictional force on the block is acting in the negative x direction trying to stop the block from moving up the incline. The magnitude f of the frictional force is directly proportional to the magnitude N of the normal force applied by the inclined surface on the block and the coefficient of friction, μ, is the constant of proportionality. Therefore, f, N, and μ are related through:

$$f = \mu N. \tag{i}$$

The unknowns in this example are P and N. Since we have an expression relating f and N, if N is known so is f. For the solution of the problem, first consider the equilibrium of the block in the y direction:

$$\sum F_y = 0: \quad \begin{aligned} N - W_y &= 0 \\ N = W_y &= W \cos \theta \end{aligned} \tag{ii}$$

Once N is determined, then:

$$f = \mu \times N = \mu \times W \cos \theta.$$

We are asked to determine the minimum force the person must apply to overcome frictional and gravitational effects to start moving the block. There is equilibrium at the instant just before the movement starts. Therefore, we can apply the equilibrium condition in the x direction:

$$\sum F_x = 0: \quad \begin{aligned} P - f - W_x &= 0 \\ P &= f + W_x \end{aligned}$$

Substitute (*i*) and (*ii*) into (*iii*) along with $W_x = W \sin \theta$:

$$P = \mu W \cos \theta + W \sin \theta.$$

This is a general solution for P in terms of W, μ, and θ. This solution is valid for any θ less than 90°, including $\theta = 0°$ which represents a flat horizontal surface (Fig. 4.40). For $\theta = 0°$, $\sin 0 = 0°$ and $\cos 0 = 0°$. Therefore, the force required to start moving the same block on a horizontal surface is

$$P = f = \mu W.$$

Fig. 4.40 Pushing a block on a horizontal surface

To have a numerical example, assume that $W = 1,000\,\text{N}$, $\mu = 0.3$, and $\theta = 15°$. Then

$$P = (0.3)(1,000)(\cos 15) + (1,000)(\sin 15) = 548.6\,\text{N}.$$

Therefore, to be able to start moving a $1,000\,\text{N}$ block up the 15° incline which has a surface friction coefficient of 0.3, the person must apply a force slightly greater than $548.6\,\text{N}$ in a direction parallel to the incline.

To start moving the same block on a horizontal surface with the same friction coefficient, the person must apply a horizontal force of

$$P = (0.3)(1,000) = 300\,\text{N}.$$

As compared to a horizontal surface, the person must apply about 83% more force on the block to start moving the block on the 15° incline, which is calculated as:

$$\frac{548.6 - 300}{300} \times 100 = 82.9$$

4.12   Center of Gravity Determination

As discussed in Chap. 2, every object may be considered to consist of an infinite number of particles that are acted upon by the force of gravity, thus forming a distributed force system. The resultant of these forces or individual weights of particles is equal to the total weight of the object, acting as a concentrated load at the *center of gravity* of the object. A concept related to the center of gravity is that of the *center of mass*, which is a point at which the entire mass of an object is assumed to be concentrated. In general, there is a difference between the centers of mass and gravity of an object. This may be worth considering if the object is large enough for the magnitude of the gravitational acceleration to vary at different parts of the object. For our applications, the centers of mass and gravity of an object refer to the same point. There is also the concept of gravity line that is used to refer to the vertical line that passes through the center of gravity.

The center of gravity of an object is such that if the object is cut into two parts by any vertical plane that goes through the center of gravity, then the weight of each part would be equal. Therefore, the object can be balanced on a knife-edge that is located directly under its center of gravity or along its gravity line. If an object has a symmetric, well-defined geometry and uniform composition, then its center of gravity is located at the geometric center of the object. There are several methods of finding the centers of gravity of irregularly shaped objects. One method is by "suspending" the object. For the sake of illustration, consider the piece of paper shown in Fig. 4.41. The center of gravity of the paper

is located at point C. Let O and Q be two points on the paper. If the paper is pinned to the wall at point O, there will be a nonzero net clockwise moment acting on the paper about point O because the center of gravity of the paper is located to the right of O. If the paper is released, the moment about O due to the weight W of the paper will cause the paper to swing in the clockwise direction first, oscillate, and finally come to rest at a position in which C lies along a vertical line *aa* (gravity line) passing through point O (Fig. 4.41b). At this position, the net moment about point O is zero because the length of the moment arm of $\underline{W}$ relative to point O is zero. If the paper is then pinned at point Q, the paper will swing in the counterclockwise direction and soon come to rest at a position in which point C is located directly under point Q, or along the vertical line *bb* passing through point Q (Fig. 4.41c). The point at which lines *aa* and *bb* intersect will indicate the center of gravity of the paper, which is point C.

Note that a piece of paper is a plane object with negligible thickness and that suspending the paper at two points is sufficient to locate its center of gravity. For a three-dimensional object, the object must be suspended at three points in two different planes.

Another method of finding the center of gravity is by "balancing" the object on a knife-edge. As illustrated in Fig. 4.42, to determine the center of gravity of a person, first balance a board on the knife-edge and then place the person supine on the board. Adjust the position of the person on the board until the board is again balanced (Fig. 4.42a). The horizontal distance between the feet of the person and the point of contact of the knife-edge and the board is the height of the center of gravity of the person.

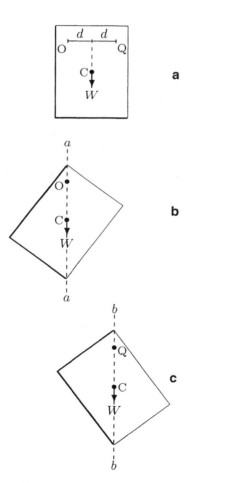

Fig. 4.41 Center of gravity of the paper is located at the intersection of lines *aa* and *bb* (suspension method)

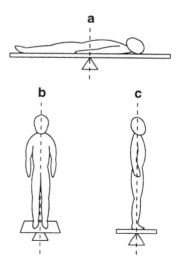

Fig. 4.42 Balance method

Consider a plane that passes through the point of contact of the knife-edge with the board that cuts the person into upper and lower portions. The center of gravity of the person lies somewhere on this plane. Note that the center of gravity of a three-dimensional body, such as a human being, has three coordinates. Therefore, the same method must be repeated in two other planes to establish the exact center of gravity. For this purpose, consider the anteroposterior balance of the person that will yield two additional planes as illustrated in (Fig. 4.42b, c). The intersection of these planes will correspond to the center of gravity of the person.

The third method of finding the center of gravity of a body involves the use of a "reaction board" with two knife-edges fixed to its undersurface (Fig. 4.43). Assume that the weight W_B of the board and the distance l between the knife-edges are known. One of the two edges (A) rests on a platform and the other edge (B) rests on a scale, such that the board is horizontal. The location of the center of gravity of a person can be determined by placing the person on the board and recording the weight indicated on the scale, which is essentially the magnitude R_B of the reaction force on the board at point B. Figure 4.43b illustrates the free-body diagram of the board. R_A is the magnitude of the reaction force at point A, W_p is the known weight of the person. The weight W_B of the board is acting at the geometric center (point C) of the board which is equidistant from points A and B. D is a point on the board directly under the center of gravity of the person. The unknown distance between points A and D is designated by x_{cg} which can be determined by

considering the rotational equilibrium of the board about point A. Assuming that clockwise moments are positive

$$\sum M_A = 0 : \quad \tfrac{l}{2} W_b + x_{cg} W_p - l R_B = 0 .$$

Solving this equation for x_{cg} will yield

$$x_{cg} = \frac{l}{W_p} \left(R_B - \frac{W_b}{2} \right).$$

If the person is placed on the board so that the feet are directly above the knife-edge at point A, x_{cg} will designate the height of the person's center of gravity as measured from the floor level.

Sometimes, we must deal with a system made up of parts with known centers of gravity where the task is to determine the center of gravity of the system as a whole. This can be achieved simply by utilizing the definition of the center of gravity. Consider the system shown in Fig. 4.44, which is composed of three spheres with weights W_1, W_2, and W_3 connected to one another through rods. Assume that the weights of the rods are negligible. Let x_1, x_2, and x_3 be the x coordinates of the centers of gravity of each sphere. The net moment about point O due to the individual weights of the spheres is

$$M_O = x_1 W_1 + x_2 W_2 + x_3 W_3.$$

The total weight of the system is $W_1 + W_2 + W_3$ which is assumed to be acting at the center of gravity of the system as a whole. If x_{cg} is the x coordinate of the center of gravity of the entire system, then the moment of the total weight of the system about point O is

$$M_O = x_{cg}(W_1 + W_2 + W_3).$$

The last two equations can be combined together so as to eliminate M_O, which will yield

$$x_{cg} = \frac{x_1 W_1 + x_2 W_2 + x_3 W_3}{W_1 + W_2 + W_3}.$$

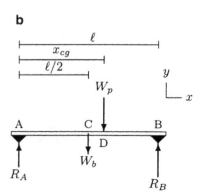

Fig. 4.43 Reaction board method

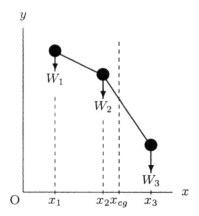

Fig. 4.44 x_{cg} is the x component of the center of gravity

This result can be generalized for any system composed of n parts:

$$x_{cg} = \frac{\sum_{i=1}^{n} x_i W_i}{\sum_{i=1}^{n} W_i}. \tag{4.12}$$

Equation (4.12) provides only the x coordinate of the center of gravity of the system. To determine the exact center, the y coordinate of the center of gravity must also be determined. For this purpose, the entire system must be rotated by an angle, preferably 90°, as illustrated in Fig. 4.45. If y_1, y_2, and y_3 correspond to the y coordinates of the centers of gravity of the spheres, then the y coordinate of the center of gravity of the system as a whole is:

$$y_{cg} = \frac{y_1 W_1 + y_2 W_2 + y_3 W_3}{W_1 + W_2 + W_3}.$$

For any system composed of n parts,

$$y_{cg} = \frac{\sum_{i=1}^{n} y_i W_i}{\sum_{i=1}^{n} W_i}. \tag{4.13}$$

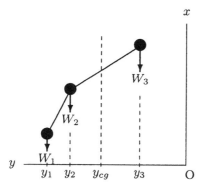

Fig. 4.45 y_{cg} is the y component of the center of gravity

In Fig. 4.46, the center of gravity of the entire system is located at the point of intersection of the perpendicular lines passing through x_{cg} and y_{cg}.

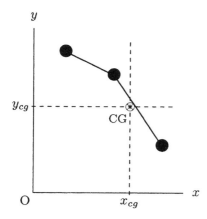

Fig. 4.46 $\otimes$ is the center of gravity of the system

Example 4.8 Consider the leg shown in Fig. 4.47, which is flexed to a right angle. The coordinates of the centers of gravity of the leg between the hip and knee joints (upper leg), the knee and ankle joints, and the foot, as measured from the floor level directly in line with the hip joint, are given in Table 4.2. The weights of the segments of the leg as percentages of the total weight W of the person are also provided in Table 4.2.

Determine the location of the center of gravity of the entire leg.

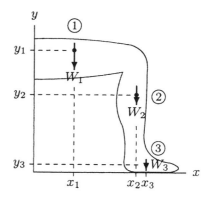

Fig. 4.47 Locating the center of gravity of a flexed leg

Table 4.2 Example 4.8

Part	x (cm)	y (cm)	W (%)
1	17.3	51.3	10.6
2	42.5	32.8	4.6
3	45.0	3.3	1.7

Solution

The coordinates (x_{cg}, y_{cg}) of the center of gravity of the entire leg can be determined by utilizing Eqs. (4.12) and (4.13). Using Eq. (4.12)

$$x_{cg} = \frac{x_1 W_1 + x_2 W_2 + x_3 W_3}{W_1 + W_2 + W_3}$$

$$x_{cg} = \frac{(17.3)(0.106W) + (42.5)(0.046W) + (45.0)(0.017W)}{0.106W + 0.046W + 0.017W}$$

$$x_{cg} = 26.9\,\text{cm}$$

To determine the y coordinate of the center of gravity of the leg, we must rotate the leg by 90°, as illustrated in Fig. 4.48, and apply Eq. (4.13):

$$y_{cg} = \frac{y_1 W_1 + y_2 W_2 + y_3 W_3}{W_1 + W_2 + W_3}$$

$$y_{cg} = \frac{(51.3)(0.106W) + (32.8)(0.046W) + (3.3)(0.017W)}{0.106W + 0.046W + 0.017W}$$

$$y_{cg} = 41.4\,\text{cm}$$

Fig. 4.48 $\otimes$ is the center of gravity of the leg

Therefore, the center of gravity of the entire lower extremity when flexed at a right angle is located at a horizontal distance of 26.9 cm from the hip joint and at a height of 41.4 cm measured from the floor level.

Remarks

- During standing with upper extremities straight, the center of gravity of a person lies within the pelvis anterior to the second sacral vertebra. If the origin of the rectangular coordinate system is placed at the center of gravity of the person, then the *xy*-plane corresponds to the *frontal, coronal*, or *longitudinal plane*, the *yz*-plane is called the *sagittal plane*, and the *xz*-plane is the *horizontal* or *transverse plane*. The frontal plane divides the body into front and back portions, the sagittal plane divides the body into right and left portions, and the horizontal plane divides it into upper and lower portions.
- The center of gravity of the entire body varies from person to person depending on build. For a given person, the position of the center of gravity can shift depending on the changes in the relative alignment of the extremities during a particular physical activity. Locations of the centers of gravity of the upper and lower extremities can also vary. For example, the center of gravity of a lower limb shifts backward as the knee is flexed. The center of gravity of the entire arm shifts forward as the elbow is flexed. These observations suggest that when the knee is flexed, the leg will tend to move forward bringing the center of gravity of the leg directly under the hip joint. When the elbow is flexed, the arm will tend to move backward bringing its center of gravity under the shoulder joint.

4.13   Exercise Problems

Problem 4.1 As illustrated in Fig. 4.49, consider an 80-kg person preparing to dive into a pool. The diving board is represented by a uniform, horizontal beam that is hinged to

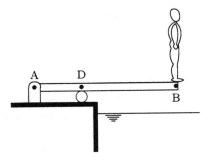

Fig. 4.49 Problem 4.1

the ground at point A and supported by a frictionless roller at point D. B is a point on the board directly under the center of gravity of the person. The distance between points A and B is $l = 6$ m and the distance between points A and D is $d = 2$ m. (Note that one-third of the board is located on the left of the roller support and two-thirds is on the right. Therefore, for the sake of force analyses, one can assume that the board consists of two boards with two different weights connected at point D.)

If the diving board has a total weight of 1,500 N, determine the reactions on the beam at points A and D.

Answers: $R_A = 2,318\,\text{N}(\downarrow)$, $R_D = 4,602\,\text{N}(\uparrow)$

Problem 4.2 The uniform, horizontal beam shown in Fig. 4.50 is hinged to the ground at point A and supported by a frictionless roller at point D. The distance between points A and B is $l = 4$ m and the distance between points A and D is $d = 3$ m. A force that makes an angle $\beta = 60°$ with the horizontal is applied at point B. The magnitude of the applied force is $P = 1,000$ N. The total weight of the beam is $W = 400$ N.

By noting that three-quarters of the beam is on the left of the roller support and one-quarter is on the right, calculate the x and y components of reaction forces on the beam at points A and D.

Answers: $R_D = 1,421\,\text{N}(\uparrow)$, $R_{Ax} = 500\,\text{N}(\leftarrow)$, $R_{Ay} = 155\,\text{N}(\downarrow)$

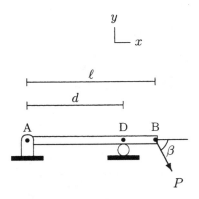

Fig. 4.50 Problem 4.2

Problem 4.3 The uniform, horizontal beam shown in Fig. 4.51 is hinged to the wall at point A and supported by a cable attached to the beam at point C. Point C also represents the center of gravity of the beam. At the other end, the cable is attached to the wall so that it makes an angle $\theta = 68°$ with the horizontal. If the length of the beam is $l = 4\,\text{m}$ and the weight of the beam is $W = 400\,\text{N}$, calculate the tension T in the cable and components of the reaction force on the beam at point A.

Answers: $T = 431\,\text{N}$, $R_{Ax} = 162\,\text{N}\,(\rightarrow)$, $R_{Ay} = 0$

Fig. 4.51 Problem 4.3

Problem 4.4 Using two different cable-pulley arrangements shown in Fig. 4.52, a block of weight W is elevated to a certain height. For each system, determine how much force is applied to the person holding the cable.

Answers: $T_1 = W/2$, $T_2 = W/4$

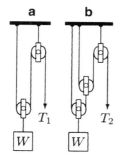

Fig. 4.52 Problem 4.4

Problem 4.5 Consider the split Russel traction device and a mechanical model of the leg shown in Fig. 4.53. The leg is held in the position shown by two weights that are connected to the leg via two cables. The combined weight of the leg and the cast is $W = 300\,\text{N}$. l is the horizontal distance between points A and B where the cables are attached to the leg. Point C is the center of gravity of the leg including the cast which is located at a distance two-thirds of l as measured from point A. The angle cable 2 makes with the horizontal is measured as $\beta = 45°$.

Determine the tensions T_1 and T_2 in the cables, weights W_1 and W_2, and angle α that cable 1 makes with the horizontal, so that the leg remains in equilibrium at the position shown.

Answers: $T_1 = W_1 = 223.6\,\text{N}$, $T_2 = W_2 = 282.8\,\text{N}$, $\alpha = 26.6°$

Fig. 4.53 Problem 4.5

Problem 4.6 Consider the uniform, horizontal cantilever beam shown in Fig. 4.54. The beam is fixed at point A and a force that makes an angle $\beta = 63°$ with the horizontal is applied at point B. The magnitude of the applied force is $P = 80\,\text{N}$. Point C is the center of gravity of the beam and the beam weighs $W = 40\,\text{N}$ and has a length $l = 2\,\text{m}$.

Determine the reactions generated at the fixed end of the beam.

Answers: $R_{Ax} = 36.3\,\text{N}\,(+x)$, $R_{Ay} = 111.3\,\text{N}\,(+y)$, $M_A = 182.3\,\text{N-m}\,(\text{ccw})$

Fig. 4.54 Problem 4.6

Problem 4.7 Consider the L-shaped beam illustrated in Fig. 4.55. The beam is welded to the wall at point A, the arm AB extends in the positive z direction, and the arm BC extends in the negative y direction. A force $\underline{P}$ is applied in the positive x direction at the free-end (point B) of the beam. The lengths of arms AB and BC are a and b, respectively, and the magnitude of the applied force is P.

Assuming that the weight of the beam is negligibly small, determine the reactions generated at the fixed end of the beam in terms of a, b, and P.

Answers: The nonzero force and moment components are $R_{Ax} = P(-x)$, $M_{Ay} = aP(-y)$, $M_{Az} = bP(-z)$.

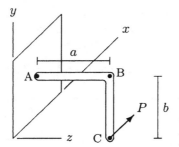

Fig. 4.55 Problems 4.7 and 4.8

Problem 4.8 Reconsider the L-shaped beam illustrated in Fig. 4.55. This time, assume that the applied force $\underline{P}$ has components in the positive x and positive z directions such that $\underline{P} = P_x \underline{i} + P_y \underline{j}$. Determine the reactions generated at the fixed end of the beam in terms of a, b, P_x, and P_y.

Answers: $R_{Ax} = P_x(-x); R_{Ay} = 0; R_{Az} = P_y(-z)$

$M_{Ax} = bP_z(+y); M_{Ay} = a; P_x(-y); M_{Az} = bP_x(-z)$

Problem 4.9 Figure 4.56 illustrates a person who is trying to pull a block on a horizontal surface using a rope. The rope makes an angle θ with the horizontal. If W is the weight of the block and μ is the coefficient of maximum friction between the bottom surface of the block and horizontal surface, show that the magnitude P of minimum force the person must apply in order to overcome the frictional and gravitational effects (to start moving the block) is

$$P = \frac{\mu W}{\cos\theta + \mu\sin\theta}.$$

Fig. 4.56 Problem 4.9

Problem 4.10 Figure 4.57 illustrates a person trying to push a block up on an inclined surface by applying a horizontal force. The weight of the block is W, the coefficient of maximum friction between the block and the incline is μ, and the incline makes an angle θ with the horizontal.

Determine the magnitude P of minimum force the person must apply in order to overcome the frictional and gravitational effects (to start moving the block) in terms of W, μ, and θ.

Answer: $P = \dfrac{\sin\theta + \mu\cos\theta}{\cos\theta + \mu\sin\theta}W.$

Fig. 4.57 Problem 4.10

Problem 4.11 As shown in Fig. 4.58, a horizontal beam is hinged to the wall at point A. The length of the beam is $l = 2\,\text{m}$ and it weighs $W = 150\,\text{N}$. Point C is the center of gravity of the beam and it is equidistant from both its ends. A cable is attached to the bean at point B making an angle $\alpha = 50°$ with the horizontal. At the other end, the cable is attached to the wall. A load that weighs $W_1 = 50\,\text{N}$ is placed on the beam such that its gravity line is passing through point C. Another load of the same weight $W_2 = 50\,\text{N}$ is attached to the beam at point B.

Determine the tension T in the cable and the reaction force at point A.

Answer: $T = 196\,\text{N}$, $R_A = 161\,\text{N}$

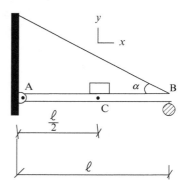

Fig. 4.58 Problem 4.11

Problem 4.12 As shown in Fig. 4.59, consider a horizontal beam hinged to the ground at point A. The length of the beam is $l = 4.5\,\text{m}$ and it weighs $W = 650\,\text{N}$. Point C represents the center of gravity of the beam and it is equidistant from point A and the free end of the beam (point B). A force $P = 850\,\text{N}$ is applied at point B making an angle $\alpha = 45°$ with the horizontal. Furthermore, a load weighing $W_1 = 125\,\text{N}$ is placed on the beam at its free end. The distance between point B and the line of gravity of the load is $l_1 = 0.3\,\text{m}$. A frictionless roller is adjusted at point D to constrain the counterclockwise rotation of the beam. The distance between points A and D is $l_2 = 3\,\text{m}$.

Calculate the reactions on the beam at points A and D.

Answer: $R_A = 729\,\text{N}$, $R_D = 239\,\text{N}$

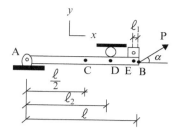

Fig. 4.59 Problem 4.12

Problem 4.13 As shown in Fig. 4.60, consider two divers preparing for sequential jumps into the pool. The divers are standing on a uniform horizontal diving board at points B and C, respectively. The diving board is hinged to the ground at point A and supported by a frictionless roller at point D. The length of the diving board is $l = 4$ m and it weighs $W = 500$ N. The distance between points A and D is $l_2 = 1.3$ m. The center of gravity of the board (point E) is equidistant from points A and B. The weight of the divers standing at points B and C is $W_1 = 680$ N and $W_2 = 710$ N, respectively, and the distance between the divers is $l_1 = 1.5$ m.

Determine the reactions on the diving board at points A and D.

Answer: $R_A = 2,337$ N($\downarrow$)

$\qquad R_D = 4,227$ N($\uparrow$)

Fig. 4.60 Problem 4.13

Problem 4.14 As illustrated in Fig. 4.61, consider a uniform horizontal beam fixed to the wall at point A. A force $F = 135$ N is applied on the beam at its free end (point B), making an angle $\alpha = 65°$ with the horizontal. The length of the beam is $l = 3$ m and it weighs $W = 150$ N. Point C is the center of gravity of the beam and it is equidistant from points A and B. Two identical boxes weighing $W_1 = W_2 = 80$ N

each are placed on the beam such that the distance between points A and B and the gravity lines of the boxes $l_1 = l_2 = 0.75$ m.

Determine the reactions at the fixed end of the beam.

Answer: $R_A = 196$ N, $M = 97.8$ N-m

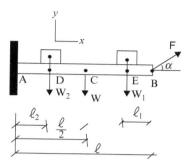

Fig. 4.61 Problem 4.14

Problem 4.15 As illustrated in Fig. 4.62, consider a simple traction device applied to the leg of a patient such that the cable of the pulley makes an angle $\alpha = 35°$ with the horizontal. The leg is in a cast and the coefficient of friction between the cast and the bed is $\mu = 0.45$. The weight of the leg is $W = 180$ N.

Determine the tension in the cable.

Answer: $T = 75$ N

Fig. 4.62 Problem 4.15

Problem 4.16 As illustrated in Fig. 4.24, consider a simple three-pulley traction system used to transmit a horizontal force to the leg of a patient. The leg is in a cast. A cable is wrapped around the pulleys such that one end of the cable is attached to the ceiling and the other end is attached to a weight pan. The weight of the weight pan is $W = 12$ N.

Determine the magnitude of the horizontal force acting on the leg.

Answer: $F = 24$ N

5.1 Skeletal Joints

The human body is rigid in the sense that it can maintain a posture, and flexible in the sense that it can change its posture and move. The flexibility of the human body is due primarily to the *joints*, or *articulations*, of the skeletal system. The primary function of joints is to provide mobility to the musculoskeletal system. In addition to providing mobility, a joint must also possess a degree of stability. Since different joints have different functions, they possess varying degrees of mobility and stability. Some joints are constructed so as to provide optimum mobility. For example, the construction of the shoulder joint (ball-and-socket) enables the arm to move in all three planes (triaxial motion). However, this high level of mobility is achieved at the expense of reduced stability, increasing the vulnerability of the joint to injuries, such as dislocations. On the other hand, the elbow joint provides movement primarily in one plane (uniaxial motion), but is more stable and less prone to injuries than the shoulder joint. The extreme case of increased stability is achieved at joints that permit no relative motion between the bones constituting the joint. The contacting surfaces of the bones in the skull are typical examples of such joints.

The joints of the human skeletal system may be classified based on their structure and/or function. *Synarthrodial joints*, such as those in the skull, are formed by two tightly fitting bones and do not allow any relative motion of the bones forming them. *Amphiarthrodial joints*, such as those between the vertebrae, allow slight relative motions, and feature an intervening substance (a cartilaginous or ligamentous tissue) whose presence eliminates direct bone-to-bone contact. The third and mechanically most significant type of articulations are called *diarthrodial joints* which permit varying degrees of relative motion and have articular cavities, ligamentous capsules, synovial membranes, and synovial fluid (Fig. 5.1). The *articular cavity* is the space between the articulating bones. The *ligamentous capsule* holds the articulating bones

together. The *synovial membrane* is the internal lining of the ligamentous capsule enclosing the *synovial fluid* which serves as a lubricant. The synovial fluid is a viscous material which functions to reduce friction, reduce wear and tear of the articulating surfaces by limiting direct contact between them, and nourish the articular cartilage lining the surfaces. The *articular cartilage*, on the other hand, is a specialized tissue designed to increase load distribution on the joints and provide a wear-resistant surface that absorbs shock. Various diarthrodial joints can be further categorized as gliding (e.g., vertebral facets), hinge (elbow and ankle), pivot (proximal radioulnar), condyloid (wrist), saddle (carpometacarpal of thumb), and ball-and-socket (shoulder and hip).

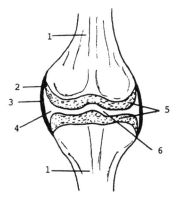

Fig. 5.1 A diarthrodial joint: (1) Bone, (2) ligamentous capsule, (3, 4) synovial membrane and fluid, (5, 6) articular cartilage and cavity

The nature of motion about a diarthrodial joint and the stability of the joint are dependent upon many factors, including the manner in which the articulating surfaces fit together, the properties of the joint capsule, the structure and length of the ligaments around the joint, and the number and orientation of the muscles crossing the joint.

N. Özkaya et al., *Fundamentals of Biomechanics: Equilibrium, Motion, and Deformation*,
DOI 10.1007/978-1-4614-1150-5_5, © Springer Science+Business Media, LLC 2012

5.2   Skeletal Muscles

In general, there are over 600 muscles in the human body, accounting for about 45° of the total body weight.

There are three types of muscles: cardiac, smooth, and skeletal. *Cardiac muscle* is the contractive tissue found in the heart that pumps the blood for circulation. *Smooth muscle* is found in the stomach, intestinal tracts, and the walls of blood vessels. *Skeletal muscle* is connected to the bones of the body and when contracted, causes body segments to move.

Movement of human body segments is achieved as a result of forces generated by skeletal muscles that convert chemical energy into mechanical work. The structural unit of skeletal muscle is the muscle fiber, which is composed of myofibrils. Myofibrils are made up of actin and myosin filaments. Muscles exhibit viscoelastic material behavior. That is, they have both solid and fluid-like material properties. Muscles are elastic in the sense that when a muscle is stretched and released it will resume its original (unstretched) size and shape. Muscles are *viscous* in the sense that there is an internal resistance to motion.

A skeletal muscle is attached, via soft tissues such as aponeuroses and/or tendons, to at least two different bones controlling the relative motion of one segment with respect to the other. When its fibers contract under the stimulation of a nerve, the muscle exerts a pulling effect on the bones to which it is attached. *Contraction* is a unique property of the muscle tissue. In engineering mechanics, contraction implies shortening under compressive forces. In muscle mechanics, contraction can occur as a result of muscle shortening or muscle lengthening, or it can occur without any change in the muscle length. Furthermore, the result of a muscle contraction is always tension: a muscle can only exert a pull. Muscles cannot exert a push.

There are various types of muscle contractions: a *concentric contraction* occurs simultaneously as the length of the muscle decreases (e.g., the biceps during flexion of the forearm); a *static contraction* occurs while muscle length remains constant (the biceps when the forearm is flexed and held without any movement); and an *eccentric contraction* occurs as the length of the muscle increases (the biceps during the extension of the forearm). A muscle can cause movement only while its length is shortening (concentric contraction). If the length of a muscle increases during a particular activity, then the tension generated by the muscle contraction is aimed at controlling the movement of the body segments associated with that muscle (excentric contraction). If a muscle contracts but there is no segmental motion, then the tension in the muscle balances the effects of applied forces such as those due to gravity (isometric contraction).

The skeletal muscles can also be named according to the functions they serve during a particular activity. For example, a muscle is called *agonist* if it causes movement through the process of its own contraction. Agonist muscles are the primary muscles responsible for generating a specific movement. An *antagonist* muscle opposes the action of another muscle. *Synergic* muscle is that which assists the agonist muscle in performing the same joint motion.

5.3   Basic Considerations

In this chapter, we want to apply the principles of statics to investigate the forces involved in various muscle groups and joints for various postural positions of the human body and its segments. Our immediate purpose is to provide answers to questions such as: what tension must the neck extensor muscles exert on the head to support the head in a specified position? When a person bends, what would be the force exerted by the erector spinae on the fifth lumbar vertebra? How does the compression at the elbow, knee, and ankle joints vary with externally applied forces and with different segmental arrangements? How does the force on the femoral head vary with loads carried in the hand? What are the forces involved in various muscle groups and joints during different exercise conditions?

The forces involved in the human body can be grouped as internal and external. Internal forces are those associated with muscles, ligaments, and tendons, and at the joints. Externally applied forces include the effect of gravitational acceleration on the body or on its segments, manually and/or mechanically applied forces on the body during exercise and stretching, and forces applied to the body by prostheses and implements. In general, the unknowns in static problems involving the musculoskeletal system are the joint reaction forces and muscle tensions. Mechanical analysis of a joint requires that we know the vector characteristics of tension in the muscle including the proper locations of muscle attachments, the weights or masses of body segments, the centers of gravity of the body segments, and the anatomical axis of rotation of the joint.

5.4   Basic Assumptions and Limitations

The complete analysis of muscle forces required to sustain various postural positions is difficult because of the complex arrangement of muscles within the human body and because of limited information. In general, the relative motion of body segments about a given joint is controlled by more than one muscle group. To be able to reduce a specific problem of biomechanics to one that is statically determinate and apply the equations of equilibrium, only the muscle group that is the primary source of control over the joint can be taken into consideration. Possible contributions of other muscle groups to the load-bearing mechanism of the joint must be ignored. Note however that approximations of the effect of other muscles may be made by considering their cross-sectional areas and their relative positions in relation to the joint. Also, if the phasic activity of muscles is known via some experiments such as the electromyography (EMG) measurements of muscle signals, then the tension in different muscle groups may be estimated.

To apply the principles of statics to analyze the mechanics of human joints, we shall adopt the following assumptions and limitations:

- The anatomical axes of rotation of joints are known.
- The locations of muscle attachments are known.
- The line of action of muscle tension is known.
- Segmental weights and their centers of gravity are known.
- Frictional factors at the joints are negligible.
- Dynamic aspects of the problems will be ignored.
- Only two-dimensional problems will be considered.

These analyses require that the anthropometric data about the segment to be analyzed must be available. For this purpose, there are tables listing anthropometric information including average weights, lengths, and centers of gravity of body segments. See Chaffin et al. (1999), Roebuck (1995), and Winter (2004) for a review of the anthropometric data available.

It is clear from this discussion that we shall analyze certain idealized problems of biomechanics. Based on the results obtained and experience gained, these models may be expanded by taking additional factors into consideration. However, a given problem will become more complex as more factors are considered.

In the following sections, the principles of statics are applied to analyze forces involved at and around the major joints of the human body. First, a brief functional anatomy of each joint and related muscles is provided, and specific biomechanical problems are constructed. For a more complete discussion about the functional anatomy of joints, see texts such as Nordin and Frankel (2012) and Thompson (1989). Next, an analogy is formed between muscles, bones, and human joints, and certain mechanical elements such as cables, beams, and mechanical joints. This enables us to construct a mechanical model of the biological system under consideration. Finally, the procedure outlined in section "Procedure to Analyze Systems in Equilibrium" of Chap. 4 is applied to analyze the mechanical model

thus constructed. See LeVeau (1992) for additional examples of the application of the principles of statics to biomechanics.

5.5 Mechanics of the Elbow

The elbow joint is composed of three separate articulations (Fig. 5.2). The *humeroulnar joint* is a hinge (ginglymus) joint formed by the articulation between the spool-shaped trochlea of the distal humerus and the concave trochlear fossa of the proximal ulna. The structure of the humeroulnar joint is such that it allows only uniaxial rotations, confining the movements about the elbow joint to flexion (movement of the forearm toward the upper arm) and extension (movement of the forearm away from the upper arm). The *humeroradial joint* is also a hinge joint formed between the capitulum of the distal humerus and the head of the radius. The *proximal radioulnar joint* is a pivot joint formed by the head of the radius and the radial notch of the proximal ulna. This articulation allows the radius and ulna to undergo relative rotation about the longitudinal axis of one or the other bone, giving rise to pronation (the movement experienced while going from the palm-up to the palm-down) or supination (the movement experienced while going from the palm-down to the palm-up).

Fig. 5.2 Bones of the elbow: (1) humerus, (2) capitulum, (3) trochlea, (4) radius, (5) ulna

The muscles coordinating and controlling the movement of the elbow joint are illustrated in Fig. 5.3. The biceps brachii muscle is a powerful flexor of the elbow joint, particularly when the elbow joint is in a supinated position. It is the most powerful supinator of the forearm. On the distal side, the biceps is attached to the tuberosity of the radius, and on the proximal side, it has attachments at the top of the coracoids process and upper lip of the glenoid fossa. Another important flexor is the brachialis muscle which, regardless of forearm orientation, has the ability to produce elbow flexion.

Fig. 5.4 Example 5.1

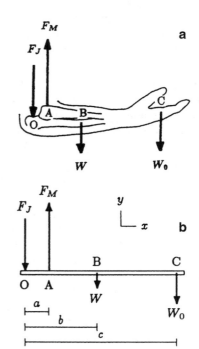

Fig. 5.5 Forces acting on the lower arm

Fig. 5.3 Muscles of the elbow: (1) biceps, (2) brachioradialis, (3) brachialis, (4) pronator teres, (5) triceps brachii, (6) anconeus, (7) supinator

It is then the strongest flexor of the elbow. It has attachments at the lower half of the anterior portion of the humerus and the coronoid process of the ulna. Since it does not inset on the radius, it cannot participate in pronation or supination. The most important muscle controlling the extension movement of the elbow is the triceps brachii muscle. It has attachments at the lower head of the glenoid cavity of the scapula, the upper half of the posterior surface of the humerus, the lower two thirds of the posterior surface of the humerus, and the olecranon process of the ulna. Pronation and supination movements of the forearm are performed by the pronator teres and supinator muscles, respectively. The pronator teres is attached to the lower part of the inner condyloid ridge of the humerus, the medial side of the ulna, and the middle third of the outer surface of the radius. The supinator muscle has attachments at the outer condyloid ridge of the humerus, the neighboring part of the ulna and the outer surface of the upper third of the radius.

Common injuries of the elbow include fractures and dislocations. Fractures usually occur at the epicondyles of the humerus and the olecranon process of the ulna. Another group of elbow injuries are associated with overuse, which causes an inflammatory process of the tendons of an elbow that has been damaged by repetitive motions. These include tennis elbow and golfer's elbow syndromes.

Example 5.1 Consider the arm shown in Fig. 5.4. The elbow is flexed to a right angle and an object is held in the hand. The forces acting on the forearm are shown in Fig. 5.5a, and the free-body diagram of the forearm is shown on a mechanical model in Fig. 5.5b. This model assumes that the biceps is the major flexor and that the line of action of the tension (line of pull) in the biceps is vertical.

Point O designates the axis of rotation of the elbow joint, which is assumed to be fixed for practical purposes. Point A is the attachment of the biceps muscle on the radius, point B is the center of gravity of the forearm, and point C is a point on the forearm that lies along a vertical line passing through the center of gravity of the weight in the hand. The distances between point O and points A, B, and C are measured as a, b, and c, respectively. W_o is the weight of the object held in the hand and W is the total weight of the forearm. F_M is the magnitude of the force exerted by the biceps on the radius, and F_J is the magnitude of the reaction force at the elbow joint. Notice that the line of action of the muscle force is assumed to be vertical. The gravitational forces are vertical as well. Therefore, for the equilibrium of the lower arm, the line of action of the joint reaction force must also be vertical (a parallel force system).

The task in this example is to determine the magnitudes of the muscle tension and the joint reaction force at the elbow.

Solution

We have a parallel force system, and the unknowns are the magnitudes F_M and F_J of the muscle and joint reaction forces. Considering the rotational equilibrium of the forearm about the elbow joint and assuming the (cw) direction is positive:

$$\sum M_o = 0.$$

That is,

$$cW_o + bW - aF_M = 0.$$

Then

$$F_M = \frac{1}{a}(bW + cW_o). \tag{i}$$

For the translational equilibrium of the forearm in the y direction,

$$\sum F_y = 0.$$

That is,

$$-F_J + F_M - W - W_o = 0.$$

Then

$$F_J = F_M - W - W_o. \tag{ii}$$

For given values of geometric parameters a, b, and c, and weights W and W_o, Eqs. (i) and (ii) can be solved for the magnitudes of the muscle and joint reaction forces. For example, assume that these parameters are given as follows: $a = 4\,\text{cm}$, $b = 15\,\text{cm}$, $c = 35\,\text{cm}$, $W = 20\,\text{N}$, and $W_o = 80\,\text{N}$. Then from Eqs. (i) and (ii):

$$F_M = \frac{1}{0.04}\,[(0.15)(20) + (0.35)(80)] = 775\,\text{N} \quad (+y).$$

$$F_J = 775 - 20 - 80 = 675\,\text{N} \quad (-y).$$

Remarks

- The numerical results indicate that the force exerted by the biceps muscle is about ten times larger than the weight of the object held in the position considered. Relative to the axis of the elbow joint, the length a of the lever arm of the muscle force is much smaller than the length c of the lever arm enjoyed by the load held in the hand. The smaller the lever arm, the greater the muscle tension required to balance the clockwise rotational effect of the load about the elbow joint. Therefore, during lifting, it is disadvantageous to have a muscle attachment close to the elbow joint. However, the closer the muscle is to the joint, the larger the range of motion of elbow flexion–extension, and the faster the distal end (hand) of the forearm can reach its goal of moving toward the upper arm or the shoulder.

- The angle between the line of action of the muscle force and the long axis of the bone upon which the muscle force is exerted is called the angle of pull and it is critical in determining the effectiveness of the muscle force. When the lower arm is flexed to a right angle, the muscle tension has only a rotational effect on the forearm about the elbow joint, because the line of action of the muscle force is at a right angle with the longitudinal axis of the forearm. For other flexed positions of the forearm, the muscle force can have a translational (stabilizing or sliding) component as well as a rotational component.

  Assume that the linkage system shown in Fig. 5.6a illustrates the position of the forearm relative to the upper arm. n designates a direction perpendicular (normal) to the long axis of the forearm and t is tangent to it. Assuming that the line of action of the muscle force remains parallel to the long axis of the humerus, $\underline{F}_M$ can be decomposed into its rectangular components $\underline{F}_{Mn}$ and $\underline{F}_{Mt}$. In this case, $\underline{F}_{Mn}$ is the *rotational* (rotatory) component of the muscle force because its primary function is to rotate the forearm about the elbow joint. The tangential component $\underline{F}_{Mt}$ of the muscle force acts to compress the elbow joint and is called the *stabilizing* component of the

Fig. 5.6 Rotational (F_{Mn}) and stabilizing or sliding (F_{Mt}) components of the muscle force

muscle force. As the angle of pull approaches 90°, the magnitude of the rotational component of the muscle force increases while its stabilizing component decreases, and less and less energy is "wasted" to compress the elbow joint. As illustrated in Fig. 5.6b, the stabilizing role of $\underline{F}_{Mt}$ changes into a *sliding* or *dislocating* role when the angle between the long axes of the forearm and upper arm becomes less than 90°.

- The elbow is a diarthrodial (synovial) joint. A ligamentous capsule encloses an articular cavity which is filled with synovial fluid. Synovial fluid is a viscous material whose primary function is to lubricate the articulating surfaces, thereby reducing the frictional forces that may develop while one articulating surface slides over the other. The synovial fluid also nourishes the articulating cartilages. A common property of fluids is that they exert pressures (force per unit area) that are distributed over the surfaces they touch. The fluid pressure always acts in a direction toward and perpendicular to the surface it touches having a compressive effect on the surface. Note that in Fig. 5.7, the small vectors indicating the fluid pressure have components in the horizontal and vertical directions. We determined that the joint reaction force at the elbow acts vertically downward on the ulna. This implies that the horizontal components of these vectors cancel out (i.e., half pointing to the left and half pointing to the right), but their vertical components (on the ulna, almost all of them are pointing downward) add up to form the resultant force F_J (shown with a dashed arrow in Fig. 5.7c). Therefore, the joint reaction force F_J corresponds to the resultant of the distributed force system (pressure) applied through the synovial fluid.

- The most critical simplification made in this example is that the biceps was assumed to be the single muscle group responsible for maintaining the flexed configuration of the forearm. The reason for making such an assumption was to reduce the system under consideration to one that is statically determinate. In reality, in addition to the biceps, the brachialis and the brachioradialis are primary elbow flexor muscles.

Consider the flexed position of the arm shown in Fig. 5.8a. The free-body diagram of the forearm is shown in Fig. 5.8b. F_{M1}, F_{M2}, and F_{M3} are the magnitudes of the forces exerted on the forearm by the biceps, the brachialis, and the brachioradialis muscles with attachments at points A_1, A_2, and A_3, respectively. Let θ_1, θ_2, and θ_3 be the angles that the biceps, the brachialis, and the brachioradialis muscles make with the long axis of the lower arm. As compared to the single-muscle system which consisted of two unknowns (F_M and F_j), the analysis of this three-muscle system is quite complex. First of all, this is not a simple parallel force system. Even if we assume that the locations of muscle attachments (A_1,A_2, and A_3), their angles of pull (θ_1, θ_2, and θ_3) and the lengths of their moment arms (a_1, a_2, and a_3) as measured from the elbow joint are known, there are still five unknowns in the problem (F_{M1}, F_{M2}, F_{M3}, F_j, and β, where the angle β is an angle between F_j and the long axes of the forearm). The total number of equations available from statics is three:

$$\sum M_o = 0: \quad a_1 F_{M1} + a_2 F_{M2} + a_3 F_{M3} = bW + cW_o. \quad \text{(iii)}$$

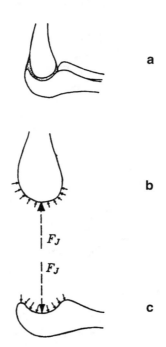

Fig. 5.7 Explaining the joint reaction force at the elbow

Fig. 5.8 Three-muscle system

$$\sum F_x = 0: \quad F_{Jx} = F_{M1x} + F_{M2x} + F_{M3x}. \quad \text{(iv)}$$

$$\sum F_y = 0: \quad F_{Jy} = F_{M1y} + F_{M2y} + F_{M3y} - W - W_o. \quad \text{(v)}$$

Note that once the muscle forces are determined, Eqs. (iv) and (v) will yield the components of the joint reaction force $\underline{F}_J$. As far as the muscle forces are concerned, we have only Eq. (iii) with three unknowns. In other words, we have a statically indeterminate problem. To obtain a unique solution, we need additional information relating F_{M1}, F_{M2}, and F_{M3}.

There may be several approaches to the solution of this problem. The criteria for estimating the force distribution among different muscle groups may be established by (1) using cross-sectional areas of muscles, (2) using EMG measurements of muscle signals, and (3) applying certain optimization techniques. It may be assumed that each muscle exerts a force proportional to its cross-sectional area. If S_1, S_2, and S_3 are the cross-sectional areas of the biceps, the brachialis, and the brachioradialis, then this criteria may be applied by expressing muscle forces in the following manner:

$$F_{M2} = k_{21} F_{M1} \quad \text{with} \quad k_{21} = \frac{S_2}{S_1}. \quad \text{(vi)}$$

$$F_{M3} = k_{31} F_{M1} \quad \text{with} \quad k_{31} = \frac{S_3}{S_1}. \quad \text{(vii)}$$

If constants k_{21} and k_{31} are known, then Eqs. (vi) and (vii) can be substituted into Eq. (iii), which can then be solved for F_{M1}:

$$F_{M1} = \frac{b \, W + c \, W_o}{a_1 + a_2 \, k_{21} + a_3 \, k_{31}}.$$

Substituting F_{M1} back into Eqs. (vi) and (vii) will then yield the magnitudes of the forces in the brachialis and the brachioradialis muscles. The values of k_{21} and k_{31} may also be estimated by using the amplitudes of muscle EMG signals.

This statically indeterminate problem may also be solved by considering some optimization techniques. If the purpose is to accomplish a certain task (static or dynamic) in the most efficient manner, then the muscles of the body must act to minimize the forces exerted, the moments about the joints (for dynamic situations), and/or the work done by the muscles. The question is, what force distribution among the various muscles facilitates the maximum efficiency?

5.6 Mechanics of the Shoulder

The bony structure and the muscles of the shoulder complex are illustrated in Figs. 5.9 and 5.10. The shoulder forms the base for all upper extremity movements. The complex

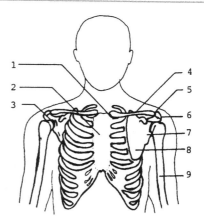

Fig. 5.9 The shoulder: (1) sternoclavicular joint, (2) sternum, (3) glenohumeral joint, (4) clavicle, (5) acromioclavicular joint, (6) acromion process, (7) glenoid fossa, (8) scapula, (9) humerus

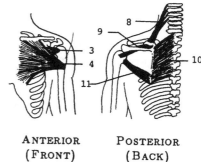

ANTERIOR (FRONT)   **POSTERIOR (BACK)**

Fig. 5.10 Shoulder muscles: (1) deltoideus, (2) pectoralis minor, (3) subscapularis, (4) pectoralis major, (5) trapezius, (6) infraspinatus and teres minor, (7) latissimus dorsi, (8) levator scapulae, (9) supraspinatus, (10) rhomboideus, (11) teres major

structure of the shoulder can be divided into two: the shoulder joint and the shoulder girdle.

The shoulder joint, also known as the *glenohumeral articulation*, is a ball-and-socket joint between the nearly hemispherical humeral head (ball) and the shallowly concave glenoid fossa (socket) of the scapula. The shallowness of the glenoid fossa allows a significant freedom of movement of the humeral head on the articulating surface of the glenoid. The movements allowed are in the sagittal plane, flexion (movement of the humerus to the front—a forward upward movement) and extension (return from flexion); in the coronal plane, abduction (horizontal upward movement of the humerus to the side) and adduction (return from abduction); and in the transverse plane, outward rotation (movement of the humerus around its long axis to the lateral side) and inward rotation (return from outward rotation). The configuration of the articulating surfaces of the shoulder joint also makes the joint more susceptible to instability and injury, such as dislocation. The stability of the joint is provided by the glenohumeral and coracohumeral ligaments, and by the muscles crossing the joint. The major muscles of the shoulder joint are deltoideus, supraspinatus, pectoralis major, coracobrachialis, latissimus dorsi, teres major, teres minor, infraspinatus, and subscapularis.

The bony structure of the shoulder girdle consists of the clavicle (collarbone) and the scapula (shoulder blade). The *acromioclavicular joint* is a small synovial articulation between the distal clavicle and the acromion process of the scapula. The stability of this joint is reinforced by the coracoclavicular ligaments. The *sternoclavicular joint* is the articulation between the manubrium of the sternum and the proximal clavicle. The stability of this joint is enhanced by the costoclavicular ligament. The acromioclavicular joint and the sternoclavicular joint both have layers of cartilage, called *menisci*, interposed between their bony surfaces.

There are four pairs of scapular movements: elevation (movement of the scapula in the frontal plane) and depression (return from elevation), upward rotation (turning the glenoid fossa upward and the lower medial border of the scapula away from the spinal column) and downward rotation (return from upward rotation), protraction (movement of the distal end of the clavicle forward) and retraction (return from protraction), and forward and backward rotation (rotation of the scapula about the shaft of the clavicle). Some of the main muscles that control and coordinate these movements are the trapezius, levator scapulae, rhomboid, pectoralis minor, serratus anterior, and subclavius.

Example 5.2 Consider a person strengthening the shoulder muscles by means of dumbbell exercises. Figure 5.11 illustrates the position of the left arm when the arm is fully abducted to horizontal. The free-body diagram of the arm is shown in Fig. 5.12 along with a mechanical model of the arm. Also in Fig. 5.12, the forces acting on the arm are resolved into their rectangular components along the horizontal and vertical directions. Point O corresponds to the axis of rotation of the shoulder joint, point A is where the deltoid muscle is attached to the humerus, point B is the center of gravity of the entire arm, and point C is the center of gravity of the dumbbell. W is the weight of the arm, W_0 is the weight of the dumbbell, F_M is the magnitude of the tension in the deltoid muscle, and F_j is the joint reaction force at the shoulder. The resultant of the deltoid muscle

Fig. 5.11 The arm is abducted to horizontal

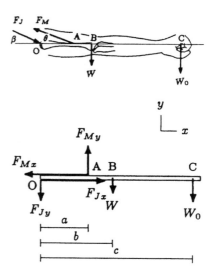

Fig. 5.12 Forces acting on the arm and a mechanical model representing the arm

force makes an angle θ with the horizontal. The distances between point O and points A, B, and C are measured as a, b, and c, respectively.

Determine the magnitude F_M of the force exerted by the deltoid muscle to hold the arm at the position shown. Also determine the magnitude and direction of the reaction force at the shoulder joint in terms of specified parameters.

Solution

With respect to the xy coordinate frame, the muscle and joint reaction forces have two components while the weights of the arm and the dumbbell act in the negative y direction. The components of the muscle force are

$$F_{Mx} = F_M \cos \theta \quad (-x). \tag{i}$$

$$F_{My} = F_M \sin \theta \quad (+y). \tag{ii}$$

Components of the joint reaction force are

$$F_{Jx} = F_J \cos \beta \quad (+x). \tag{iii}$$

$$F_{Jy} = F_J \sin \beta \quad (-y) \tag{iv}$$

β is the angle that the joint reaction force makes with the horizontal. The line of action and direction (in terms of θ) of the force exerted by the muscle on the arm are known. However, the magnitude F_M of the muscle force, the magnitude F_j, and the direction (β) of the joint reaction force are unknowns. We have a total of three unknowns, F_M, F_j, and β (or F_M, F_{Jx}, and F_{Jy}). To be able to solve this two-dimensional problem, we have to utilize all three equilibrium equations.

First, consider the rotational equilibrium of the arm about the shoulder joint at point O. The joint reaction force produces no torque about point O because its line of action passes through it. For practical purposes, we can neglect the possible contribution of the horizontal component of the muscle force to the moment generated about point O by assuming that its line of action also passes through point O. Note that this is not a critical or necessary assumption to solve this problem. If we knew the length of its moment arm (i.e., the vertical distance between O and A), we could easily incorporate the torque generated by F_{Mx} about point O into the analysis. Under these considerations, there are only three moment producing forces about point O. For the rotational equilibrium of the arm, the net moment about point O must be equal to zero. Taking counterclockwise moments to be positive:

$$\sum M_o = 0: \quad a F_{My} - b W - c W_o = 0$$

$$F_{My} = \frac{1}{a}(b W + c W_o). \tag{v}$$

For given a, b, c, W, and W_o, Eq. (v) can be used to determine the vertical component of the force exerted by the deltoid muscle. Equation, (ii) can now be used to determine the total force exerted by the muscle:

$$F_M = \frac{F_{My}}{\sin \theta}. \tag{vi}$$

Knowing F_M, Eq. (i) will yield the horizontal component of the tension in the muscle:

$$F_{Mx} = F_M \cos \theta. \tag{vii}$$

The components of the joint reaction force can be determined by considering the translational equilibrium of the arm in the horizontal and vertical directions:

$$\sum F_x = 0 \quad \text{that is}, F_{Jx} - F_{Mx} = 0, \text{ then } F_{Jx} = F_{Mx}. \tag{viii}$$

$$\sum F_y = 0 \quad \text{that is}, -F_{Jy} + F_{My} - W - W_o = 0,$$

$$\text{then } F_{Jy} = F_{My} - W - W_o. \tag{ix}$$

Knowing the rectangular components of the joint reaction force enables us to compute the magnitude of the force itself and the angle its line of action makes with the horizontal:

$$F_J = \sqrt{(F_{Jx})^2 + (F_{Jy})^2}. \tag{x}$$

$$\beta = \tan^{-1}\left(\frac{F_{Jy}}{F_{Jx}}\right). \tag{xi}$$

Now consider that $a = 15$ cm, $b = 30$ cm, $c = 60$ cm, $\theta = 15°$, $W = 40$ N, and $W_o = 60$ N. Then,

$$F_{My} = \frac{1}{0.15}[(0.30)(40) + (0.60)(60)] = 320 \text{ N} \quad (+y).$$

$$F_M = \frac{320}{\sin 15°} = 1{,}236 \text{ N}.$$

$$F_{Mx} = (1{,}236)(\cos 15°) = 1{,}194 \text{ N} \quad (-x).$$

$$F_{Jx} = 1{,}194 \text{ N} \quad (+x).$$

$$F_{Jy} = 320 - 40 - 60 = 220 \text{ N} \quad (-y).$$

$$F_J = \sqrt{1{,}194^2 + 220^2} = 1{,}214 \text{ N}.$$

$$\beta = \tan^{-1}\left(\frac{220}{1{,}194}\right) = 10°.$$

Remarks

- F_{Mx} is the stabilizing component and F_{My} is the rotational component of the deltoid muscle. F_{Mx} is approximately four times larger than F_{My}. A large stabilizing component suggests that the horizontal position of the arm is not stable, and that the muscle needs to exert a high horizontal force to stabilize it.

- The human shoulder is very susceptible to injuries. The most common injuries are dislocations of the shoulder joint and the fracture of the humerus. Since the socket of the glenohumeral joint is shallow, the head of the humerus is relatively free to rotate about the articulating surface of the glenoid fossa. This freedom of movement is achieved, however, by reduced joint stability. The humeral head may be displaced in various ways, depending on the strength or weakness of the muscular and ligamentous structure of the shoulder, and depending on the physical activity. Humeral fractures are another common type of injuries. The humerus is particularly vulnerable to injuries because of its unprotected configuration.

- Average ranges of motion of the arm about the shoulder joint are 230° during flexion–extension, and 170° in both abduction–adduction and inward–outward rotation.

5.7   Mechanics of the Spinal Column

The human spinal column is the most complex part of the human musculoskeletal system. The principal functions of the spinal column are to protect the spinal cord; to support the head, neck, and upper extremities; to transfer loads from the head and trunk to the pelvis; and to permit a variety of movements. The spinal column consists of the cervical (neck), thoracic (chest), lumbar (lower back), sacral, and coccygeal regions. The thoracic and lumbar sections of the spinal column make up the trunk. The sacral and coccygeal regions are united with the pelvis and can be considered parts of the pelvic girdle.

The vertebral column consists of 24 intricate and complex vertebrae (Fig. 5.13). The articulations between the vertebrae are amphiarthrodial joints. A fibrocartilaginous disk is interposed between each pair of vertebrae. The primary functions of these intervertebral disks are to sustain loads transmitted from segments above, act as shock absorbers, eliminate bone-to-bone contact, and reduce the effects of impact forces by preventing direct contact between the bony structures of the vertabrae. The articulations of each vertebra with the adjacent vertabrae permit movement in three planes, and the entire spine functions like a single

ball-and-socket joint. The structure of the spine allows a wide variety of movements including flexion–extension, lateral flexion, and rotation.

Two particularly important joints of the spinal column are those with the head (occiput bone of the skull) and the first cervical vertebrae, atlas, and the atlas and the second vertebrae, the axis. The *atlantooccipital joint* is the union between the first cervical vertebra (the atlas) and the occipital bone of the head. This is a double condyloid joint and permits movements of the head in the sagittal and frontal planes. The *atlantoaxial joint* is the union between the atlas and the odontoid process of the head. It is a pivot joint, enabling the head to rotate in the transverse plane. The muscle groups providing, controlling, and coordinating the movement of the head and the neck are the prevertebrals (anterior), hyoids (anterior), sternocleidomastoid (anterior–lateral), scalene (lateral), levator scapulae (lateral), suboccipitals (posterior), and spleni (posterior).

The spine gains its stability from the intervertebral discs and from the surrounding ligaments and muscles (Fig. 5.14). The discs and ligaments provide intrinsic stability, and the muscles supply extrinsic support. The muscles of the spine exist in pairs. The anterior portion of the spine contains the abdominal muscles: the rectus abdominis, transverse abdominus, external obliques, and internal obliques. These muscles provide the necessary force for trunk flexion and maintain the internal organs in proper position. There are three layers of posterior trunk muscles: erector spinae, semispinalis, and the deep posterior spinal muscle groups. The primary function of the muscles located at the posterior portion of the spine is to provide trunk extension. These muscles also support the spine against the effects of gravity. The quadratus lumborum muscle is important in lateral trunk flexion. It also stabilizes the pelvis and lumbar spine.

Fig. 5.13 The spinal column: (1) cervical vertebrae, (2) thoracic vertebrae, (3) lumbar vertabrae, (4) sacrum

Fig. 5.14 Selected muscles of the neck and spine: (1) splenius, (2) sternocleidomastoid, (3) hyoid, (4) levator scapula, (5) erector spinae, (6) obliques, (7) rectus abdominis, (8) transversus abdominis

The lateral flexion of the trunk results from the actions of the abdominal and posterior muscles. The rotational movement of the trunk is controlled by the simultaneous action of anterior and posterior muscles.

The spinal column is vulnerable to various injuries. The most severe injury involves the spinal cord, which is immersed in fluid and protected by the bony structure. Other critical injuries include fractured vertebrae and herniated intervertebral disks. Lower back pain may also result from strains in the lower regions of the spine.

Example 5.3 Consider the position of the head and the neck shown in Fig. 5.15. Also shown are the forces acting on the head. The head weighs $W = 50\,\text{N}$ and its center of gravity is located at point C. F_M is the magnitude of the resultant force exerted by the neck extensor muscles, which is applied on the skull at point A. The atlantooccipital joint center is located at point B. For this flexed position of the head, it is estimated that the line of action of the neck muscle force makes an angle $\theta = 30°$ and the line of action of the joint reaction force makes an angle $\beta = 60°$ with the horizontal.

What tension must the neck extensor muscles exert to support the head? What is the compressive force applied on the first cervical vertebra at the atlantooccipital joint?

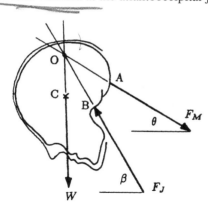

Fig. 5.15 Forces on the skull form a concurrent system

Solution

We have a three-force system with two unknowns: magnitudes F_M and F_j of the muscle and joint reaction forces. Since the problem has a relatively complicated geometry, it is convenient to utilize the condition that for a body to be in equilibrium the force system acting on it must be either concurrent or parallel. In this case, it is clear that the forces involved do not form a parallel force system. Therefore, the system of forces under consideration must be concurrent. Recall that a system of forces is concurrent if the lines of action of all forces have a common point of intersection.

In Fig. 5.15, the lines of action of all three forces acting on the head are extended to meet at point O. In Fig. 5.16,

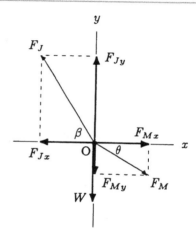

Fig. 5.16 Components of the forces acting on the head

the forces $\underline{W}$, $\underline{F}_M$, and $\underline{F}_J$ acting on the skull are translated to point O, which is also chosen to be the origin of the xy coordinate frame. The rectangular components of the muscle and joint reaction forces in the x and y directions are

$$F_{Mx} = F_M\,\cos\theta. \tag{i}$$

$$F_{My} = F_M\,\sin\theta. \tag{ii}$$

$$F_{Jx} = F_J\,\cos\beta. \tag{iii}$$

$$F_{Jy} = F_J\,\sin\beta. \tag{iv}$$

The translational equilibrium conditions in the x and y directions will yield

$$\sum F_x = 0 \quad \text{that is}: \; -F_{Jx} + F_{Mx} = 0,$$
$$\text{then } F_{Jx} = F_{Mx}. \tag{v}$$

$$\sum F_y = 0 \quad \text{that is}: \; -W - F_{My} + F_{Jy} = 0,$$
$$\text{then } F_{Jy} = W + F_{My}. \tag{vi}$$

Substitute Eqs. (*i*) and (*iii*) into Eq. (*v*):

$$F_J\,\cos\beta = F_M\,\cos\theta. \tag{vii}$$

Substitute Eqs. (*ii*) and (*iv*) into Eq. (*vi*):

$$F_J\,\sin\beta = W + F_M\,\sin\theta \tag{viii}$$

Substitute this equation into Eq. (*viii*), that is,

$$\frac{F_M \times \cos\theta}{\cos\beta} \times \sin\beta = W + F_M\,\sin\theta,$$

$$F_M \times \cos\theta \ \tan\beta = W + F_M \ \sin\theta,$$

then

$$\tan\beta = \frac{W + F_M \ \sin\theta}{F_M \ \cos\theta}. \qquad (ix)$$

Equation (ix) can now be solved for the unknown muscle force F_M:

$$F_M \ \cos\theta \ \tan\beta = W + F_M \ \sin\theta$$

$$F_M (\cos\theta \ \tan\beta - \sin\theta) = W$$

$$F_M = \frac{W}{\cos\theta \ \tan\beta - \sin\theta}. \qquad (x)$$

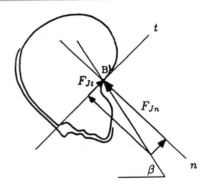

Fig. 5.17 Normal and shear components of the joint reaction force

Equation (x) gives the tension in the muscle as a function of the weight W of the head and the angles θ and β that the lines of action of the muscle and joint reaction forces make with the horizontal. Substituting the numerical values of W, θ, and β will yield

$$F_M = \frac{50}{(\cos 30°)(\tan 60°) - (\sin 30°)} = 50\,\text{N}.$$

From Eqs. (i) and (ii),

$$F_{Mx} = (50)(\cos 30°) = 43\,\text{N} \quad (+x)$$
$$F_{My} = (50)(\sin 30°) = 25\,\text{N} \quad (-y).$$

From Eqs. (v) and (vi),

$$F_{Jx} = 43\,\text{N} \qquad (-x)$$
$$F_{Jy} = 50 + 25 = 75\,\text{N} \quad (+y).$$

The resultant of the joint reaction force can be computed either from Eqs. (iii) or (iv). Using Eq. (iii)

$$F_J = \frac{F_{Jx}}{\cos\beta} = \frac{43}{\cos 60°} = 86\,\text{N}.$$

Remarks

- The extensor muscles of the head must apply a force of 50 N to support the head in the position considered. The reaction force developed at the atlantooccipital joint is about 86 N.
- The joint reaction force can be resolved into two rectangular components, as shown in Fig. 5.17. F_{Jn} is the magnitude of the normal component of $\underline{F}_J$ compressing the articulating joint surface, and F_{Jt} is the magnitude of its tangential component having a shearing effect on the

joint surfaces. Forces in the muscles and ligaments of the neck operate in a manner to counterbalance this shearing effect.

Example 5.4 Consider the weight lifter illustrated in Fig. 5.18, who is bent forward and lifting a weight W_0. At the position shown, the athlete's trunk is flexed by an angle θ as measured from the upright (vertical) position.

Fig. 5.18 A weight lifter

The forces acting on the lower portion of the athlete's body are shown in Fig. 5.19 by considering a section passing through the fifth lumbar vertebra. A mechanical model of the athlete's lower body (the pelvis and legs) is illustrated in Fig. 5.20 along with the geometric parameters of the problem under consideration. W is the total weight of the athlete, W_1 is the weight of the legs including the pelvis, $(W + W_0)$ is the total ground reaction force applied to the athlete through the feet (at point C), F_M is the magnitude of the resultant force exerted by the erector spinae muscles supporting the trunk, and F_J is the magnitude of the compressive force generated at the union (point O) of the sacrum

Fig. 5.19 Forces acting on the lower body of the athlete

Fig. 5.20 Free-body diagram

and the fifth lumbar vertebra. The center of gravity of the legs including the pelvis is located at point B. Relative to point O, the lengths of the lever arms of the muscle force,

lower body weight, and ground reaction force are measured as a, b, and c, respectively.

Assuming that the line of pull of the resultant muscle force exerted by the erector spinae muscles is parallel to the trunk (i.e., making an angle θ with the vertical), determine F_M and F_J in terms of, b, c, θ, W_0, W_1, and W.

Solution

In this case, there are three unknowns: F_M, F_{Jx}, and F_{Jy}. The lengths of the lever arms of the muscle force, ground reaction force, and the gravitational force of the legs including the pelvis are given as measured from point O. Therefore, we can apply the rotational equilibrium condition about point O to determine the magnitude F_M of the resultant force exerted by the erector spinae muscles. Considering clockwise moments to be positive

$$\sum M_o = 0 : \quad aF_M + bW_1 - c(W + W_0) = 0.$$

Solving this equation for F_M will yield

$$F_M = \frac{c(W + W_0) - bW_1}{a}. \tag{i}$$

For given numerical values of, b, c, θ, W_0, W_1, and W, Eq. (i) can be used to determine the magnitude of the resultant muscle force. Once F_M is calculated, its components in the x and y directions can be determined using

$$F_{Mx} = F_M \sin\theta. \tag{ii}$$

$$F_{My} = F_M \cos\theta. \tag{iii}$$

The horizontal and vertical components of the reaction force developed at the sacrum can now be determined by utilizing the translational equilibrium conditions of the lower body of the athlete in the x and y directions:

$$\sum F_x = 0 \quad \text{That is, } F_{Mx} - F_{Jx} = 0, \text{ then } F_{Jx} = F_{Mx}. \tag{iv}$$

$$\sum F_y = 0 \quad \text{That is, } F_{My} - F_{Jy} - W_1 + (W + W_0) = 0,$$
$$\text{then } F_{Jy} = F_{My} + W + W_0 - W_1.$$
$$\tag{v}$$

Assume that at an instant the athlete is bent so that his trunk makes an angle $\theta = 45°$ with the vertical, and that the lengths of the lever arms are measured in terms of the height h of the athlete and the weights are given in terms of the weight W of the athlete as $a = 0.02h$, $b = 0.08h$, $c = 0.12h$, $W_0 = W$, and $W_1 = 0.4W$. Using Eq. (i)

$$F_M = \frac{(0.12h)(W + W) - (0.08h)(0.4W)}{0.02h} = 10.4\,\text{W}.$$

From Eqs. (*ii*) and (*iii*)

$$F_{Mx} = (10.4\,\text{W})\,(\sin 45°) = 7.4\,\text{W}$$
$$F_{My} = (10.4\,\text{W})\,(\cos 45°) = 7.4\,\text{W}.$$

From Eqs. (*iv*) and (*v*)

$$F_{Jx} = 7.4\,W$$
$$F_{Jy} = 7.4\,W + W + W - 0.4\,W = 9.0\,W.$$

Therefore, the magnitude of the resultant force on the sacrum is

$$F_J = \sqrt{(F_{Jx})^2 + (F_{Jy})^2} = 11.7\,\text{W}.$$

A remark

- The results obtained are quite significant. While the athlete is bent forward by 45° and lifting a weight with magnitude equal to his own body weight, the erector spinae muscles exert a force more than ten times the weight of the athlete and the force applied to the union of the sacrum and the fifth lumbar vertebra is about 12 times that of the body weight.

5.8   Mechanics of the Hip

The articulation between the head of the femur and the acetabulum of the pelvis (Fig. 5.21) forms a diarthrodial joint. The stability of the hip joint is provided by its relatively rigid ball-and-socket type of configuration, its ligaments, and by the large and strong muscles crossing it. The femoral head fits well into the deep socket of the acetabulum. The ligaments of the hip joint, as well as the labrum (a flat rim of fibrocartilage), support and hold the femoral head in the acetabulum as the femoral head moves. The construction of the hip joint is such that it is very stable and has a great deal of mobility, thereby allowing a wide range of motion required for activities such as walking, sitting, and squatting. Movements of the femur about the hip joint include flexion and extension, abduction and adduction, and inward and outward rotation. In some instances, the extent of these movements is constrained by ligaments, muscles, and/or the bony structure of the hip. The articulating surfaces of the femoral head and the acetabulum are lined with hyaline cartilage. Derangements of the hip can produce altered force distributions in the joint cartilage, leading to degenerative arthritis.

The pelvis consists of the ilium, ischium, and pubis bones, and the sacrum. At birth and during growth, the

Fig. 5.21 Pelvis and the hip: (1) ilium, (2) sacrum, (3) acetabulum, (4) ischium, (5) greater trochanter, (6) lesser trochanter, (7) femur

bones of the pelvis are distinct. In adults, the bones of the pelvis are fused and form synarthrodial joints that allow no movement. The pelvis is located between the spine and the two femurs. The position of the pelvis makes it relatively less stable. Movements of the pelvis occur primarily for the purpose of facilitating the movements of the spine or the femurs. There are no muscles whose primary purpose is to move the pelvis. Movements of the pelvis are caused by the muscles of the trunk and the hip.

Based on their primary actions, the muscles of the hip joint can be divided into several groups (Fig. 5.22). The psoas, iliacus, rectus femoris, pectineus, and tensor fascia latae are the primary hip flexors. They are also used to carry out activities such as running or kicking. The gluteus maximus and the hamstring muscles (the biceps femoris, semitendinosus, and semimembranosus) are hip extensors. The hamstring muscles also function as knee flexors. The gluteus medius and gluteus minimus are hip abductor muscles providing for the inward rotation of the femur. The gluteus medius is also the primary muscle group stabilizing the pelvis in the frontal plane. The adductor longus, adductor brevis, adductor magnus, and gracilis muscles are the hip adductors. There are also small, deeply placed muscles (outward rotators) that provide for the outward rotation of the femur.

The hip muscles predominantly suffer contusions and strains occurring in the pelvis region.

Example 5.5 During walking and running, we momentarily put all of our body weight on one leg (the right leg in Fig. 5.23). The forces acting on the leg carrying the total body weight are shown in Fig. 5.24 during such a single-leg stance. F_M is the magnitude of the resultant force exerted by the hip abductor muscles, F_J is the magnitude of the joint reaction force applied by the pelvis on the femur, W_1 is the weight of the leg, W is the total weight of the body applied as a normal force by the ground on the leg. The angle between the line of action of the resultant muscle force and the horizontal is designated by θ.

Fig. 5.22 Muscles of the hip (**a**) anterior and (**b**) posterior views: (1) psoas, (2) iliacus, (3) tensor fascia latae, (4) rectus femoris, (5) sartorius, (6) gracilis, (7) gluteus minimis, (8) pectineus, (9) adductors, (10), (11) gluteus maximus & medius, (12) lateral rotators, (13) biceps femoris, (14) semitendinosus, (15) semimembranosus

Fig. 5.23 Single-leg stance

Fig. 5.24 Forces acting on the right leg carrying the entire weight of the body

A mechanical model of the leg, rectangular components of the forces acting on it, and the parameters necessary to define the geometry of the problem are shown in Fig. 5.25. O is a point along the instantaneous axis of rotation of the hip joint, point A is where the hip abductor muscles are attached to the femur, point B is the center of gravity of the leg, and point C is where the ground reaction force is applied on the foot. The distances between point A and points O, B, and C are specified as a, b, and c, respectively. α is the angle of inclination of the femoral neck to the horizontal, and β is the angle that the long axis of the femoral shaft makes with the horizontal. Therefore, $\alpha + \beta$ is approximately equal to the total neck-to-shaft angle of the femur.

Determine the force exerted by the hip abductor muscles and the joint reaction force at the hip to support the leg and the hip in the position shown.

Solution 1: Utilizing the Free-Body Diagram of the Leg
For the solution of the problem, we can utilize the free-body diagram of the right leg supporting the entire weight of the person. In Fig. 5.25a, the muscle and joint reaction forces are shown in terms of their components in the x and y directions. The resultant muscle force has a line of action that makes an angle θ with the horizontal. Therefore,

$$F_{Mx} = F_M \cos \theta. \tag{i}$$

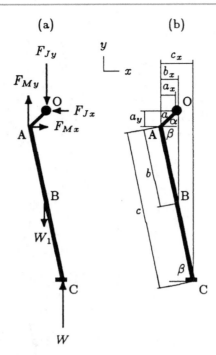

Fig. 5.25 Free-body diagram of the leg (**a**) and the geometric parameters (**b**)

$$F_{My} = F_M \sin \theta. \qquad (ii)$$

Since angle θ is specified (given as a measured quantity), the only unknown for the muscle force is its magnitude F_M. For the joint reaction force, neither the magnitude nor the direction is known. With respect to the axis of the hip joint located at point O, a_x in Fig. 5.25b is the moment arm of the vertical component F_{My} of the muscle force, and a_y is the moment arm of the horizontal component of the muscle force F_{Mx}. Similarly, $(b_x - a_x)$ is the moment arm for W_1 and $(c_x - a_x)$ is the moment arm for the force W applied by the ground on the leg.

From the geometry of the problem

$$a_x = a \cos \alpha. \qquad (iii)$$

$$a_y = a \sin \alpha. \qquad (iv)$$

$$b_x = b \cos \beta. \qquad (v)$$

$$c_x = c \cos \beta. \qquad (vi)$$

Now that the horizontal and vertical components of all forces involved, and their moment arms with respect to point O are established, the condition for the rotational equilibrium of the leg about point O can be utilized to determine the

magnitude of the resultant muscle force applied at point A. Assuming that the clockwise moments are positive:

$$\sum M_o = 0 : \quad a_x F_{My} - a_y F_{Mx} - (c_x - a_x)W$$
$$+ (b_x - a_x)W_1 = 0.$$

Substituting Eqs. (i) through (vi) into the above equation:

$$(a \cos \alpha)(F_M \sin \theta) - (a \sin \alpha)(F_M \cos \theta)$$
$$- (c \cos \beta - a \cos \alpha)W + (b \cos \beta - a \cos \alpha)W_1 = 0.$$

Solving this equation for the muscle force:

$$F_M = \frac{(cW - bW_1)\cos \beta - a(W - W_1)\cos \alpha}{a(\cos \alpha \sin \theta - \sin \alpha \cos \theta)}. \qquad (vii)$$

Notice that the denominator of Eq. (vii) can be simplified as $a \sin(\theta - \alpha)$. To determine the components of the joint reaction force, we can utilize the horizontal and vertical equilibrium conditions of the leg:

$$\sum F_x = 0 : \quad F_{Jx} = F_{Mx} = F_M \cos \theta. \qquad (viii)$$

$$\sum F_y = 0 : \quad F_{Jy} = F_{My} + W - W_1$$
$$F_{Jy} = F_M \sin \theta + W - W_1. \qquad (ix)$$

Therefore, the resultant force acting at the hip joint is

$$F_J = \sqrt{(F_{Jx})^2 + (F_{Jy})^2}. \qquad (x)$$

Assume that the geometric parameters of the problem and the weight of the leg are measured in terms of the person's height h and total weight W as follows: $a = 0.05h$, $b = 0.20h$, $c = 0.52h$, $\alpha = 45°$, $\beta = 80°$, $\theta = 70°$, and $W_1 = 0.17 W$. The solution of the above equations for the muscle and joint reaction forces will yield $F_M = 2.6 W$ and $F_J = 3.4 W$, the joint reaction force making an angle $\varphi = \tan^{-1}(F_{Jy}/F_{Jx}) = 74.8°$ with the horizontal.

Solution 2: Utilizing the Free-Body Diagram of the Upper Body Here, we have an alternative approach to the solution of the same problem. In this case, instead of the free-body diagram of the right leg, the free-body diagram of the upper body (including the left leg) is utilized. The forces acting on the upper body are shown in Figs. 5.26 and 5.27. F_M is the magnitude of the resultant force exerted by the hip abductor muscles applied on the pelvis at point D. θ is again the angle between the line of action of the resultant muscle force and the horizontal. F_J is the magnitude of the reaction force applied by the head of the femur on the hip joint at

Fig. 5.26 Forces acting on the pelvis during a single-leg (right leg) stance

Fig. 5.27 Forces involved form a concurrent system

point E. $W_2 = W - W_1$ (total body weight minus the weight of the right leg) is the weight of the upper body and the left leg acting as a concentrated force at point G. Note that point G is

not the center of gravity of the entire body. Since the right leg is not included in the free-body, the left-hand side of the body is "heavier" than the right-hand side, and point G is located to the left of the original center of gravity (a point along the vertical dashed line in Fig. 5.27) of the person. The location of point G can be determined utilizing the method provided in Sect. 4.12.

By combining the individual weights of the segments constituting the body under consideration, the problem is reduced to a three-force system. It is clear from the geometry of the problem that the forces involved do not form a parallel system. Therefore, for the equilibrium of the body, they have to form a concurrent system of forces. This implies that the lines of action of the forces must have a common point of intersection (point Q in Fig. 5.27), which can be obtained by extending the lines of action of $\underline{W}_2$ and $\underline{F}_M$. A line passing through points Q and E designates the line of action of the joint reaction force $\underline{F}_j$. The angle φ that $\underline{F}_j$ makes with the horizontal can now be measured from the geometry of the problem. Since the direction of $\underline{F}_j$ is determined through certain geometric considerations, the number of unknowns is reduced by one. As illustrated in Fig. 5.28, the unknown magnitudes F_M and F_J of the muscle and joint reaction forces can now be determined simply by translating $\underline{W}_2$, $\underline{F}_M$, and $\underline{F}_j$ to point Q, and decomposing them into their components along the horizontal (x) and vertical (y) directions:

$$
\begin{aligned}
F_{Mx} &= F_M \cos\theta, \\
F_{My} &= F_M \sin\theta, \\
F_{Jx} &= F_J \cos\varphi, \\
F_{Jy} &= F_J \sin\varphi.
\end{aligned}
\tag{xi}
$$

For the translational equilibrium in the x and y directions

$$\sum F_x = 0 \qquad \text{That is, } -F_{Mx} + F_{Jx} = 0,$$

$$\text{then } F_{Jx} = F_{Mx}.$$

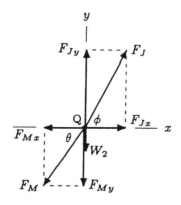

Fig. 5.28 Resolution of the forces into their components

$$\sum F_y = 0 \qquad \text{That is, } F_{Jy} - W_2 - F_{My} = 0,$$
$$\text{then } F_{Jy} = F_{My} + W_2.$$

Considering Eq. (xi)

$$F_J \ \cos\varphi = F_M \ \cos\theta, \quad \text{and} \qquad (xii)$$

$$F_J \ \sin\varphi = F_M \ \sin\theta + W_2. \qquad (xiii)$$

From Eq. (xii)

$$F_J = \frac{F_M \ \cos\theta}{\cos\varphi}.$$

Substituting this equation into Eq. (xiii) will yield

$$\frac{F_M \ \cos\theta}{\cos\varphi} \ \sin\varphi = F_M \ \sin\theta + W_2,$$

that is

$$\frac{F_M \ \cos\theta}{\cos\varphi} \ \sin\varphi - F_M \ \sin\theta = W_2,$$

$$F_M\left(\frac{\cos\theta \ \sin\varphi - \sin\theta \ \cos\varphi}{\cos\varphi}\right) = W_2,$$

then

$$F_M = \frac{\cos\varphi \ W_2}{\sin(\varphi - \theta)},$$

and

$$F_J = \frac{\cos\theta \ W_2}{\sin(\varphi - \theta)}.$$

For example, if $\theta = 70°$, $\varphi = 74.8°$, and $W_2 = 0.83\,W$ (W is the total weight of the person), then the last two equations will yield $F_M = 2.6\,W$ and $F_J = 3.4\,W$.

How would the muscle and hip joint reaction forces vary if the person is carrying a load of W_0 in each hand during single-leg stance (Fig. 5.29)?

The free-body diagram of the upper body while the person is carrying a load of W_0 in each hand is shown in Fig. 5.30. The system to be analyzed consists of the upper body of the person (including the left leg) and the loads carried in each hand. To counter-balance both the rotational and translational (downward) effects of the extra loads, the hip abductor muscles will exert additional forces, and there will be larger compressive forces generated at the hip joint.

In this case, the number of forces is five. The gravitational pull on the upper body (W_2) and on the masses carried in the hands (W_0) form a parallel force system. If these parallel forces can be replaced by a single resultant force, then the

Fig. 5.29 Carrying a load in each hand

Fig. 5.30 Forces acting on the upper body

number of forces can be reduced to three, and the problem can be solved by applying the same technique explained above (Solution 2). For this purpose, consider the force system shown in Fig. 5.31. Points M and N correspond to the right and left hands of the person where external forces of equal magnitude (W_0) are applied. Point G is the center of gravity of the upper body including the left leg. The vertical

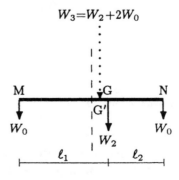

Fig. 5.31 $\underline{W}_3$ is the resultant of the three-force system

dashed line shows the symmetry axis (midline) of the person in the frontal plane, and point G is located to the left of this axis. Note that the distance l_1 between points M and G is greater than the distance l_2 between points N and G. If l_1, l_2, W_2, and W_0 are given, then a new center of gravity (point G′) can be determined by applying the technique of finding the center of gravity of a system composed of a number of parts whose centers of gravity are known (see Sect. 4.12). By intuition, point G′ is located somewhere between the symmetry axis and point G. In other words, G′ is closer to the right hip joint, and therefore, the length of the moment arm of the total weight as measured from the right hip joint is shorter as compared to the case when there is no load carried in the hands. On the other hand, the magnitude of the resultant gravitational force is $W_3 = W_2 + 2W_0$, which overcompensates for the advantage gained by the reduction of the moment arm.

Once the new center of gravity of the upper body is determined, including the left leg and the loads carried in each hand, Eqs. (xi) and (xii) can be utilized to calculate the resultant force exerted by the hip abductor muscles and the reaction force generated at the hip joint:

$$F_M = \frac{\cos \varphi' \, (W_2 + 2W_0)}{\cos \theta \, \sin \varphi' - \sin \theta \, \cos \varphi'}.$$

$$F_j = \frac{\cos \theta \, (W_2 + 2W_0)}{\cos \theta \, \sin \varphi' - \sin \theta \, \cos \varphi'}.$$

Here, Eqs. (xi) and (xii) are modified by replacing the weight W_2 of the upper body with the new total weight $W_3 = W_2 + 2W_0$, and by replacing the angle φ that the line of action of the joint reaction force makes with the horizontal with the new angle φ' (Fig. 5.32). φ' is slightly larger than φ because of the shift of the center of gravity from point G to point G′ toward the right of the person. Also, it is assumed that the angle θ between the line of action of the muscle force and the horizontal remains unchanged.

What happens if the person is carrying a load of W_0 in the left hand during a right-leg stance (Fig. 5.33)?

Fig. 5.33 Carrying a load in one hand

Assuming that the system we are analyzing consists of the upper body, left leg, and the load in hand, the extra load W_0 carried in the left hand will shift the center of gravity of the system from point G to point G″ toward the left of the person. Consequently, the length of the lever arm of the total gravitational force $W_4 = W_2 + W_0$ as measured from the right hip joint (Fig. 5.34) will increase. This will require larger hip abductor muscle forces to counter-balance the clockwise rotational effect of W_4 and also increase the compressive forces at the right hip joint.

Fig. 5.34 Forces acting on the upper body

Fig. 5.32 The problem is reduced to a three-force concurrent system

It can be observed from the geometry of the system analyzed that a shift in the center of gravity from point G to point G″ toward the left of the person will decrease the angle between the line of action of the joint reaction force and the horizontal from φ to φ''. For the new configuration of the free-body shown in Fig. 5.34, Eqs. (xi) and (xii) can again be utilized to calculate the required hip abductor muscle force and joint reaction force produced at the right hip (opposite to the side where the load is carried):

$$F_M = \frac{\cos \varphi'' \ (W_2 + W_0)}{\cos \theta \sin \varphi'' - \sin \theta \cos \varphi''}.$$

$$F_J = \frac{\cos \theta \ (W_2 + W_0)}{\cos \theta \sin \varphi'' - \sin \theta \cos \varphi''}.$$

Remarks

- When the body weight is supported equally on both feet, half of the supra-femoral weight falls on each hip joint. During walking and running, the entire mass of the body is momentarily supported by one joint, and we have analyzed some of these cases.

- The above analyses indicate that the supporting forces required at the hip joint are greater when a load is carried on the opposite side of the body as compared to the forces required to carry the load when it is distributed on either side. Carrying loads by using both hands and by bringing the loads closer to the midline of the body is effective in reducing required musculoskeletal forces.

- While carrying a load on one side, people tend to lean toward the other side. This brings the center of gravity of the upper body and the load being carried in the hand closer to the midline of the body, thereby reducing the length of the moment arm of the resultant gravitational force as measured from the hip joint distal to the load.

- People with weak hip abductor muscles and/or painful hip joints usually lean toward the weaker side and walk with a so-called abductor gait. Leaning the trunk sideways toward the affected hip shifts the center of gravity of the body closer to that hip joint, and consequently reduces the rotational action of the moment of the body weight about the hip joint by reducing its moment arm. This in return reduces the magnitude of the forces exerted by the hip abductor muscles required to stabilize the pelvis.

- Abductor gait can be corrected more effectively with a cane held in the hand opposite to the weak hip, as compared to the cane held in the hand on the same side as the weak hip.

5.9 Mechanics of the Knee

The knee is the largest joint in the body. It is a modified hinge joint. In addition to flexion and extension action of the leg in the sagittal plane, the knee joint permits some automatic inward and outward rotation. The knee joint is designed to sustain large loads. It is an essential component of the linkage system responsible for human locomotion. The knee is extremely vulnerable to injuries.

The knee is a two-joint structure composed of the tibiofemoral joint and the patellofemoral joint (Fig. 5.35). The *tibiofemoral joint* has two distinct articulations between the medial and lateral condyles of the femur and the tibia. These articulations are separated by layers of cartilage, called *menisci*. The lateral and medial menisci eliminate bone-to-bone contact between the femur and the tibia, and function as shock absorbers. The *patellofemoral joint* is the articulation between the patella and the anterior end of the femoral condyles. The patella is a "floating" bone kept in position by the quadriceps tendon and the patellar ligament. It increases the mechanical advantage of the quadriceps muscle, improving its pulling effect on the tibia via the patellar tendon. The stability of the knee is provided by an intricate ligamentous structure, the menisci and the muscles crossing the joint. Most knee injuries are characterized by ligament and cartilage damage occurring on the medial side.

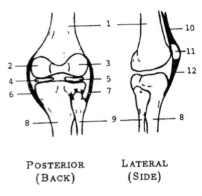

POSTERIOR   **LATERAL**
(BACK)   **(SIDE)**

Fig. 5.35 The knee: (1) femur, (2), medial condyle, (3) lateral condyle, (4) medial meniscus, (5) lateral meniscus, (6) tibial collateral ligament, (7) fibular collateral ligament, (8) tibia, (9) fibula, (10) quadriceps tendon, (11) patella, (12) patellar ligament

The muscles crossing the knee protect it, provide internal forces for movement, and/or control its movement. The muscular control of the knee is produced primarily by the quadriceps muscles and the hamstring muscle group (Fig. 5.36). The quadriceps muscle group is composed of the rectus femoris, vastus lateralis, vastus medialis, and vastus intermedius muscles. The rectus femoris muscle has attachments at the anterior–inferior iliac spine and the patella, and its primary actions are the flexion of the hip and the extension of the knee. The vastus lateralis, medialis, and intermedius muscles connect the femur and tibia through the patella, and they are all knee extensors. The biceps femoris, semitendinosus, and semimembranosus muscles make up the hamstring muscle group, which help control the extension of the hip, flexion of the knee, and some inward–outward rotation of the tibia. Semitendinosus and semimembranosus muscles have proximal attachments on the pelvic bone and distal attachments on the tibia.

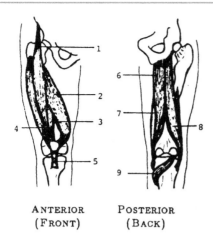

ANTERIOR
(FRONT)

POSTERIOR
(BACK)

Fig. 5.36 Muscles of the knee: (1) rectus femoris, (2) vastus medialis, (3) vastus intermedius, (4) vastus lateralis, (5) patellar ligament, (6) semitendinosus, (7) semimembranosus, (8) biceps femoris, (9) gastrocnemius

The biceps femoris has proximal attachments on the pelvic bone and the femur, and distal attachments on the fibula. There is also the popliteus muscle that has attachments on the femur and tibia. The primary function of this muscle is knee flexion. The other muscles of the knee are sartorius, gracilis, gastrocnemius, and plantaris.

Example 5.6 Consider a person wearing a weight boot, and from a sitting position, doing lower leg flexion/extension exercises to strengthen the quadriceps muscles (Fig. 5.37).

Fig. 5.37 Exercising the muscles around the knee joint

Forces acting on the lower leg and a simple mechanical model of the leg are illustrated in Fig. 5.38. W_1 is the weight

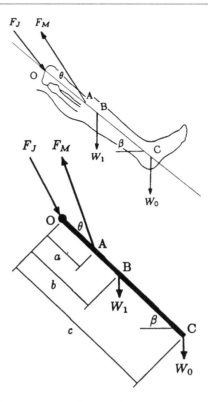

Fig. 5.38 Forces acting on the lower leg

of the lower leg, W_0 is the weight of the boot, F_M is the magnitude of the tensile force exerted by the quadriceps muscle on the tibia through the patellar tendon, and F_J is the magnitude of the tibiofemoral joint reaction force applied by the femur on the tibial plateau. The tibiofemoral joint center is located at point O, the patellar tendon is attached to the tibia at point A, the center of gravity of the lower leg is located at point B, and the center of gravity of the weight boot is located at point C. The distances between point O and points A, B, and C are measured as a, b, and c, respectively. For the position of the lower leg shown, the long axis of the tibia makes an angle β with the horizontal, and the line of action of the quadriceps muscle force makes an angle θ with the long axis of the tibia.

Assuming that points O, A, B, and C all lie along a straight line, determine F_M and F_J in terms of a, b, c, θ, β, W_1, and W_0.

Solution Horizontal (x) and vertical (y) components of the forces acting on the leg and their lever arms as measured from the knee joint located at point O are shown in Fig. 5.39. The components of the muscle force are

$$F_{Mx} = F_M \cos(\theta + \beta) \qquad \text{(i)}$$

$$F_{My} = F_M \sin(\theta + \beta) \qquad \text{(ii)}$$

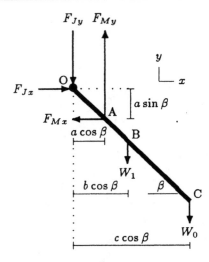

Fig. 5.39 Force components, and their lever arms

There are three unknowns, namely F_M, F_{Jx}, and F_{Jy}. For the solution of this two-dimensional (plane) problem, all three equilibrium conditions must be utilized. Assuming that the counter-clockwise moments are positive, consider the rotational equilibrium of the lower leg about point O:

$$\sum M_0 = 0: \quad (a \cos \beta)F_{My} - (a \sin \beta)F_{Mx}$$
$$- (b \cos \beta)W_1 - (c \cos \beta) W_0 = 0$$

Substituting Eqs. (*i*) and (*ii*) into the above equation, and solving it for F_M will yield

$$F_M = \frac{(bW_1 + cW_0) \cos \beta}{a[\cos \beta \sin(\theta + \beta) - \sin \beta \cos(\theta + \beta)]} \quad \text{(iii)}$$

Note that this equation can be simplified by considering that $[\cos \beta \, \sin(\theta + \beta) - \sin \beta \, \cos(\theta + \beta)] = \sin \theta$, that is

$$F_M = \frac{bW_1 + cW_0 \cos \beta}{a \sin \theta}$$

Equation (*iii*) yields the magnitude of the force that must be exerted by the quadriceps muscles to support the leg when it is extended forward making an angle β with the horizontal. Once F_M is determined, the components of the reaction force developed at the knee joint along the horizontal and vertical directions can also be evaluated by considering the translational equilibrium of the lower leg in the x and y directions:

$$\sum F_x = 0: \quad F_{Jx} = F_{Mx} = F_M \cos(\theta + \beta)$$

$$\sum F_y = 0: \quad F_{Jy} = F_{My} - W_0 - W_1$$
$$F_{Jy} = F_M \sin(\theta + \beta) - W_0 - W_1$$

The magnitude of the resultant compressive force applied on the tibial plateau at the knee joint is

$$F_J = \sqrt{(F_{jx})^2 + (F_{jy})^2} \quad \text{(iv)}$$

Furthermore, once F_{jx} and F_{jy} are determined, an angle φ that the joint reaction force F_J makes with the horizontal can also be determined:

$$\varphi = arc \, \tan\left(\frac{F_{Jy}}{F_{Jx}}\right)$$

Assume that the geometric parameters and the weights involved are given as $a = 12 \, \text{cm}$, $b = 22 \, \text{cm}$, $c = 50 \, \text{cm}$, $W_1 = 150 \, \text{N}$, $W_0 = 100 \, \text{N}$, $\theta = 15°$, and $\beta = 45°$, then by using Eqs. (*iii*) and (*iv*):

$$F_M = 1,893 \, \text{N}, \quad F_J = 1,681 \, \text{N}, \quad \varphi = 55.7°$$

Remarks

- The force $\underline{F}_M$ exerted by the quadriceps muscle on the tibia through the patellar tendon can be expressed in terms of two components normal and tangential to the long axis of the tibia (Fig. 5.40). The primary function of the normal component $\underline{F}_{Mn}$ of the muscle force is to rotate the tibia about the knee joint, while its tangential component $\underline{F}_{Mt}$ tends to translate the lower leg in a direction collinear with the long axis of the tibia and applies a compressive force on the articulating surfaces of the tibiofemoral joint. Since the normal component of $\underline{F}_M$ is a sine function of angle θ, a larger angle between the patellar tendon and the long axis of the tibia indicates a larger rotational effect of the muscle exertion. This implies that for large θ, less muscle force is wasted to compress the knee joint, and a larger portion of the muscle tension is utilized to rotate the lower leg about the knee joint.

Fig. 5.40 Rotational and translatory components of $\underline{F}_M$

- One of the most important biomechanical functions of the patella is to provide anterior displacement of the quadriceps and patellar tendons, thus lengthening the lever arm of the knee extensor muscle forces with respect to the center of rotation of the knee by increasing angle θ (Fig. 5.41a). Surgical removal of the patella brings the patellar tendon closer to the center of rotation of the knee joint (Fig. 5.41b), which causes the length of the lever arm of the muscle force to decrease ($d_2 < d_1$). Losing the advantage of having a relatively long lever arm, the quadriceps muscle has to exert more force than normal to rotate the lower leg about the knee joint.

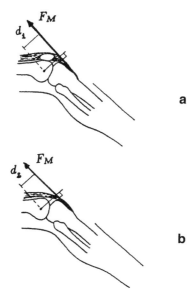

Fig. 5.41 Patella increases the length of the lever arm

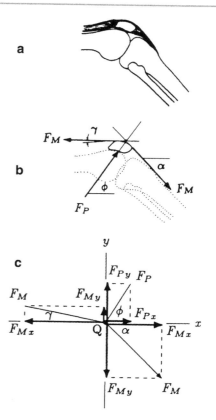

Fig. 5.42 Static analysis of the forces acting on the patella

- The human knee has a two-joint structure composed of the tibiofemoral and patellofemoral joints. Notice that the quadriceps muscle goes over the patella, and the patella and the muscle form a pulley–rope arrangement. The higher the tension in the muscle, the larger the compressive force (pressure) the patella exerts on the patellofemoral joint.

We have analyzed the forces involved around the tibiofemoral joint by considering the free-body diagram of the lower leg. Having determined the tension in the patellar tendon, and assuming that the tension is uniform throughout the quadriceps, we can calculate the compressive force applied on the patellofemoral joint by considering the free-body diagram of the patella (Fig. 5.42). Let F_M be the uniform magnitude of the tensile force in the patellar and quadriceps tendons, F_p be the magnitude of the force exerted on the patellofemoral joint, α be the angle between the patellar tendon and the horizontal, γ be the angle between the quadriceps tendon and the horizontal, and φ be the unknown angle between the line of action of the compressive reaction force at the joint (Fig. 5.42b) and the horizontal. We have a three-force system and for the equilibrium of the patella it has to be concurrent.

We can first determine the common point of intersection Q by extending the lines of action of patellar and quadriceps tendon forces. A line connecting point Q and the point of application of $\underline{F}_p$ will correspond to the line of action of $\underline{F}_p$. The forces can then be translated to point Q (Fig. 5.42c), and the equilibrium equations can be applied. For the equilibrium of the patella in the x and y directions:

$$\sum F_x = 0 : \quad F_p \cos\varphi = F_M(\cos\gamma - \cos\alpha) \quad (v)$$

$$\sum F_y = 0 : \quad F_p \sin\varphi = F_M(\sin\alpha - \sin\gamma) \quad (vi)$$

These equations can be solved simultaneously for angle φ and the magnitude F_p of the compressive force applied by the femur on the patella at the patellofemoral joint:

From Eq. (v)

$$F_p = \frac{F_M(\cos\gamma - \cos\gamma)}{\cos\varphi};$$

From Eq. (vi)

$$F_p = \frac{F_M(\sin\gamma - \sin\gamma)}{\sin\varphi}, \text{ that is}$$

$$\frac{F_M(\cos\gamma - \cos\gamma)}{\cos\varphi} = \frac{F_M(\sin\gamma - \sin\gamma)}{\sin\varphi}, \text{ then:}$$

$$\sin\varphi(\cos\gamma - \cos\gamma) = \cos\varphi(\sin\gamma - \sin\gamma),$$

$$\tan\varphi = \frac{\sin\gamma - \sin\gamma}{\cos\gamma - \cos\gamma} \text{ and}$$

$$\varphi = \tan^{-1}\left(\frac{\sin\gamma - \sin\gamma}{\cos\gamma - \cos\gamma}\right)$$

Once angle φ is determined, then the magnitude of force exerted on the patellofemoral joint F_p can also be determined:

$$F_p = F_M\left(\frac{\cos\gamma - \cos\gamma}{\cos\varphi}\right).$$

5.10 Mechanics of the Ankle

The ankle is the union of three bones: the tibia, fibula, and the talus of the foot (Fig. 5.43). Like other major joints in the lower extremity, the ankle is responsible for load-bearing and kinematic functions. The ankle joint is inherently more stable than the knee joint which requires ligamentous and muscular restraints for its stability.

POSTERIOR   SUPERIOR
(BACK)   (TOP)

Fig. 5.43 The ankle and foot: (1) tibia, (2) fibula, (3) medial malleolus, (4) lateral malleolus, (5) talus, (6) calcaneus

The ankle joint complex consists of the tibiotalar, fibulotalar, and distal tibiofibular articulations. The *ankle (tibiotalar) joint* is a hinge or ginglymus-type articulation between the spool-like convex surface of the trochlea of the talus and the concave distal end of the tibia. Being a hinge joint, the ankle permits only flexion–extension (dorsiflexion–plantar flexion) movement of the foot in the sagittal

plane. Other foot movements include inversion and eversion, inward and outward rotation, and pronation and supination. These movements occur about the foot joints such as the subtalar joint between the talus and calcaneus and the transverse tarsal joints, talonavicular and calcaneocuboid.

The ankle mortise is maintained by the shape of the three articulations, and the ligaments and muscles crossing the joint. The integrity of the ankle joint is improved by the medial (deltoid) and lateral collateral ligament systems, and the interosseous ligaments. There are numerous muscle groups crossing the ankle. The most important ankle plantar flexors are the gastrocnemius and soleus muscles (Fig. 5.44). Both the gastrocnemius and soleus muscles are located in the posterior compartment of the leg and have attachments to the posterior surface of the calcaneus via the Achilles tendon. The gastrocnemius crosses the knee and ankle joints and has functions in both. In the knee, it collaborates with knee flexion and in the ankle it is the main plantar flexor. The plantar extensors or dorsiflexors are anterior muscles. They are the tibialis anterior, extensor digitorum longus, extensor hallucis longus and peroneous tertius muscles. The primary function of the lateral muscles (the peroneus longus and peroneus brevis) is to exert and plantarflex the ankle.

Fig. 5.44 Ankle muscles, (**a**) posterior, (**b**) anterior and (**c**) lateral views: (1) gastrocnemius, (2) soleus, (3) Achilles tendon, (4) tibialis anterior, (5) extensor digitorum longus, (6) extensor hallucis longus, (7) peroneus longus, (8) peroneus brevis

The ankle joint responds poorly to small changes in its anatomical configuration. Loss of kinematic and structural restraints due to severe sprains can seriously affect ankle stability and can produce malalignment of the ankle joint surfaces. The most common ankle injury, inversion sprain, occurs when the body weight is forcefully transmitted to the ankle while the foot is inverted (the sole of the foot facing inward).

Example 5.7 Consider a person standing on tiptoe on one foot (a strenuous position illustrated). The forces acting on the foot during this instant are shown in Fig. 5.45. W is the person's weight applied on the foot as the ground reaction force, F_M is the magnitude of the tensile force exerted by the gastrocnemius and soleus muscles on the calcaneus through the Achilles tendon, and F_J is the magnitude of the ankle joint reaction force applied by the tibia on the dome of the talus. The weight of the foot is small compared to the weight of the body and is therefore ignored. The Achilles tendon is attached to the calcaneus at point A, the ankle joint center is located at point B, and the ground reaction force is applied on the foot at point C. For this position of the foot, it is estimated that the line of action of the tensile force in the Achilles tendon makes an angle θ with the horizontal, and the line of action of the ankle joint reaction force makes an angle β with the horizontal.

Fig. 5.45 Forces acting on the foot form a concurrent system of forces

Assuming that the relative positions of points A, B, and C are known, determine expressions for the tension in the Achilles tendon and the magnitude of the reaction force at the ankle joint.

Solution

We have a three-force system composed of muscle force $\underline{F}_M$, joint reaction force $\underline{F}_J$, and the ground reaction force $\underline{W}$. From the geometry of the problem, it is obvious that for the position of the foot shown, the forces acting on the foot do not form a parallel force system. Therefore, the force system must be a concurrent one. The common point of intersection (point O in Fig. 5.45) of these forces can be determined by extending the lines of action of $\underline{W}$ and $\underline{F}_M$. A straight line passing through both points O and B represents the line of action of the joint reaction force. Assuming that the relative positions of points A, B, and C are known (as stated in the problem), the angle (say β) of the line of action of the joint reaction force can be measured.

Once the line of action of the joint reaction force is determined by graphical means, the magnitudes of the joint reaction and muscle forces can be calculated by translating all three forces involved to the common point of intersection at O (Fig. 5.46). The two unknowns F_M and F_J can now be determined by applying the translational equilibrium conditions in the horizontal (x) and vertical (y) directions. For this purpose, the joint reaction and muscle forces must be decomposed into their rectangular components first:

$$F_{Mx} = F_M \cos \theta$$
$$F_{My} = F_M \sin \theta$$
$$F_{Jx} = F_J \cos \beta$$
$$F_{Jy} = F_J \sin \beta.$$

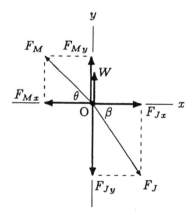

Fig. 5.46 Components of the forces acting on the foot

For the translational equilibrium of the foot in the horizontal and vertical directions:

$$\sum F_x = 0: \quad F_{Jx} = F_{Mx}, \quad \text{that is, } F_J \cos\beta = F_M \cos\theta$$

$$\sum F_y = 0: \quad F_{Jy} = F_{My} + W, \quad \text{that is, } F_J \sin\beta = F_M \sin\theta + W$$

Simultaneous solutions of these equations will yield:

$$F_M = \frac{W \cos\beta}{\cos\theta \, \sin\beta - \sin\theta \, \cos\beta},$$

that is, $F_M = \dfrac{W \cos\beta}{\sin(\beta - \theta)}$

$$F_J = \frac{W \cos\theta}{\cos\theta \, \sin\beta - \sin\theta \, \cos\beta},$$

that is, $F_j = \dfrac{W \cos\theta}{\sin(\beta - \theta)}$

For example, assume that $\theta = 45°$ and $\beta = 60°$. Then

$$F_M = 1.93\,W \quad F_J = 2.73\,W.$$

6.1 Dynamics

Dynamics is the study of bodies in motion. Dynamics is concerned with describing motion and explaining its causes. The general field of dynamics consists of two major areas: kinematics and kinetics. Each of these areas can be further divided to describe and explain linear, angular, or general motion of bodies. The fundamental concepts in dynamics are space (relative position or displacement), time, mass, and force. Other important concepts include velocity, acceleration, torque, moment, work, energy, power, impulse, and momentum.

The broad definitions of basic terms and concepts in dynamics will be introduced in this chapter. The details of kinematic and kinetic characteristics of moving bodies will be covered in the following chapters.

6.2 Kinematics and Kinetics

The field of *kinematics* is concerned with the description of geometric and time-dependent aspects of motion without dealing with the forces causing the motion. Kinematic analyses are based on the relationships between displacement, velocity, and acceleration vectors. These relationships appear in the form of differential and integral equations.

The field of *kinetics* is based on kinematics, and it incorporates into the analyses the effects of forces and torques that cause the motion. Kinetic analyses utilize Newton's second law of motion that can take various mathematical forms. There are a number of different approaches to the solutions of problems in kinetics. These approaches are based on the equations of motion, work and energy methods, and impulse and momentum methods. Different methods may be applied to different situations or depending on what is required to be determined. For example, the equations of motion are used for problems requiring the analysis of acceleration. Energy methods are suitable when a problem

requires the analysis of forces related to changes in velocity. Momentum methods are applied if the forces involved are impulsive, which is the case during impact and collision.

There is also the field of *kinesiology* that is related to the study of human motion characteristics, joint and muscle forces, and neurological and other factors that may be important in studying human motion. The term "kinesiology" is not a mechanical but a medical term. It is commonly used to refer to the biomechanics of human motion.

6.3 Linear, Angular, and General Motions

To study both kinematics and kinetics in an organized manner, it is a common practice to divide them into branches according to whether the motion is translational, rotational, or general. *Translational* or *linear motion* occurs if all parts of a body move the same distance at the same time and in the same direction. For example, if a block is pushed on a horizontal surface, the block will undergo translational motion only (Fig. 6.1). Another typical example of translational motion is the vertical motion of an elevator in a shaft. It should be noted, however, that linear motion does not imply movement along a straight line. In a given time interval, an object may translate in one direction, and it may translate in a different direction during a different time interval.

Fig. 6.1 An object subjected to externally applied forces

N. Özkaya et al., *Fundamentals of Biomechanics: Equilibrium, Motion, and Deformation*,
DOI 10.1007/978-1-4614-1150-5_6, © Springer Science+Business Media, LLC 2012

Rotational or *angular motion* occurs when a body moves in a circular path such that all parts of the body move in the same direction through the same angle at the same time. The angular motion occurs about a central line known as the *axis of rotation*, which lies perpendicular to the plane of motion. For example, for a gymnast doing giant circles, the center of gravity of the gymnast may undergo rotational motion with the centerline of the bar acting as the axis of rotation of the motion (Fig. 6.2).

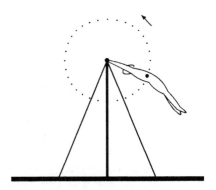

Fig. 6.2 Method of sections

The third class of motion is called *general motion*, which occurs if a body undergoes translational and rotational motions simultaneously. It is more complex to analyze motions composed of both translation and rotation as compared to a pure translational or a pure rotational motion. The diver illustrated in Fig. 6.3 is an example of a body undergoing general motion. Most human body segmental motions are of the general type. For example, while walking, the lower extremities both translate and rotate.

Fig. 6.3 Internal forces and moments

The branch of kinematics that deals with the description of translational motion is known as *linear kinematics* and the branch that deals with rotational motion is *angular kinematics*. Similarly, the field of kinetics can be divided into *linear* and *angular kinetics*.

Linear movements are direct consequences of applied forces. The linear motion of an object occurs in the direction of the net force acting on the object. On the other hand, angular movements are due to the rotational effects of applied forces, which are known as torques. There are linear and angular quantities defined to analyze linear and angular motions, respectively. For example, there are linear and angular displacements, linear and angular velocities, and linear and angular accelerations. It is important to note, however, that linear and angular quantities are not mutually independent. That is, if angular quantities are known, then linear quantities can also be determined, and vice versa.

6.4 Distance and Displacement

In mechanics, *distance* is defined as the total length of the path followed while moving from one point to another, and *displacement* is the length of the straight line joining the two points along with some indication of direction involved. Distance is a scalar quantity (has only a magnitude) and displacement is a vector quantity (has both a magnitude and a direction).

To understand the differences between distance and displacement, consider a person who lives in an apartment building located at the corner of Third Avenue and 18th Street, and walks to work in a building located at the corner of Second Avenue and 17th Street in New York City. In Fig. 6.4, A represents the corner of Third Avenue and 18th Street, B represents the corner of Second Avenue and 18th Street, and C represents the corner of Second Avenue and 17th Street. Every morning this person walks toward the east from A to B, and then towards the south from B to C. Assume that the length of the straight line between A and B is 100 m, and between B and C is 50 m. Therefore, the total distance the person walks every morning is 150 m.

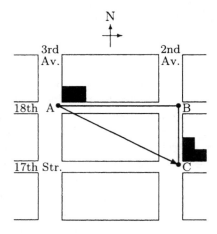

Fig. 6.4 Distance vs. displacement

On the other hand, the door-to-door southeasterly displacement of the person is equal to the length of the straight line joining A and C, which is $\sqrt{(100)^2 + (50)^2} = 112\,\text{m}$.

6.5 Speed and Velocity

While the terms speed and velocity are used interchangeably in ordinary language, they have distinctly different meanings in mechanics. *Velocity* is defined as the time rate of change of position. Velocity is a vector quantity having both a magnitude and a direction. *Speed* is a scalar quantity equal to the magnitude of the velocity vector.

6.6 Acceleration

Acceleration is defined as the time rate of change of velocity and is a vector quantity. Although the term "acceleration" is more commonly used to describe situations where speed increases over time and the term "deceleration" is used to indicate decreasing speed over time, the mathematical definitions of the two are the same.

6.7 Inertia and Momentum

Inertia is the tendency of an object to maintain its state of rest or uniform motion. Inertia can also be defined as the resistance to the change in motion of an object. The more inertia an object has, the more difficult it is to start moving it from rest or to change its state of motion. The greater the mass of the object, the greater its inertia. For example, a truck has a greater inertia than a passenger car because of the difference in their mass. If both of them are traveling at the same speed, it is always more difficult to stop the truck as compared to the car.

Like inertia, *momentum* is a tendency to resist changes in the existing state of motion and is defined as the product of mass and velocity. Only moving objects have momentum, whereas every object—stationary or moving—has an inertia. If two moving objects having the same mass are considered, then the one with higher speed has the greater momentum. If two moving objects having the same speed are considered, then the one with higher mass has the greater momentum.

6.8 Degree of Freedom

Degree of freedom is an expression that describes the ability of an object to move in space. A completely unrestrained object, such as a ball, has six degrees of freedom (three related to translational motion along three mutually perpendicular axes and three related to rotational motion about the same axes). The human hip joint has three degrees of freedom because it enables the lower extremity to rotate about one axis and undergo angular movements in two planes. On the other hand, the elbow and forearm system has two degrees of freedom because it allows the lower arm to rotate about one axis and undergo angular movement in one plane.

6.9 Particle Concept

The "particle" concept in mechanics is rather a hypothetical one. It undermines the size and shape of the object under consideration, and assumes that the object is a particle with a mass equal to the total mass of the object and located at the center of gravity of the object. In some problems, the shape of the object under investigation may not be pertinent to the discussion of certain aspects of its motion. This is particularly true if the object is undergoing a translational motion only. For example, what is significant for a person pushing a wheelchair is the total mass of the wheelchair, not its size or shape. Therefore, the wheelchair may be treated as a particle with a mass equal to the total mass of the wheelchair, and proceed with relatively simple analyses. The size and shape of the object may become important if the object undergoes a rotational motion.

6.10 Reference Frames and Coordinate Systems

To be able to describe the motion of a body properly, a *reference frame* must be adopted. The *rectangular* or *Cartesian coordinate system* that is composed of three mutually perpendicular directions is the most suitable reference frame for describing linear movements. The axes of this system are commonly labeled with x, y, and z. For two-dimensional problems, the number of axes may be reduced to two by eliminating the z axis (Fig. 6.5).

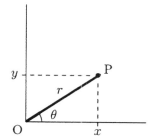

Fig. 6.5 Rectangular (x, y) and polar (r, θ) coordinates of a point

Another commonly used reference frame is the *polar coordinate system*, which is better suited for analyzing angular motions. As shown in Fig. 6.5, the polar coordinates of a point P are defined by parameters r and θ. r is the distance between the origin O of the coordinate frame and point P, and θ is the angle line OP makes with the horizontal. The details of polar coordinates will be provided in later chapters.

6.11   Prerequisites for Dynamic Analysis

The prerequisites for dynamic analysis are vector algebra, differential calculus, and integral calculus. Vector algebra is reviewed in Appendix B. The principles of differential and integral calculus are provided in Appendix C, along with the definitions and properties of commonly encountered functions that form the basis of calculus. Appendices B and C must be reviewed before proceeding to the following chapters. Also important in dynamic analyses are the properties of force and torque vectors as covered in Chaps. 2 and 3, respectively. It should be noted that the static analyses covered in Chaps. 4 and 5 are specific cases of dynamic analyses for which acceleration is zero.

6.12   Topics to Be Covered

Chapters 6 through 11 constitute the second part of this textbook, which is devoted to the analyses of moving systems. In Chap. 7, mathematical definitions of displacement, velocity, and acceleration vectors are introduced, kinematic relationships between linear quantities are defined, uniaxial and biaxial motion analyses are discussed, and the concepts introduced are applied to problems of sports biomechanics. Linear kinetics is studied in Chap. 8. Solving problems in kinetics using the equations of motion and work and energy methods are discussed in Chap. 8. Angular kinematics and kinetics are covered in Chaps. 9 and 10, respectively, and the concepts and procedures introduced are applied to investigate some of the problems of biomechanics. Topics such as impulse, momentum, impact, and collision are covered in Chap. 11.

7.1 Uniaxial Motion

Uniaxial motion is the one in which the motion occurs only in one direction, and it is the simplest form of linear or translational motion. A car traveling on a straight highway, an elevator going up and down in a shaft, and a sprinter running a 100-m race are examples of uniaxial motion.

Kinematic analyses utilize the relationships between the position, velocity, and acceleration vectors. For uniaxial motion analyses, it is usually more practical to define a direction, such as x, to coincide with the direction of motion, to define kinematic parameters in that direction, and to carry out the analyses as if displacement, velocity, and acceleration are scalar quantities.

7.2 Position, Displacement, Velocity, and Acceleration

Consider the car illustrated in Fig. 7.1. Assume that the car is initially stationary and located at 0. At time t_0, the car starts moving to the right on a straight horizontal path. At some time t_1, the car is observed to be at 1 and at a later time t_2 it is located at 2; 0, 1, and 2 represent *positions* of the car at different times, and 0 also represents the *initial position* of

the car. It is a common practice to start measuring time beginning with the instant when the motion starts, in which case $t_0 = 0$.

The position of the car at different times must be measured with respect to a point in space. Let x be a measure of horizontal distances relative to the initial position of the car. If x_0 represents the initial position of the car, then $x_1 = 0$. If 1 and 2 are located at x_1 and x_2 distances away from 0, then x_1 and x_2 define the relative positions of the object at times t_1 and t_2, respectively. Since the relative position of the car is changing with time, x is a function of time t, or $x = f(t)$. In the time interval between t_1 and t_2, the position of the car changed by an amount $\Delta x = x_2 - x_1$, where Δ(capital delta) implies change. This change in position is the displacement of the car in the time interval $\Delta t = t_2 - t_1$.

During a uniaxial horizontal motion, the car may be located on the right or the left of the origin 0 of the x axis. Assuming that the positive x axis is toward the right, the position of the car is positive if it is located on the right of 0 and negative if it is on the left of 0. Similarly, the displacement of the car is positive if it is moving toward the right, and it is negative if the car is moving toward the left.

Velocity is defined as the time rate of change of relative position. If the position of an object moving in the x direction is known as a function of time, then the instantaneous velocity, v, of the object can be determined by considering the derivative of x with respect to t:

$$v = \frac{dx}{dt}. \qquad (7.1)$$

If required, the average velocity, $\bar{v}$, of the object in any time interval can be determined by considering the ratio of change in position (displacement) of the object and the time it takes to make that change. For example, the average velocity of the car in Fig. 7.1 in the time interval between t_1 and t_2 is:

$$\bar{v} = \frac{\Delta v}{\Delta t} = \frac{x_2 - x_1}{t_2 - t_1}. \qquad (7.2)$$

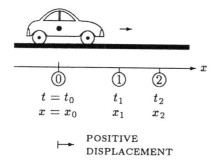

POSITIVE DISPLACEMENT

Fig. 7.1 The car is located at positions 0, 1, and 2 at times t_0, t_1, and t_2, respectively

N. Özkaya et al., *Fundamentals of Biomechanics: Equilibrium, Motion, and Deformation*, DOI 10.1007/978-1-4614-1150-5_7, © Springer Science+Business Media, LLC 2012

In Eq. (7.2), the "bar" over v indicates average, and x_1 and x_2 are the relative positions of the car at times t_1 and t_2, respectively.

Velocity is a vector quantity and may take positive and negative values, indicating the direction of motion. The velocity is positive if the object is moving away from the origin in the positive x direction, and it is negative if the object is moving in the negative x direction. The magnitude of the velocity vector is called *speed*, which is always a positive quantity.

The instantaneous velocity of an object may vary during a particular motion. In other words, velocity may be a function of time or $v = f(t)$. Acceleration is defined as the time rate of change of velocity. If the velocity of an object is known as a function of time, then its instantaneous acceleration, a, can be determined by considering the derivative of v with respect to t:

$$a = \frac{dv}{dt}. \qquad (7.3)$$

In general, the acceleration of a moving object may vary with time. In other words, acceleration may be a function of time or $a = f(t)$.

There is also *average acceleration*, $\bar{a}$, that can be determined by considering the ratio of the change in velocity of the object and the time elapsed during that change. For example, if the instantaneous velocities v_1 and v_2 of the car in Fig. 7.2 at times t_1 and t_2 are known, then the average acceleration of the car in the time interval between t_1 and t_2 can be calculated:

$$\bar{a} = \frac{\Delta v}{\Delta t} = \frac{v_2 - v_1}{t_2 - t_1}. \qquad (7.4)$$

Fig. 7.2 v_0 is the initial velocity, and v_1 and v_2 are the velocities of the car at times t_1 and t_2, respectively

Acceleration is a vector quantity and may be positive or negative. Positive acceleration does not always mean that the object is speeding up and negative acceleration does not always imply that the object is slowing down. At a given instant, if the velocity and acceleration are both positive or

negative, then the object is said to be speeding up or accelerating. For a uniaxial motion in the x direction, if both the velocity and acceleration are positive, then the object is moving in the positive x direction with an increasing speed. If both the velocity and acceleration are negative, then the object is moving in the negative x direction with an increasing speed. On the other hand, if the velocity and acceleration have opposite signs, then the object is slowing down or decelerating. For example, for a uniaxial motion in the x direction, if the velocity is positive and acceleration is negative, then the object is moving in the positive x direction with a decreasing speed. If the velocity is negative and acceleration is positive, then the object is moving in the negative x direction with a decreasing speed. Finally, if the acceleration is zero, then the object is said to have a constant or uniform velocity. All of these possibilities are summarized in Table 7.1.

Table 7.1 Acceleration, deceleration, and constant velocity conditions

	V	a
Increasing speed Or acceleration	+ −	+ −
Decreasing speed Or deceleration	+ −	− +
Constant speed	±	0

Acceleration is derived from velocity, which is itself derived from position. Therefore, there must be a way to relate acceleration and position directly. This relationship can be derived by substituting Eq. (7.1) into Eq. (7.3):

$$v = \frac{dx}{dt} = \dot{x},$$

$$a = \frac{dv}{dt} = \frac{d}{dt}\left(\frac{dx}{dt}\right) = \frac{d^2x}{dt^2} = \ddot{x}.$$

The "dots" over x in the above equations indicate differentiation with respect to time. One dot signifies the first derivative with respect to time and two dots imply the second derivative.

7.3   Dimensions and Units

The relative position is measured in units of length. By definition, displacement is equal to the change of position, velocity is the time rate of change of relative position, and acceleration is the time rate of change of velocity. Therefore, relative position and displacement have the dimension of length, velocity has the dimension of length divided by time, and acceleration has the dimension of velocity divided by time:

$$[\text{POSITION}] = L,$$

$$[\text{DISPLACEMENT}] = L,$$

$$[\text{VELOCITY}] = \frac{[\text{DISPLACEMENT}]}{[\text{TIME}]} = \frac{L}{T},$$

$$[\text{ACCELERATION}] = \frac{[\text{VELOCITY}]}{[\text{TIME}]} = \frac{(L/T)}{T} = \frac{L}{T^2}.$$

Based on these dimensions, the units of displacement, velocity, and acceleration in different unit systems can be determined. Some of these units are listed in Table 7.2.

Table 7.2 Units of displacement, velocity, and acceleration

Unit system	Displacement	Velocity	Acceleration
SI	meter (m)	m/s	m/s^2
CGS	centimeter (cm)	cm/s	cm/s^2
British	foot (ft)	ft/s	ft/s^2

7.4 Measured and Derived Quantities

In practice, it is possible to measure position, velocity, and acceleration over time. From any one of the three, the other two quantities can be determined by employing proper differentiation and/or integration, or through the use of graphical and numerical techniques. If the position of an object undergoing uniaxial motion in the x direction is measured and recorded, then the position can be expressed as a function of time, $x = f(t)$. Once the function representing the position of the object is established, the velocity and acceleration of the object at different times can be calculated using:

$$v = \frac{dx}{dt}, \tag{7.5}$$

$$a = \frac{dv}{dt} = \frac{d^2x}{dt^2}. \tag{7.6}$$

If the velocity of an object undergoing uniaxial motion in the x direction is measured and expressed as a function of time, $v = f(t)$, then the position of the object relative to its initial position and instantaneous acceleration of the object can be calculated using:

$$x = x_0 + \int_{t_0}^{t} v \, dt, \tag{7.7}$$

$$a = \frac{dv}{dt}. \tag{7.8}$$

The lower limit of integration, t_0, in Eq. (7.7) corresponds to the time at which the first measurements are taken, and the upper limit corresponds to any time t. x_0 is the initial position of the object at time t_0. For practical purposes, t_0 can be taken to be zero. This would mean that all time measurements are made relative to the instant when the motion began. Also, $x_0 = 0$ if all position measurements are made relative to the initial position of the object.

If the acceleration of an object is measured and expressed as a function of time, $a = f(t)$, then the instantaneous velocity and position of the object relative to its initial velocity and position can be calculated using:

$$v = v_0 + \int_{t_0}^{t} a \, dt, \tag{7.9}$$

$$x = x_0 + \int_{t_0}^{t} v \, dt. \tag{7.10}$$

In Eqs. (7.9) and (7.10), x_0 and v_0 correspond to the initial position and initial velocity of the object at time t_0. Note that these equations relate change of position and velocity relative to the initial position and velocity of the moving object. However, these equations are valid relative to the position and velocity of the object at any time. For example, if x_1 and v_1 represent the position and velocity of the object at time t_1, then Eqs. (7.9) and (7.10) can also be expressed as:

$$v = v_1 + \int_{t_1}^{t} a \, dt,$$

$$x = x_1 + \int_{t_1}^{t} v \, dt.$$

7.5 Uniaxial Motion with Constant Acceleration

A common type of uniaxial motion occurs when the acceleration is constant. If a_0 represents the constant acceleration of an object, v_0 is its initial velocity, and x_0 is its initial position at time $t_0 = 0$, then Eqs. (7.9) and (7.10) will yield:

$$v = v_1 + a_0 t, \tag{7.11}$$

$$x = x_0 + \int_{t_0}^{t} (v_0 + a_0 t) dt,$$

$$= x_0 + \int_{t_0}^{t} v_0 dt + \int_{t_0}^{t} a_0 t \, dt,$$

$$= x_0 + v_0 t + \frac{1}{2} a_0 t^2,$$

$$x = x_0 + v_0 t + \frac{1}{2} a_0 t^2. \tag{7.12}$$

For a given initial position, initial velocity, and constant acceleration of an object undergoing uniaxial motion in the x direction, Eqs. (7.11) and (7.12) can be used to determine the velocity and position of the object as a function of time relative to its initial velocity and position. Note that Eqs. (7.11) and (7.12) can be expressed relative to any other time and position. For example, if x_1 and v_1 represent the known position and velocity of the object at time t_1, then:

$$v = v_1 + a_0(t - t_1),$$

$$x = x_1 + v_1(t - t_1) + \frac{1}{2}a_0(t - t_1)^2.$$

Figure 7.3 shows an acceleration versus time graph for an object moving with constant acceleration, a_0. According to Eq. (7.11), velocity is a linear function of time. As illustrated in Fig. 7.4, the velocity versus time graph is a straight line with constant slope that is equal to the magnitude of the constant acceleration. This is consistent with the fact that the slope of a function is equal to the derivative of that function, and that the derivative of velocity with respect to time is equal to acceleration. In Eq. (7.12), displacement is a quadratic function of time, and as illustrated in Fig. 7.5, the graph of this function is a parabola. At any given time, the slope of this function is equal to the velocity of the object at that instant.

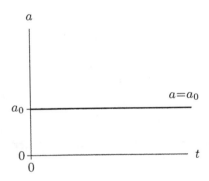

Fig. 7.3 Constant (uniform) acceleration

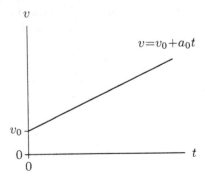

Fig. 7.4 When acceleration is constant, velocity is a linear function of time

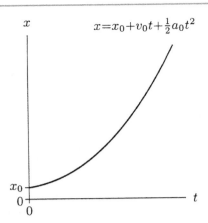

Fig. 7.5 When acceleration is constant, change of position is a quadratic function of time

For a uniaxial motion with constant acceleration, it is also possible to derive an expression between velocity, displacement, and time by solving Eq. (7.11) for a_0 and substituting it into Eq. (7.12). This will yield:

$$x = x_0 + \frac{1}{2}(v + v_0)t. \tag{7.13}$$

Similarly, an expression between velocity, displacement, and acceleration can be derived by solving Eq. (7.11) for t and substituting it into Eq. (7.12):

$$v^2 = v_0{}^2 + 2a_0(x - x_0). \tag{7.14}$$

Caution Equations (7.11) through (7.14) are valid if the acceleration is constant. Furthermore, the direction of the parameters involved must be handled properly. For example, if the direction of acceleration is opposite to that of the positive x direction, then the "plus" sign in front of the terms carrying acceleration must be changed to a "minus" sign.

7.6   Examples of Uniaxial Motion

The following examples are aimed to demonstrate the use of the kinematic equations (7.5) through (7.10).

Example 7.1 The short distance runner illustrated in Fig. 7.6 completed a 100-m race in 10s. The time it took for the runner to reach the first 10m and each successive 10m mark was recorded by ten observers using stopwatches. The data collected were then plotted to obtain the position versus time graph shown in Fig. 7.6. It is suggested that the data may be represented with the following function:

$$x = 0.46t^{7/3}.$$

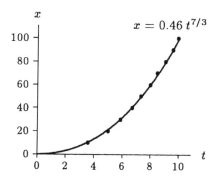

Fig. 7.6 Relative position, x, measured in meters versus time, t, measured in seconds

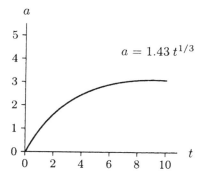

Fig. 7.7 Speed, v (m/s), and acceleration, a (m/s^2), versus time, t (s), curves for the runner

Here, change of position x is measured in meters, and time t is measured in seconds.

Determine the velocity and acceleration of the runner as a function of time, and the instantaneous velocity and acceleration of the runner 5 s after the start.

Solution

Since the function representing the position of the runner is known, it can be differentiated with respect to time once to determine the velocity and twice to determine the acceleration:

$$v = \frac{dx}{dt} = \frac{d}{dt}\left(0.46t^{7/3}\right) = 1.07t^{4/3},$$

$$a = \frac{dv}{dt} = \frac{d}{dt}\left(1.07t^{4/3}\right) = 1.43t^{1/3}.$$

The graphs of these functions are shown in Fig. 7.7.

To evaluate the velocity and acceleration of the runner 5s after the start, substitute $t = 5$s in the above equations and carry out the calculations. This will yield:

$$v = 9.15 \text{ m/s},$$

$$a = 2.45 \text{ m/s}^2.$$

Example 7.2 The speedometer reading of a car driven on a straight highway is recorded for a total time interval of 3 min. The data collected are represented with a speed versus time diagram shown in Fig. 7.8. The dotted curve in Fig. 7.8

represents the actual measurements that are approximated by three straight lines (solid lines in Fig. 7.8). According to the information presented in Fig. 7.8, the speed of the car increases linearly from $v_0 = 0$ to $v_1 = 72$ km/h between times $t_0 = 0$ and $t_1 = 30$s. Between times $t_1 = 30$s and $t_2 = 120$s, the speed of the car is constant at 72 km/h. Beginning at time $t_2 = 120$s, the driver applies the brakes, decreases the speed of the car linearly with time, and brings the car to a stop in 60s.

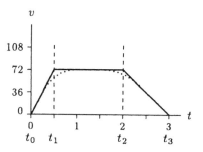

Fig. 7.8 Speed, v (km/h), versus time, t (min), diagram for the car

Determine expressions for the speed, displacement, and acceleration of the car as a function of time. Calculate the total distance traveled by the car in 3 min.

Solution

The speed measurements were made in kilometers per hour (km/h) that need to be converted to meters per second (m/s). This can be achieved by noting that 1 km is equal to 1,000m and that there are 3,600s in 1 h. Therefore, 72 km/h is equal to 20 m/s, which is calculated as:

$$72 \text{ km/h} = 72 \times \frac{1,000}{3,600} = 20 \text{ m/s}.$$

The speed versus time graph in Fig. 7.8 is redrawn in Fig. 7.9, in which speed is expressed in meters per second and time in seconds.

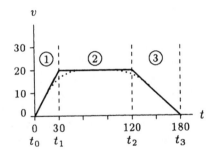

Fig. 7.9 Speed, v (m/s), versus time, t (s), diagram for the car

Because of the approximations made, the speed versus time graph in Fig. 7.9 has three distinct regions and there is not a single function that can represent the entire graph. Therefore, this problem should be analyzed in three phases.

Phase 1

Between $t_0 = 0$ and $t_1 = 30$s, the speed of the car increased linearly with time from 0 to 20 m/s. As discussed in Appendix C, all linear functions can be represented as $X = A + BY$. In this expression, Y is the independent variable, X is the dependent variable, and A and B are some constant coefficients. In this case, we have time as the independent variable and speed as the dependent variable. Since the relationship between the speed of the car and time in phase 1 is linear, we can write:

$$v = A + Bt. \tag{i}$$

The function given in Eq. (*i*) is a general expression between v and t because coefficients A and B are not yet determined. We need two conditions to calculate A and B (two unknowns). These conditions can be obtained from Fig. 7.9. When the car first started to move, $t = 0$ and $v = 0$, and $v = 20$ m/s when $t = 30$s. Substituting the initial condition ($v = 0$ when $t = 0$) into Eq. (*i*) will yield $A = 0$, and substituting the second condition ($v = 20$ m/s when $t = 30$s) will yield $B = 0.667$. Substituting $A = 0$ and $B = 0.667$ back into Eq. (*i*) will yield the function relating the speed of the car and time in phase 1:

$$v = 0.667t. \tag{ii}$$

Note that since we already converted speed measurements into meter per second and time into seconds, the speed in Eq. (*ii*) is in meters per second and time is in seconds.

Now, Eqs. (7.7) and (7.8) can be utilized to determine the displacement and acceleration of the car in phase 1. If we measure displacements relative to the starting point, then the initial position of the car was $x_0 = 0$. Therefore:

$$x = x_0 + \int_0^t v \, dt = \int_0^t (0.667t)dt$$
$$= 0.667 \left[\frac{t^2}{2} \right]_0^t = 0.333t^2, \tag{iii}$$

$$a = \frac{dv}{dt} = \frac{d}{dt}(0.67t) = 0.67. \tag{iv}$$

From Eq. (*iv*), the acceleration of the car in phase 1 was constant at 0.667 m/s^2. The total distance traveled by the car at the end of phase 1 can be determined by substituting $t = 30$s into Eq. (*iii*). This will yield:

$$x_1 = 0.333t^2 = 0.333(30)^2 = 300 \text{ m}.$$

Phase 2

Phase 2 starts when time is $t_1 = 30$s and ends when it is $t_2 = 120$s. In phase 2, the speed of the car was constant at 20 m/s. Therefore, the function representing the speed in phase 2 is:

$$v = 20. \tag{v}$$

The total distance traveled by the car in phase 1 was computed as $x_1 = 300$m. $x_1 = 300$m also represents the initial position of the car in phase 2. Phase 1 ended when time was $t_1 = 30$s. Therefore, phase 2 began when time was $t_1 = 30$s. We can now write Eq. (7.7) relative to t_1 and x_1:

$$x = x_1 + \int_{t_1}^t v dt = 300 + \int_{30}^t 20 dt = 300 + 20[t]_{30}^t,$$

$$x = -300 + 20t. \tag{vi}$$

The acceleration of the car in phase 2 can be determined using Eq. (7.8):

$$a = \frac{dv}{dt} = 0. \tag{vii}$$

From Eq. (*vii*), the acceleration of the car in phase 2 is zero. The total distance traveled by the car at the end of phase 2 can be determined by substituting $t = 120s$ into Eq. (*vi*). This will yield:

$$x_2 = -300 + 20(120) = 1,200 \text{ m.}$$

Phase 3

Between $t_2 = 120s$ and $t_3 = 180s$, the speed of the car decreased linearly with time and to zero in 60s. The function representing the relationship between speed of the car and time in phase 3 can be determined using Eq. (*i*). The coefficients A and B in Eq. (*i*) can be calculated by taking into consideration two conditions related to phase 3. For example, $v = 20m/s$ when $t = 120s$ and $v = 0$ when $t = 180s$. Substituting the second condition in Eq. (*i*) will yield $A + 180B = 0$ or $A = -180B$. Substituting the first condition and $A = -180B$ into Eq. (*i*) will yield $B = -0.333$. Since $A = -180B$ and $B = -0.333, A = 60$. Therefore, the function that relates the speed of the car and time in phase 3 is:

$$v = 60 - 0.333t. \tag{viii}$$

Phase 3 begins when time is $t_2 = 120s$ and the initial position of the car at phase 3 is $x_2 = 2,100m$. Using Eq. (7.7):

$$x_3 = x_2 + \int_{t_2}^{t} v_3 dt = 2,100 + \int_{120}^{t} (60 - 0.333t)dt,$$
$$= 2,100 + [60t - 0.167t^2]_{120}^{t},$$
$$= 2,100 + [60t - 0.167t^2] - [60(120) - 0.167(120)^2],$$
$$x_3 = -2,700 + 60t - 0.167t^2. \tag{ix}$$

Using Eq. (7.8):

$$a = \frac{dv}{dt} = \frac{d}{dt}(60 - 0.333t) = -0.333. \tag{x}$$

The total distance traveled by the car can be determined by substituting $t = 180s$ into Eq. (*ix*). This will yield:

$$x_3 = -2,700 + 60(180) - 0.167(180)^2 = 2,690 \text{ m.}$$

In Fig. 7.10, the functions derived for the displacement and acceleration of the car in different phases are used to plot displacement and acceleration versus time graphs (solid curves). In all phases, the car is moving in the positive x direction. Therefore, the displacement of the car is positive throughout. In phase 1, the acceleration of the car is positive, indicating increasing speed in the positive x direction. In phase 2, the acceleration of the car is zero and the speed is constant. In phase 3, the acceleration of the car is negative, indicating deceleration in the positive x direction.

Fig. 7.10 Displacement, x (m), and acceleration, a (m/s^2), versus time, t (s), graphs for the car

Note that the displacement and acceleration versus time graphs (solid curves) in Fig. 7.10 are not continuous. For example, the slope of the x versus t curve at the end of phase 1 is not necessarily equal to the slope of the x versus t curve at the beginning of phase 2. The discontinuity is more significant for the a versus t graph. This is due to the fact that we approximated the actual speed versus time graph of the car with three straight lines in three regions. In reality, as illustrated by the dotted curves in Fig. 7.10, the variations in the slopes of these curves would be less marked and more continuous.

Example 7.3 Consider the skier illustrated in Fig. 7.11 descending a straight slope. Assume that the skier is moving down the slope at a constant acceleration of 2 m/s^2 and that the speed of the skier at position 0 is observed to be 10 m/s.

Calculate the speed v_1 of the skier when the skier is at position 1, which is at a distance $l = 100m$ from position

0 measured parallel to the slope. Also, calculate the time t_1 it took for the skier to cover the distance between positions 0 and 1.

Fig. 7.11 A skier is moving down the slope with a constant acceleration of $a_0 = 2 \text{ m/s}^2$

Solution

Since the skier is moving with a constant acceleration, Eqs. (7.11) and (7.12) can be used to analyze this problem. In Fig. 7.12, the direction parallel to the slope or the direction in which the skier is moving is identified by the x axis. For the sake of simplicity, the origin of the x axis is placed to coincide with position 0 so that $x_0 = 0$. Furthermore, we can make all time measurements relative to the instant when the skier was at position 0. That is, $t_0 = 0$. As indicated in Fig. 7.12, the speed of the skier at position 0 is $v_0 = 10$ m/s. What we know about position 1 is the fact that it is located at a distance $l = 100$m from position 0. Therefore, the position of the skier at 1 is $x_1 = l = 100$m. The time t_1 it took for the skier to cover the distance between positions 0 and 1, and the speed v_1 of the skier at position 1 are unknowns to be determined.

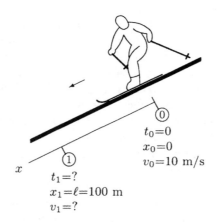

Fig. 7.12 Conditions at positions 0 and 1

Use Eq. (7.12) first. Writing this equation between positions 0 and 1:

$$x_1 = x_0 + v_0 t_1 + \frac{1}{2} a_0 t_1^2,$$

Here, $x_0 = 0$, $v_0 = 10$ m/s, and $a_0 = 2$ m/s² is the constant acceleration of the skier. Substituting these parameters into the above equation and rearranging the order of terms will yield:

$$t_1^2 + 10t_1 - 100 = 0.$$

Note that this is a quadratic equation. Solutions of quadratic equations are discussed in Appendix C.5. This equation has two solutions for t_1, one positive and one negative. Since negative time does not make any sense, the positive solution must be adopted, which is $t_1 = 6.18$s. (For the validity of $t_1 = 6.18$s, substitute it back into the quadratic equation and check whether the equilibrium is satisfied.) In other words, it took 6.18s for the skier to cover the distance between positions 0 and 1.

Now, we can use Eq. (7.11) to calculate the speed of the skier at position 1:

$$v_1 = v_0 + a_0 t_1.$$

Substituting $v_0 = 10$ m/s, $a_0 = 2$ m/s², and $t_1 = 6.18$s into the above equation and carrying out the calculations will yield $v_1 = 22.36$ m/s.

Example 7.4 One of the most common examples of uniformly accelerated motion is that of an object allowed to fall vertically downward, which is called free fall. Free fall is a consequence of the effect of gravitational acceleration on the mass of the object. If the possible effects of air resistance are ignored (assuming that the motion occurs in vacuum), then the object released from a height would move downward with a constant acceleration equal to the magnitude of the gravitational acceleration, which is about 9.8 m/s².

As illustrated in Fig. 7.13, consider a person holding a ball at a height $h = 1.5$m above the ground level. If the ball is released to descend, how much time it would take for the ball to hit the ground and what would be its impact velocity?

Fig. 7.13 Free fall

Solution

This is another example of uniaxial motion with constant acceleration. Once the ball is released, it moves downward with constant acceleration $a_0 = 9.8$ m/s^2, which is the magnitude of the gravitational acceleration. In Fig. 7.14, the direction of motion of the ball is identified with the y axis such that positive y direction is downward. Since the direction of gravitational acceleration is also downward, the acceleration of the ball is positive.

Fig. 7.14 Conditions at positions 0 and 1

In Fig. 7.14, the origin of the y axis is chosen to coincide with the initial position, 0, of the ball. Therefore, the initial position of the ball is $y_0 = 0$. The initial time $t_0 = 0$ because all time measurements are made relative to the instant when the ball was released. The initial speed of the ball is $v_0 = 0$ because the ball was initially at rest. The ground level is identified as position 1, which is at a vertical distance $y_1 = h = 1.5$m away from position 0. The task is to determine the time t_1 it took for the ball to cover the vertical distance between positions 0 and 1, and the impact speed v_1 of the ball.

Since the acceleration is constant at $a_0 = 9.8$ m/s^2, we can use Eqs. (7.11) and (7.12). However, we must replace x in these equations with y. Writing Eq. (7.12) between positions 0 and 1:

$$y_1 = y_0 + v_0 t_1 + \frac{1}{2} a_0 t_1{}^2.$$

Substituting $y_1 = 1.5$m, $y_0 = 0$, $v_0 = 0$, and $a_0 = 9.8$ m/s^2 into this equation and solving it for t_1 will yield $t_1 = 0.55$s. That is, in the absence of air resistance, it would take only 0.55s for the ball to hit the ground when it is released from a height of 1.5m from the ground level.

The impact speed can be calculated using Eq. (7.11):

$$v_1 = v_0 + a_0 t_1.$$

Substituting $v_0 = 0$, $a_0 = 9.8$ m/s^2, and $t_1 = 0.55$s into the above equation and solving it for v_1 will yield $v_1 = 5.39$ m/s.

7.7 Biaxial Motion

Biaxial or two-dimensional motion is one in which the movement occurs on a plane surface. One-dimensional linear motion characteristics of an object are completely known if, for example, the position of the object in the direction of motion is known as a function of time. The concepts introduced earlier for uniaxial motion analysis can be expanded to analyze two- and three-dimensional linear movements. This can be achieved by considering the properties of displacement, velocity, and acceleration as vector quantities.

7.8 Position, Velocity, and Acceleration Vectors

For one-dimensional problems, the position of an object is defined by using a single coordinate axis. For plane problems, two coordinates must be specified to define the position of an object uniquely. As shown in Fig. 7.15, let x and y represent the usual Cartesian (rectangular) coordinate directions with unit vectors $\underline{i}$ and $\underline{j}$ indicating positive x and y directions, respectively. The origin of the coordinate system is located at O. The position vector $\underline{r}$ of a point P in this xy-plane is a vector drawn from O toward P. The position vector can be represented in terms of x and y coordinates of point P:

$$\underline{r} = x\underline{i} + y\underline{j}. \tag{7.15}$$

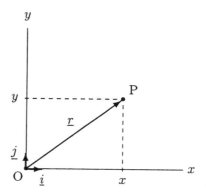

Fig. 7.15 $\underline{r}$ is the position vector of point P

The magnitude r of the position vector is equal to the length of the line connecting points O and P, which can be calculated as:

$$r = \sqrt{x^2 + y^2}. \tag{7.16}$$

If point P represents the position of a moving object at some time t, then $\underline{r}$ represents the *instantaneous position* of that object at that time. This implies that $\underline{r}$ can change with time or x and y coordinates of the moving object are functions of time.

By definition, velocity is the time rate of change of position. Therefore, the *velocity vector* is equal to the derivative of the position vector with respect to time:

$$\underline{v} = \frac{d}{dt}(\underline{r}) = \frac{d}{dt}\left(x\underline{i} + y\underline{j}\right) = \frac{dx}{dt}\underline{i} + \frac{dy}{dt}\underline{j}, \quad (7.17)$$

For two-dimensional problems, the velocity vector may have up to two components (Fig. 7.16). If v_x and v_y refer to the scalar components of $\underline{v}$ in the x and y directions, respectively, then the velocity vector can also be expressed as:

$$\underline{v} = v_x\underline{i} + v_y\underline{j}. \quad (7.18)$$

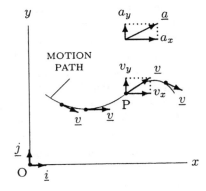

Fig. 7.16 Velocity vector is always tangent to the path of motion

By comparing Eqs. (7.17) and (7.18), we can conclude that:

$$v_x = \frac{dx}{dt} = \dot{x}, \quad (7.19)$$

$$v_y = \frac{dy}{dt} = \dot{y}, \quad (7.20)$$

v_x and v_y are also known as the *rectangular components* of $\underline{v}$, and they indicate how fast the object is moving in the x and y directions, respectively. If the components v_x and v_y of the velocity vector are known, then the magnitude v of their resultant can also be determined:

$$v = \sqrt{v_x{}^2 + v_y{}^2}. \quad (7.21)$$

Note that v is a scalar quantity also known as the *speed*. As illustrated in Fig. 7.16, it is very important to remember that the direction of the velocity vector is always tangent to the path of the motion and pointing in the direction of motion.

The direction of the velocity vector can also be determined if its scalar components in the horizontal and vertical directions are known: $\tan\alpha = \sigma_y/\sigma_x$, then $\alpha = \tan^{-1}(\sigma_y/\sigma_x)$, where α is an angle that the velocity vector makes with the horizontal axis.

By definition, acceleration is the time rate of change of velocity. Therefore, if the velocity vector of an object is known as a function of time, then its acceleration vector $\underline{a}$ can also be determined by considering the derivative of $\underline{v}$ with respect to time:

$$\underline{a} = \frac{d}{dt}(\underline{v}) = \frac{d}{dt}\left(v_x\underline{i} + v_y\underline{j}\right) = \frac{dv_x}{dt}\underline{i} + \frac{dv_y}{dt}\underline{j}. \quad (7.22)$$

The acceleration vector can also be expressed in terms of its components in the x and y directions (Fig. 7.16):

$$\underline{a} = a_x\underline{i} + a_y\underline{j}. \quad (7.23)$$

By comparing Eqs. (7.22) and (7.23), the rectangular components of the acceleration vector can alternatively be written as:

$$a_x = \frac{dv_x}{dt} = \frac{d^2x}{dt^2} = \ddot{x}, \quad (7.24)$$

$$a_y = \frac{dv_y}{dt} = \frac{d^2y}{dt^2} = \ddot{y}. \quad (7.25)$$

If needed, the magnitude a of the acceleration vector can be calculated as:

$$a = \sqrt{a_x{}^2 + a_y{}^2}. \quad (7.26)$$

The extension of these concepts to analyze three-dimensional movements is straightforward. For three-dimensional motion analyses, there is a need for a third dimension, namely z. For example, the position vector of a point in space can be expressed as:

$$\underline{r} = x\underline{i} + y\underline{j} + z\underline{k}.$$

Here, $\underline{k}$ is the unit vector indicating the positive z direction. Similarly, the velocity and acceleration vectors in space can be expressed as:

$$\underline{v} = v_x\underline{i} + v_y\underline{j} + v_z\underline{k},$$

$$\underline{a} = a_x\underline{i} + a_y\underline{j} + a_z\underline{k}.$$

For example, v_z is the speed of the object in the z direction and is equal to the time rate of change of position in the z direction, and a_z is the scalar component of the acceleration vector in the z direction and is equal to the time rate of change of the speed of the object in the z direction.

7.9 Biaxial Motion with Constant Acceleration

Two-dimensional linear motion of an object in the xy-plane can be analyzed in two stages by first considering its motion in the x and y directions separately and then combining the results using the vectorial properties of the parameters involved. The parameters defining the motion in the x direction are x, its first time derivative v_x, and its second time derivative a_x. Similarly, y, v_y, and a_y are the parameters that define the motion of the object in the y direction. If the acceleration of an object undergoing two-dimensional linear motion is constant, then a_x and a_y must be constants. The details of uniaxial motion with constant acceleration were analyzed in the previous sections. The results of these analyses can readily be adopted to analyze two-dimensional motions with constant acceleration.

In the x direction, Eqs. (7.11) and (7.12) can be rewritten in the following more specific forms:

$$v_x = v_{x_0} + a_{x_0}t, \tag{7.27}$$

$$x = x_0 + v_{x_0}t + \frac{1}{2}a_{x_0}t^2. \tag{7.28}$$

Similarly, in the y direction:

$$v_y = v_{y_0} + a_{y_0}t, \tag{7.29}$$

$$y = y_0 + v_{y_0}t + \frac{1}{2}a_{y_0}t^2. \tag{7.30}$$

Here, x_0 and y_0 are the initial coordinates of the object, v_{x_0} and v_{y_0} are the initial velocity components in the x and y directions, and a_{x_0} and a_{y_0} are the constant components of the acceleration vector in the x and y directions, respectively. For given x_0, y_0, v_{x_0}, v_{y_0}, a_{x_0}, and a_{y_0}, Eqs. (7.27) through (7.30) can be used to calculate the relative position of the moving object and its velocity components at any time t.

Furthermore, these equations are valid if acceleration is constant and its direction coincides with the direction of motion. If the direction of acceleration is opposite the direction of motion, then the sign in front of the terms carrying the acceleration must be changed into minus $(-)$. In the x direction, Eqs. (7.11) and (7.12) can be rewritten in the following forms:

$$v_x = v_{x_0} - a_{x_0}t,$$

$$x = x_0 + v_{x_0}t - \frac{1}{2}a_{x_0}t^2.$$

Similarly, in the y direction:

$$v_y = v_{y_0} - a_{y_0}t,$$

$$y = y_0 + v_{y_0}t - \frac{1}{2}a_{y_0}t^2.$$

7.10 Projectile Motion

When an object is thrown into the air in any direction other than the vertical, it will move in a curved path under the influence of gravity and air resistance. The gravity of Earth will pull the object downward with a constant gravitational acceleration of about 9.8 m/s^2, and the air resistance will retard its motion in a direction opposite to the direction of motion. This very common form of motion, called projectile motion, is relatively simple to analyze once the effect of air resistance is ignored.

Projectile motion is a particular form of two-dimensional linear motion with constant acceleration. To be able to define the basic parameters involved in all projectile motions, consider the motion of a cannonball fired into the air (Fig. 7.17). Assume that the cannonball leaves the barrel and lands on the ground at the same elevation. As illustrated in Fig. 7.18, the cannonball ascends, reaches a peak, starts

Fig. 7.17 A cannonball fired into the air will undergo a projectile motion

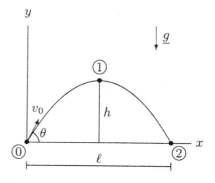

Fig. 7.18 Projectile motion

descending, and finally lands on the ground. The curved flight path of the cannonball is called the trajectory of motion. 0 represents the initial position of the cannonball, 1 is the peak it reaches, and 2 is the location of landing. v_0 is the magnitude of the initial velocity of the cannonball, which is called the *speed of release* or *takeoff speed*. θ is the angle the initial velocity vector makes with the horizontal and is called the *angle of release*. The vertical distance h between 0 and 1 is the maximum height that the cannonball reaches, and the horizontal distance l between 0 and 2 is called the *horizontal range of motion*. The total time the cannon ball remains in the air is called the *time of flight*.

The equations necessary to analyze projectile motions can be derived from Eqs. (7.27) through (7.30). For example, if the speed and angle of release of the projectile are known, then components of the velocity vector along the horizontal (x) and vertical (y) directions can be calculated:

$$v_{x_0} = v_0 \cos \theta,$$

$$v_{y_0} = v_0 \sin \theta.$$

Assuming that the air resistance on the cannonball is negligible, the acceleration of the cannonball in the x direction is zero throughout the motion. That is, $a_{x_0} = 0$. The gravitational acceleration g acts downward. Assuming that the y axis is positive upward, the gravitational acceleration acts in the negative y direction. To account for the negative direction of gravitational acceleration, the plus signs in front of the terms carrying a_{y_0} in Eqs. (7.29) and (7.30) must be changed to minus. Under these considerations, Eqs. (7.27) through (7.30) take the following special forms for projectile motion:

$$x = x_0 + (v_0 \cos \theta)t, \tag{7.31}$$

$$y = y_0 + (v_0 \sin \theta)t - \frac{1}{2}gt^2, \tag{7.32}$$

$$v_x = v_0 \cos \theta, \tag{7.33}$$

$$v_y = v_0 \sin \theta - gt. \tag{7.34}$$

Here, if the origin of the xy coordinate frame is chosen to coincide with the initial position of the ball, then $x_0 = 0$ and $y_0 = 0$. Notice from Eq. (7.33) that the magnitude of the horizontal component of the velocity vector is not a function of time. Therefore, $v_x = v_{x_0}$ remains constant throughout the projectile motion (Fig. 7.19). The magnitude of the vertical component of the velocity vector is a linear function of time. It is positive (upward) initially, decreases in time as the object ascends, and drops to zero at the peak. After reaching the peak, the vertical component of the velocity vector

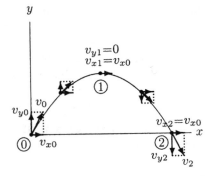

Fig. 7.19 Horizontal component of the velocity remains constant through the motion. Velocity vector is always tangent to the trajectory of motion

changes its direction from upward to downward, while its magnitude increases until it lands on the ground. At any instant during the flight, the resultant velocity vector is tangent to the trajectory of the projectile motion.

Another important aspect of the projectile motion is that if both the takeoff and landing occur at the same elevation, then the motion is symmetric with respect to a plane that passes through the peak and cuts the plane of motion at right angles. In other words, the time it takes for the object to ascend is equal to the time it takes to descend.

In some cases, the objective of the projectile motion may be to increase the horizontal range of motion to a maximum. This is particularly true for a ski jumper, for example. Other situations may require a control over the height, which is the case for a high-jumper. Therefore, it may be useful to derive some expressions for the horizontal range and maximum height of the projectile motion. Such derivations will be performed within the context of the following example problem.

Example 7.5 Consider Fig. 7.20 which shows the trajectory of a projectile motion along with some of the parameters involved. $x_0 = 0$ and $y_0 = 0$ because the origin of the xy coordinate frame is chosen to coincide with the initial position of the object. Let t_1 be the time it takes to reach the peak, and t_2 be the total time of flight. At the peak, $y_1 = h$, $v_{y_1} = 0$, and $x_1 = l/2$ from the symmetry of the motion. At the location where the object lands, $x_2 = l$ and $y_2 = 0$. The maximum height h reached by the object can be determined by noting that $v_{y_1} = 0$ at the peak. Writing Eq. (7.34) between 0 and 1:

$$v_{y_1} = v_0 \sin \theta - gt_1,$$

$$0 = v_0 \sin \theta - gt_1.$$

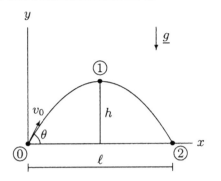

Fig. 7.20 h is the maximum elevation and is the horizontal range of motion

Solving this equation for t_1 will yield:

$$t_1 = \frac{v_0 \sin \theta}{g}. \tag{i}$$

Between 0 and 1, Eq. (7.32) will take the following form:

$$h = 0 + (v_0 \sin \theta)t_1 - \frac{1}{2}gt_1^2.$$

Substituting Eq. (i) into this equation will yield:

$$h = \frac{v_0^2 \sin^2\theta}{2g}. \tag{ii}$$

Again from the symmetry of motion, the time lapses during the ascent must be equal to the time lapses during the descent. In other words, $t_2 = 2t_1$. This can be proven by writing Eq. (7.32) between 0 and 2, and solving it for t_2:

$$y_2 = y_0 + (v_0 \sin \theta)t_2 - \frac{1}{2}gt_2^2,$$

$$0 = 0 + (v_0 \sin \theta)t_2 - \frac{1}{2}gt_2^2,$$

$$0 = v_0 \sin \theta - \frac{1}{2}gt_2.$$

Solving this equation for t_2 will yield:

$$t_2 = \frac{2v_0 \sin \theta}{g}. \tag{iii}$$

By comparing Eqs. (iii) with (i), one can conclude that:

$$t_2 = 2t_1.$$

Between 0 and 2, Eq. (7.31) can be written as:

$$l = 0 + (v_0 \cos \theta)t_2.$$

Substituting Eq. (iii) into the above equation, and noting that $2 \cos \theta \sin \theta = \sin(2\theta)$ (see Appendix C.4):

$$l = \frac{v_0^2 \sin(2\theta)}{g}. \tag{iv}$$

Equations (i) through (iv) are special forms of more general equations for projectile motions as given in Eqs. (7.31) through (7.34). From Eqs. (ii) and (iv), it is clear that for a given v_0, the maximum height and the horizontal range of motion of the projectile are functions of the angle of release θ. To see the variations of h and l with θ, $\sin^2\theta$ and $\sin(2\theta)$ are computed for θ between 0° and 90°. The values calculated are listed in Table 7.3. In Fig. 7.21, a value of 10 m/s is assigned to v_0 and corresponding trajectories are calculated for $\theta = 30°$, 45°, and 60°. The significance of these results is that, for given v_0, the range of motion l is maximum when $\theta = 45°$. Therefore, if the purpose is to maximize the horizontal range of motion of the projectile, then the angle of takeoff must be close to 45°.

Table 7.3 Variations of $\sin^2\theta$ and $\sin(2\theta)$ for $0° \leq \theta \leq 90°$

θ	$\sin^2\theta$	$\sin(2\theta)$
0°	0.000	0.000
15°	0.067	0.500
30°	0.250	0.866
45°	0.500	1.000
60°	0.750	0.866
75°	0.933	0.500
90°	1.000	0.000

The maximum height h of the projectile is a function of $\sin^2\theta$, and range l depends on $\sin(2\theta)$

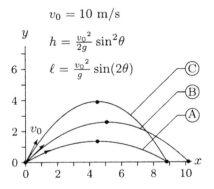

Fig. 7.21 Trajectories for $v_0 = 10$ m/s and $\theta = 30$, 45, 60° represented as A, B, and C (both x and y are in meters)

Sometimes it is easier to measure the range of motion l and the maximum height h of the projectile. In such cases, the unknowns are the takeoff speed v_0 and the angle of release θ. The relationship between these parameters can be derived as:

$$\theta = \arctan\left(\frac{4h}{l}\right), \tag{v}$$

$$v_0 = \frac{\sqrt{2gh}}{\sin\theta}. \tag{vi}$$

Note that the results obtained in this example are valid for cases in which takeoff and landing occur at the same elevation. Also, these results are not meant to be "memorized." One of the objectives of this example was to demonstrate how Eqs. (7.31) through (7.34) that govern projectile motions could be manipulated.

7.11 Applications to Athletics

The concept of projectile motion may have many applications in athletics and sports mechanics. These applications include the motion analyses of athletes doing long jumping, high jumping, ski jumping, diving, and gymnastics, and the motion analyses of the discus, javelin, shot, baseball, basketball, football, and golf ball. The following examples are aimed to illustrate some of these applications. It should be reiterated here that we are ignoring the fundamental approximation in projectile motion analyses that possible effects of air resistance can influence the motion characteristics.

Example 7.6 Based on the assumption that the air resistance is negligible, it is suggested that the overall motion characteristic of a long jumper may be analyzed by assuming that the center of gravity of the athlete undergoes a projectile motion (Fig. 7.22).

Consider an athlete who jumps a horizontal distance of

Fig. 7.22 Long jumper

8m. If the athlete was airborne for 1s, calculate the takeoff speed, angle of release, and the maximum height the athlete's center of gravity reached.

Solution

Figure 7.23 illustrates the motion of the center of gravity of the long jumper. 0 represents takeoff, 1 represents the peak, and 2 is landing. Note that the origin of the xy coordinate system is placed at 0 so that all horizontal distance and elevation measurements are relative to the position of the center of gravity of the athlete at takeoff. $x_2 = l = 8$m is the horizontal range of motion and $t_2 = 1$s is the total time the athlete was airborne. The task is to calculate the takeoff speed v_0, takeoff angle θ, and maximum height h. Note that since t_2 is known and the motion is symmetric with respect to the peak, the time it took for the athlete to reach to the peak is $t_1 = t_2/2 = 0.5$s.

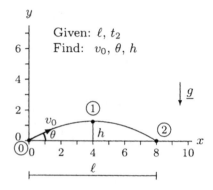

Fig. 7.23 Trajectory of the center of gravity of the athlete (both x and y are in meters)

We know l and t_2. The equation that governs the displacement in the horizontal (x) direction is Eq. (7.31). Writing this equation for position 2 relative to position 0 and substituting the known parameters:

$$x_2 = x_0 + v_{x_0}t_2,$$

$$l = 0 + v_{x_0}t_2,$$

$$8 = v_{x_0}(1).$$

Solving this equation for the unknown parameter will yield the horizontal component of the takeoff velocity, $v_{x_0} = 8.0\,\text{m/s}$.

We also know that at landing the elevation of the center of gravity is zero. That is, $y_2 = 0$. The equation that governs the displacement in the vertical (y) direction is Eq. (7.32). Writing this equation for position 2 relative to position 0 and substituting the known parameters:

$$y_2 = y_0 + v_{y_0}t_2 - \frac{1}{2}g\,t_2^{\,2},$$

$$0 = 0 + v_{y_0}(1) - \frac{1}{2}(9.8)(1)^2,$$

$$0 = v_{y_0} - 4.9.$$

Solving this equation for the vertical component of the takeoff velocity will yield $v_{y_0} = 4.9$ m/s.

Now that we know the magnitudes of the horizontal and vertical components of the takeoff velocity, we can calculate the takeoff speed:

$$v_0 = \sqrt{v_{x_0}^2 + v_{y_0}^2} = \sqrt{(8.0)^2 + (4.9)^2} = 9.4 \text{ m/s}.$$

Note that $v_{x_0} = v_0 \cos\theta$. Since we know v_{x_0} and v_0, we can calculate the takeoff angle:

$$\theta = \cos^{-1}\left(\frac{v_{x_0}}{v_0}\right) = \cos^{-1}\left(\frac{8.0}{9.4}\right) = \cos^{-1}(0.85) = 31.7°.$$

To calculate the maximum height reached, we can again utilize Eq. (7.32). Writing this equation for position 1 relative to 0 and solving it for h will yield:

$$y_1 = y_0 + v_{y_0}t_1 - \frac{1}{2}g\, t_1^2,$$

$$h = 0 + (4.9)(0.5) - \frac{1}{2}(9.8)(0.5)^2,$$

$$h = 1.2\text{m}.$$

Therefore, at the peak, the center of gravity of the athlete was 1.2m above the level it was at the takeoff.

Example 7.7 During a practice, a shot-putter puts the shot at a distance $l = 6$m. At the instant the athlete releases the shot, the elevation of the shot is $h_0 = 1.8$m as measured from the ground level, and the angle of release is $\theta = 30°$(Fig. 7.24).

Fig. 7.24 Shot-putter

Determine the speed at which the athlete released the shot, the landing speed of the shot, and the total time the shot was in the air.

Solution
Equations (7.31) and (7.32) can be utilized to solve this problem. In Fig. 7.25, the origin of the xy coordinate frame

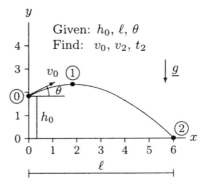

Fig. 7.25 Trajectory of the shot (x and y are in meters)

is located at the ground level directly under the point of release which is designated as 0. The shot ascends, reaches a peak at 1, and lands on the field at 2. With respect to the coordinate frame adopted, the initial and landing coordinates of the shot are: $x_0 = 0$, $y_0 = h_0$, $x_2 = l$, and $y_2 = 0$. If t_2 refers to the total time the shot was in the air, then Eq. (7.31) can be written between 0 and 2 as:

$$x_2 = x_0 + (v_0 \cos\theta)t_2,$$

$$l = 0 + (v_0 \cos\theta)t_2.$$

Solving this equation for t_2 and substituting the known parameters:

$$t_2 = \frac{l}{v_0 \cos\theta} = \frac{6}{v_0 \cos 30°} = \frac{6.93}{v_0}. \tag{i}$$

Similarly, writing Eq. (7.32) between 0 and 2 and substituting the known parameters:

$$y_2 = y_0 + (v_0 \sin\theta)t_2 - \frac{1}{2}g\, t_2^2,$$

$$0 = h_0 + (v_0 \sin\theta)t_2 - \frac{1}{2}g\, t_2^2,$$

$$0 = 1.8 + (v_0 \sin 30°)t_2 - \frac{1}{2}(9.8)t_2^2,$$

$$0 = 1.8 + 0.5v_0 t_2 = 4.9t_2^2. \tag{ii}$$

Substituting Eq. (*i*) into Eq. (*ii*):

$$0 = 1.8 + 0.5v_0\left(\frac{6.93}{v_0}\right) - 4.9\left(\frac{6.93}{v_0}\right)^2.$$

Simplifying the second term on the right-hand side of this equation by eliminating v_0, carrying out the calculations, and solving this equation for v_0 will yield:

$$v_0 = 6.69 \text{ m/s}.$$

Knowing v_0, time t_2 can be calculated using (i). This will yield $t_2 = 1.04$s.

We can utilize Eqs. (7.33) and (7.34) to calculate the landing speed v_2 of the shot. Since the horizontal component of the velocity vector is constant throughout the projectile motion:

$$v_{x_2} = v_{x_0} = v_0 \cos \theta = (6.69)(\cos 30°) = 5.79 \text{ m/s}.$$

Writing Eq. (7.34) between 0 and 2:

$$v_{y_2} = v_{y_0} - g\, t_2 = v_0 \sin \theta - g\, t_2.$$

Substituting the known parameters and carrying out the calculations will yield:

$$v_{y_2} = (6.69)(\sin 30°) - (9.8)(1.04) = -6.85 \text{ m/s}.$$

Note that we obtained a negative value for a scalar quantity, which is not permitted. Here, the negative sign implies direction. We adopted the upward direction to be positive for the y axis. The negative value calculated above indicates that the direction of the vertical component of the landing velocity is downward (opposite to that of positive y axis). Now, we can rewrite v_{y_2} as:

$$v_{y_2} = 6.85 \text{ m/s } (\downarrow).$$

Knowing the magnitudes of the horizontal and vertical components of the landing velocity enables us to calculate the landing speed:

$$v_2 = \sqrt{v_{x_2}^2 + v_{y_2}^2} = \sqrt{(5.79)^2 + (6.85)^2} = 8.97 \text{ m/s}.$$

Example 7.8 The diver illustrated in Fig. 7.26 undergoes both translational and rotational, or general motion. The overall general motion of the diver can be analyzed by

Fig. 7.26 A diver

observing the trajectory of the diver's center of gravity which can be assumed to undergo a projectile motion.

Consider a case in which a diver takes off from a diving board located at a height $h_0 = 10$m above the water level and enters the water at a horizontal distance $l = 5$m from the end of the board. If the total time the diver remains in the air is $t_2 = 2.5$s, calculate the speed and angle of takeoff of the diver's center of gravity.

Solution
The trajectory of the center of gravity of the diver is shown in Fig. 7.27. In this case, speed and angle of takeoff (v_0 and θ) are not known, but are to be determined. In Fig. 7.27, 0, 1, and 2 represent the takeoff, peak, and entry into the water stages of motion, respectively. The origin of the xy coordinate frame is located at the water level directly under 0. Therefore, the coordinates of position 0 are: $x_0 = 0$ and $y_0 = h_0 = 10$m. We know the coordinates of position 2 as well: $x_2 = l = 5$m and $y_2 = 0$. We know another parameter associated with position 2, which is $t_2 = 2.5$s.

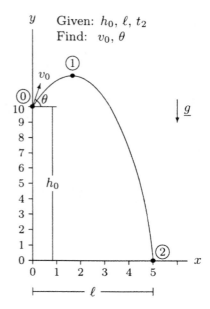

Fig. 7.27 Trajectory of the center of gravity of the diver (both x and y are in meters)

We can utilize Eqs. (7.31) and (7.32). Writing Eq. (7.31) between 0 and 2, and substituting the known parameters:

$$x_2 = x_0 + (v_0 \cos \theta)t_2,$$
$$5 = 0 + (v_0 \cos \theta)(2.5).$$

Solving this equation for $v_0 \cos \theta$:

$$v_0 \cos \theta = \frac{5}{2.5} = 2. \tag{i}$$

Similarly, writing Eq. (7.32) between 0 and 2, and substituting the known parameters:

$$y_2 = y_0 + (v_0 \cos \theta)t_2 - \frac{1}{2}g \, t_2^2,$$

$$0 = 10 + (v_0 \sin \theta)(2.5) - \frac{1}{2}(9.8)(2.5)^2.$$

Solving this equation for $v_0 \sin \theta$:

$$v_0 \sin \theta = 8.25. \qquad (ii)$$

Noting that $v_0 \sin \theta$ over $v_0 \cos \theta$ is equal to $\tan \theta$, divide Eq. (*ii*) by Eq. (*i*):

$$\tan \theta = \frac{8.25}{2} = 4.125.$$

Considering the inverse tangent of the value calculated above will yield:

$$\theta = 76.4°.$$

The speed of takeoff can now be determined from Eq. (*i*):

$$v_0 = \frac{2}{\cos \theta} = \frac{2}{\cos 76.4°} = 8.5 \text{ m/s}.$$

7.12 Exercise Problems

Problem 7.1 Consider a person throwing a ball upward into the air with an initial speed of $v_0 = 10$ m/s (Fig. 7.28). Assume that at the instant when the ball is released, the person's hand is at a height $h_0 = 1.5$m above the ground level.

Fig. 7.28 Problem 7.1

Neglecting the possible effects of air resistance, determine the maximum height h_1 that the ball reached, the total time t_2 it took for the ball to ascend and descend, and the speed v_2 of the ball just before it hit the ground.

Note that this problem must be handled in two phases: ascent and descent. Also note that the speed of the ball at the peak was zero.

Answers: $h_1 = 6.6$m, $t_2 = 1.16$s, $v_2 = 11.4$ m/s.

Problem 7.2 Consider the car shown in Fig. 7.29. At position 0, the car is stationary on a hill that makes an angle θ with the horizontal. Assume that the gear of the car is at "neutral" and that at time $t_0 = 0$ the brakes of the car are released. Under the effect of the gravitational acceleration g, the car will start moving down the hill. After some time, the car will be at position 1, which is at a distance d from position 0 measured parallel to the hill.

Fig. 7.29 Problem 7.2

Show that time t_1 to cover the distance between positions 0 and 1, and speed v_1 of the car at position 1 can be expressed as:

$$t_1 = \sqrt{\frac{2d}{g \sin \theta}},$$

$$v_1 = \sqrt{2gd \sin \theta}.$$

Problem 7.3 Based on the assumption that the air resistance is negligible, it is suggested that the overall motion characteristic of a long jumper may be analyzed by assuming that the center of gravity of the athlete undergoes a projectile motion (Fig. 7.30).

Fig. 7.30 Problem 7.3

Consider an athlete who jumps a horizontal distance of 9m after reaching a maximum height of 1.5m. What was the takeoff speed v_0 of the athlete? Discuss how the athlete can improve his/her performance.

Answer: $v_0 = 9.8$ m/s.

Problem 7.4 The ski jumper shown in Fig. 7.31 leaves the ramp with a horizontal speed of v_0 and lands on a slope that makes an angle $\beta = 45°$ with the horizontal.

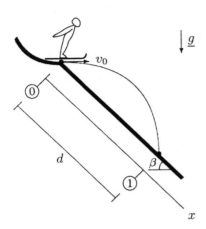

Fig. 7.31 Problem 7.4

Neglecting air resistance (the effect of which may be quite significant), determine the takeoff speed v_0, landing speed v_1, and the total time t_1 that the ski jumper was airborne if the skier touched down at a distance $d = 50$m from the ramp measured parallel to the slope.

Answers: $v_0 = 13.2$ m/s, $v_1 = 29.6$ m/s, $t_1 = 2.7$s.

Problem 7.5 Assume that in another trial, the ski jumper in the previous problem manages to maintain the takeoff speed at $v_0 = 13.2$ m/s, but leaves the ramp at an angle $\theta = 10°$ with the horizontal (Fig. 7.32).

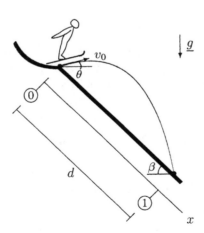

Fig. 7.32 Problem 7.5

How far from the ramp the ski jumper would land on the slope? Discuss whether the ski jumper improved his/her performance as compared to the trial in Problem 7.4. If yes, by how much?

Answers: $d = 57$m, 14% improvement.

Problem 7.6 Figure 7.33 illustrates the trajectory of a cannonball. Assume that the cannonball was fired into the air with an initial speed of $v_0 = 100$ m/s at position 0. The cannonball landed at position 2 that is at a horizontal distance $l = 1,000$m measured from position 0.

Fig. 7.33 Problem 7.6

Calculate the angle of takeoff θ, time of flight t_2, and maximum height h that the cannonball reached.

Note that $2\sin\theta\cos\theta = \sin(2\theta)$ and take $g = 10$ m/s^2.

Answers: $\theta = 45°$, $t_2 = 14$s, $h = 250$m.

8.1 Overview

As studied in the previous chapter, kinematic analyses are concerned with the description of the geometric and time-dependent aspects of motion in terms of displacement, velocity, and acceleration without dealing with the factors causing the motion. The field of *kinetics*, on the other hand, is based on kinematics and incorporates into the analysis the effects of forces that cause the motion.

Based on the type of motion involved, the field of kinetics can be divided into linear (translational) and angular (rotational) kinetics. Translation is caused by the net force applied on an object, whereas rotation is the consequence of the net torque. An object will translate and rotate simultaneously (undergo a general motion) if there is both a net force and a net moment acting on it. In addition to classifying a motion as translational, rotational, or general, the field of kinetics can be further distinguished as the *kinetics of particles* and the *kinetics of rigid bodies*. Particle kinetics is easier to implement than rigid body kinetics that introduces the size and shape of the bodies into the analyses. The distinction between a particle and a rigid body is particularly important if the object is undergoing a rotational motion. If the object is sufficiently small or it is undergoing translational motion only, then the geometric characteristics of the object may be ignored and the object can be treated as a particle located at its center of gravity with a mass equal to the total mass of the object. For example, what is significant for a person pushing a block on a flat surface is the total mass of the block, not its size or shape.

Kinetic analyses utilize Newton's second law of motion which can be formulated in various ways. One way of representing Newton's second law of motion is in terms of the equations of motion, which are particularly suitable for solving problems requiring the analysis of acceleration. The use of the equations of motion for linear kinetics will be discussed next. Another way of formulating Newton's second law is through work and energy methods that will also be discussed in this chapter within the context of linear kinetics. There are also methods based on impulse and momentum that will be presented in Chap. 11.

8.2 Equations of Motion

A body accelerates if there is a non-zero net force acting on it. Newton's second law of motion states that the magnitude of the acceleration of a body is directly proportional to the magnitude of the resultant force and inversely proportional to its mass. The direction of the acceleration is the same as the direction of the resultant force.

Consider a particle of mass m that is acted upon by a force $\underline{F}$ and let $\underline{a}$ be the resulting acceleration of the particle (Fig. 8.1). Newton's second law of motion can be expressed as:

$$\underline{F} = m\underline{a}. \tag{8.1}$$

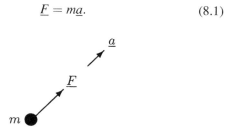

Fig. 8.1 An object will accelerate in the direction of the applied force

If there is more than one force acting on the particle (Fig. 8.2), then $\underline{F}$ in Eq. (8.1) must be replaced by the net or the resultant of all forces acting on it. The resultant of a system of forces can be determined by considering the vector sum of all forces. Therefore:

$$\sum \underline{F} = m\underline{a}. \tag{8.2}$$

N. Özkaya et al., *Fundamentals of Biomechanics: Equilibrium, Motion, and Deformation*,
DOI 10.1007/978-1-4614-1150-5_8, © Springer Science+Business Media, LLC 2012

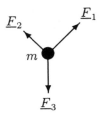

Fig. 8.2 The net force is the vector sum of all forces acting on the object

This is known as the *equation of motion*. Since force and acceleration are vector quantities, they can be expressed in terms of their components in reference to a chosen coordinate frame. For translational motion analyses, it is best to use the Cartesian (rectangular) coordinate system that consists of the x, y, and z axes with $\underline{i}$, $\underline{j}$, and $\underline{k}$ unit vectors indicating the positive x, y, and z directions, respectively (Fig. 8.3). If there is a single force acting on an object, then the force vector and the resulting acceleration vector can be expressed in terms of their components along the rectangular coordinate directions:

$$\underline{F} = F_x\underline{i} + F_y\underline{j} + F_z\underline{k}, \tag{8.3}$$

$$\underline{a} = a_x\underline{i} + a_y\underline{j} + a_z\underline{k}. \tag{8.4}$$

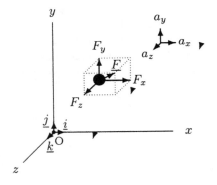

Fig. 8.3 Rectangular components of the force and acceleration vectors

Substituting Eqs. (8.3) and (8.4) into Eq. (8.1):

$$F_x\underline{i} + F_y\underline{j} + F_z\underline{k} = ma_x\underline{i} + ma_y\underline{j} + ma_z\underline{k}. \tag{8.5}$$

This vector equation is valid if the following conditions are satisfied:

$$F_x = ma_x,$$
$$F_y = ma_y, \tag{8.6}$$
$$F_z = ma_z.$$

If there is more than one force acting on the object, then F_x, F_y, and F_z must be replaced by the sum of all forces acting in the x, y, and z directions, respectively:

$$\sum F_x = ma_x,$$
$$\sum F_y = ma_y, \tag{8.7}$$
$$\sum F_z = ma_z.$$

Equations (8.7) state that the sum of all forces acting in one direction is equal to the mass times the acceleration of the body in that direction. Note that for one-dimensional motion analysis, only one of these equations need to be considered. For a two-dimensional case, two of the above equations are sufficient to analyze the problem.

8.3   Special Cases of Translational Motion

A force can be applied in various ways. For example, an applied force may be constant or it may vary over time. Applied forces can be measured in various ways as well. The magnitude of a force vector can be measured as a function of time, as a function of the relative position of the object upon which it is applied, or as a function of velocity. Some of these cases will be discussed next. To illustrate the methods of handling these cases in a concise manner, it will be assumed that the motion is along a straight line in the x direction and under the effect of only one applied force. Using the vectorial properties of the parameters involved, these methods can be easily expanded to analyze two- and three-dimensional translational motions under the action of more than one force.

Note that the derivations provided in this section are aimed to demonstrate that different cases can be handled through proper mathematical manipulations. The mathematics involved for the cases in which the applied force is a function of displacement may be beyond the scope of this text, and can be omitted without losing the continuity of the topics to be covered in the following sections.

8.3.1   Force is Constant

If a force applied on an object has a constant magnitude and direction, the object will move with a constant acceleration in the direction of the applied force. Assume that a force with magnitude F_x is applied on an object with mass m. The magnitude a_x of the constant acceleration of the object in the x direction can be calculated using the equation of motion in the x direction:

$$a_x = \frac{F_x}{m} = \text{constant}.$$

Once the acceleration of the object is determined, the kinematic equations can be utilized to calculate the velocity and displacement of the object as well:

$$v_x = v_{x_0} + \int_{t_0}^{t} a_x dt = v_{x_0} + \frac{F_x}{m} t, \qquad (8.8)$$

$$x = x_0 + \int_{t_0}^{t} v_x dt = x_0 + v_{x_0} t + \frac{1}{2} \frac{F_x}{m} t^2. \qquad (8.9)$$

Here, v_{x_0} and x_0 are the initial speed and displacement of the object in the x direction at time $t = 0$. Note that these results can be used for a situation in which there is a second force with constant magnitude F_y acting on the object in the y direction simply by replacing x with y throughout the equations.

8.3.2 Force Is a Function of Time

If the magnitude of a force applied on an object is a function of time, then $F_x = F_x(t)$. The resulting acceleration of the object is also a function of time:

$$a_x(t) = \frac{F_x(t)}{m}.$$

The velocity and displacement of the object can now be determined using the kinematic relationships:

$$v_x = v_{x_0} + \int_{t_0}^{t} a_x(t) dt, \qquad (8.10)$$

$$x = x_0 + \int_{t_0}^{t} v_x(t) dt. \qquad (8.11)$$

The function $F_x(t)$ must be provided so that the integral in Eq. (8.10) can be evaluated.

8.3.3 Force Is a Function of Displacement

Sometimes it is more convenient to express force as a function of displacement, in which case $F_x = F_x(x)$. By definition, acceleration is equal to the time rate of change of velocity. Therefore, the equation of motion in the x direction can be expressed as:

$$\frac{dv_x}{dt} = \frac{F_x(x)}{m}.$$

Employing the chain rule of differentiation (see Appendix C.2.6), the time derivative of velocity can be expressed as:

$$\frac{dv_x}{dt} = \frac{dv_x}{dx} \frac{dx}{dt} = \frac{dv_x}{dx} v_x.$$

Therefore, the equation of motion in the x direction is:

$$v_x \frac{dv_x}{dx} = \frac{F_x(x)}{m}.$$

Multiplying both sides by dx:

$$v_x dv_x = \frac{F_x(x)}{m} dx.$$

The left-hand side of this equation is a function of v_x only and can be integrated with respect to v_x, and the right-hand side is a function of x only and can be integrated with respect to x:

$$\int_{v_{x_0}}^{v_x} v_x dv_x = \int_{x_0}^{x} \frac{F_x(x)}{m} dx.$$

Evaluating the integral on the left-hand side:

$$\frac{1}{2} \left(v_x^2 - v_{x_0}^2 \right) = \frac{1}{m} \int_{x_0}^{x} F_x(x) dx.$$

Rearranging the order of terms:

$$v_x^2 = v_{x_0}^2 + \frac{2}{m} \int_{x_0}^{x} F_x(x) dx. \qquad (8.12)$$

F_x must be provided as a function of x, so that the integral in Eq. (8.12) can be evaluated. For given $F_x(x)$, Eq. (8.12) will yield v_x as a function of x. Once F_x is known, the acceleration of the object can be determined using:

$$a_x = v_x \frac{dv_x}{dx}. \qquad (8.13)$$

8.4 Procedure for Problem Solving in Kinetics

The procedure for analyzing kinetic characteristics of objects undergoing translational motion using the equations of motion can be outlined as follows:

- Draw a simple, neat diagram of the system to be analyzed.
- Isolate the bodies of interest from their surroundings and draw their free-body diagrams by showing all external

forces acting on them. Indicate the correct directions for the known forces. If the direction of a force vector is not known, assume a positive direction for it. If that force appears to have a negative value in the solution, it would mean that the assumed direction for the force vector was incorrect.

- Designate the direction of motion of each object on the sidelines (not as parts of the free-body diagrams). It is particularly important to be consistent with the assumed direction of the motion throughout the analyses.
- Choose a convenient coordinate system. For two-dimensional cases, rectangular coordinates x and y are usually the most convenient.
- Apply the equations of motion. For two-dimensional motion analysis there are two governing equations, and therefore, the number of unknowns to be determined cannot be more than two. In linear kinetics, the unknowns are either forces or accelerations.
- Include the correct directions of forces and accelerations in the solution, along with their units.
- The kinematic relations between position, velocity, and acceleration can also be utilized if the information about the velocity and/or position of the object analyzed is given or required.

Example 8.1 As illustrated in Fig. 8.4, consider a block of mass $m = 50$ kg which is being pulled on a rough, horizontal surface by a person using a rope. Assume that the person is applying a constant force of $T = 150$ N on the block, the rope makes an angle $\theta = 30°$ with the horizontal, and the coefficient of kinetic friction between the block and the horizontal surface is $\mu = 0.2$.

Fig. 8.4 A block is being pulled on a horizontal surface

Determine the acceleration of the block if the bottom surface of the block remains in full contact with the floor throughout the motion.

Solution
The free-body diagram of the block is shown in Fig. 8.5. The positive x direction is chosen in the direction of motion of the block. W is the weight of the block, f is the magnitude of the frictional force acting in the direction opposite to the direction of motion, and N is the magnitude of the reaction force applied by the floor on the block. $\underline{T}$ is the force exerted by the person which is transmitted to the

Fig. 8.5 The free-body diagram of the block

block through the rope. The rope makes an angle $\theta = 30°$ with the horizontal. Therefore, $\underline{T}$ has components in the x and y directions:

$$T_x = T \cos \theta \quad (\rightarrow),$$

$$T_y = T \sin \theta \quad (\uparrow).$$

The weight of the block is due to the gravitational effect of Earth on the mass of the block, and can be expressed as:

$$W = mg \quad (\downarrow).$$

The magnitude of the frictional force is proportional to the magnitude of the normal force, and they are related through the coefficient of friction between the surfaces in contact:

$$f = \mu N \quad (\leftarrow). \tag{i}$$

Equations of motion in the x and y directions can now be applied to determine an expression for the acceleration of the block. The block has no motion in the y direction, and therefore, the acceleration of the block in the y direction is zero ($a_y = 0$). The equation of motion in the y direction is:

$$\sum F_y = 0: \quad N + T_y - W = 0.$$

Solving this equilibrium equation for force N will yield:

$$N = W - T_y = mg - T \sin \theta. \tag{ii}$$

Substituting Eq. (*ii*) into Eq. (*i*) will yield:

$$f = \mu(mg - T \sin \theta). \tag{iii}$$

Now, the equation of motion in the x direction can be considered:

$$\sum F_x = ma_x: \quad T_x - f = ma_x.$$

Solving this equation for a_x will yield

$$a_x = \frac{1}{m}(T_x - f). \tag{iv}$$

Substituting Eq. (*iii*) and $T_x = T \cos \theta$ into Eq. (*iv*):

$$a_x = \frac{1}{m}[T \cos \theta - \mu(mg - T \sin \theta)].$$

Substituting the numerical values $m = 50$ kg, $T = 150$ N, $\theta = 30°$, $\mu = 0.2$, and $g = 9.8$ m/s, and carrying out the calculations will yield $a_x = 0.94$ m/s^2 ($\rightarrow$).

8.5 Work and Energy Methods

The fundamental method of analyzing the kinetic characteristics of bodies is based on the equations of motion which are mathematical representations of Newton's second law of motion. Using the equations of motion, one can determine accelerations. In some cases, particularly when the forces involved are not constant, the solution of equations of motion may be difficult. To handle such situations, alternative methods are developed that are based on the concepts of work and energy. These methods are also derived from Newton's laws, and can be applied to analyze the forces, velocities, and displacements involved in relatively complex systems without resorting to the equations of motion.

8.6 Mechanical Work

By definition, *mechanical work* is the product of force and corresponding displacement. Work is a scalar quantity. There is no direction associated with work.

8.6.1 Work Done by a Constant Force

To explore the definition of work, consider the block in Fig. 8.6. Assume that a constant, horizontal force $\underline{F}$ is applied on the block so as to move it from position 1 to position 2, which are s distance apart. The work done, W, by force $\underline{F}$ on the block to move the block from position 1 to 2 is

Fig. 8.6 A constant force applied on the block displaces it from position 1 to position 2

equal to the magnitude of the force vector times the displacement:

$$W = Fs. \tag{8.14}$$

Consider the same block which is pulled from position 1 to 2 by another constant force $\underline{F}$ that makes an angle θ with the horizontal (Fig. 8.7). The work done by $\underline{F}$ on the block is equal to the magnitude of the force component in the direction of displacement times the displacement itself. Since the component of $\underline{F}$ along the horizontal is $F_x = F \cos \theta$, the work done by $\underline{F}$ to move the block from position 1 to 2 is:

$$W = F_x s = Fs \cos \theta. \tag{8.15}$$

Fig. 8.7 A constant force that makes an angle θ with the horizontal is applied on the block

Note that Eqs. (8.14) and (8.15) are consistent with each other since $\cos \theta = 1$ when $\theta = 0°$.

For a force to do work, the body on which the force is applied must undergo a displacement and the force vector must have a non-zero component in the direction of displacement. For example, the vertical component, $F_y = F \sin \theta$, of the force vector in Fig. 8.7, does no work because the block is not displaced in the vertical direction.

Work done can be positive or negative. The work done by a force is positive if the force is applied in the same direction as the displacement. If the applied force and displacement have opposite directions, then the work done by that force is negative. A typical example of negative work is the one done by a frictional force. As illustrated in Fig. 8.8, assume that a

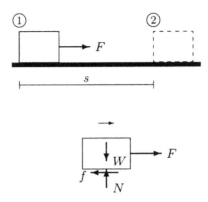

Fig. 8.8 Frictional forces do negative work

block is pulled by a force $\underline{F}$ toward the right to displace the block by a distance s. The work done W_f by the frictional force $\underline{f}$ on the block while the block was displaced by a distance s is:

$$W_f = -fs. \tag{8.16}$$

If there is more than one external force acting on a body in motion, then there is one work done for each force. The net work done is the algebraic sum of work done by individual forces. For example, the net work done for the case illustrated in Fig. 8.8 is:

$$W = F_s - fs.$$

8.6.2 Work Done by a Varying Force

Equation (8.14) can only be used to calculate the work done by a constant force. If an applied force is a function of displacement, then the work done can be calculated by considering the integral of the force over the distance it is applied.

As illustrated in Fig. 8.9, consider a block pulled along the x direction by a force F_x that varies with the displacement of the block in the x direction. That is, $F_x = F_x(x)$. Assume that the block that was originally located at position 1 moves to position 2, which are s distance apart. Let x_1 and x_2 represent the initial and final positions of the block, respectively. If the variation of F_x with respect to x is

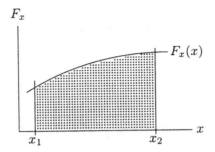

Fig. 8.9 Work is equal to the area under the force versus displacement curve

known, then the work done by F_x to move the block from position 1 to 2 can be determined using:

$$W = \int_{x_1}^{x_2} F_x \, dx. \tag{8.17}$$

Note that the evaluation of the definite integral in Eq. (8.17) will yield the total area under the force versus position curve, the x axis, and vertical lines passing through $x = x_1$ and $x = x_2$ (shaded area in Fig. 8.9). Also note that if the force F_x is constant, then the integration in Eq. (8.17) will yield $W = F_x(x_2 - x_1) = F_x s$, which is consistent with Eq. (8.14).

8.6.3 Work as a Scalar Product

For some applications, it may be convenient to utilize the definition of work as the dot (scalar) product of the force and displacement vectors. As discussed in Appendix B.14, the dot product of any two vectors is a scalar quantity equal to the product of magnitudes of the two vectors multiplied by the cosine of the smaller angle between the two. In the case of work done by a constant force $\underline{F}$ on a body whose displacement vector is given by $\underline{s}$:

$$W = \underline{F} \cdot \underline{s}. \tag{8.18}$$

If θ is the smaller angle between vectors $\underline{F}$ and $\underline{s}$, then:

$$W = \underline{F} \cdot \underline{s} = Fs \cos \theta. \tag{8.19}$$

Force and displacement vectors can be expressed in terms of their rectangular components:

$$\underline{F} = F_x \underline{i} + F_y \underline{j} + F_z \underline{k}, \tag{8.20}$$

$$\underline{s} = x\underline{i} + y\underline{j} + z\underline{k}. \tag{8.21}$$

The dot product of unit vectors are such that $\underline{i} \cdot \underline{i} = \underline{j} \cdot \underline{j} = \underline{k} \cdot \underline{k} = 1$ and $\underline{i} \cdot \underline{j} = \underline{j} \cdot \underline{k} = \underline{k} \cdot \underline{i} = 0$. Substituting Eqs. (8.20) and (8.21) into Eq. (8.18) and carrying out the dot products of unit vectors will yield:

$$W = F_x x + F_y y + F_z z. \tag{8.22}$$

Equation (8.22) is significant in that it represents the total work done by the components of the force vector in the x, y, and z directions. For example, the work done in the x direction is equal to the magnitude of the force component in the x direction times the displacement in the same direction. Note that for a biaxial motion in the xy-plane, Eq. (8.22) reduces to $W = F_x x + F_y y$ and $W = F_x x$ for a uniaxial motion in the x direction.

8.7 Mechanical Energy

The term *energy* is used to describe the capacity of a system to do work on another system. Energy can take various forms such as mechanical, thermal, chemical, and nuclear. The field of mechanics is primarily concerned with the mechanical form of energy. Mechanical energy can be categorized as potential energy and kinetic energy. Energy is also a scalar quantity.

8.7.1 Potential Energy

The *potential energy* of a system is associated with its position or elevation. It is the energy stored in the system that can be converted into kinetic energy. The concept of potential energy comes from the perception that an object located at a height can do useful work if it is allowed to descend. The potential of an object to do work due to the relative height of its center of gravity is defined as *gravitational potential energy*. Consider the object with weight $W = mg$ shown in Fig. 8.10. The object is at position 1 which is located at a height h measured relative to position 2. The gravitational potential energy, $\mathcal{E}_P$, of the object at position 1 relative to position 2 is:

$$\mathcal{E}_P = Wh = mgh. \tag{8.23}$$

Fig. 8.10 Gravitational potential energy

Notice that Wh is essentially the work that the force of gravity would do on the object to move it from position 1 to position 2, which are h distance apart.

8.7.2 Kinetic Energy

Kinetic energy is associated with motion. Every moving object has a kinetic energy. The kinetic energy, $\mathcal{E}_K$, of an object with mass m moving with a speed v is equal to the product of one-half of the mass and the square of the speed of the object:

$$\mathcal{E}_K = \frac{1}{2}mv^2. \tag{8.24}$$

8.8 Work–Energy Theorem

There is a relationship between the kinetic energy and the work done. The net work done, W_{12}, on an object to displace the object from position 1 to position 2 is equal to the change in kinetic energy, $\Delta\mathcal{E}_K$, of the object between positions 1 and 2. This is known as the *work–energy theorem* and can be expressed as:

$$W_{12} = \Delta\mathcal{E}_K = \mathcal{E}_{K2} - \mathcal{E}_{K1}. \tag{8.25}$$

8.9 Conservation of Energy Principle

Forces may be conservative and nonconservative. A force is conservative if the work done by that force to move an object between two positions is independent of the path taken. A typical example of conservative forces is the gravitational force. The frictional force, on the other hand, is a nonconservative force. Nonconservative forces dissipate energy as heat.

The net work done on a system by conservative forces is converted into kinetic and potential energies in such a manner that the total energy of the system (sum of kinetic and potential energies) remains constant throughout the motion. This is known as the *principle of conservation of mechanical energy*, and between any two positions 1 and 2 it can be stated as:

$$\mathcal{E}_{K1} + \mathcal{E}_{P1} = \mathcal{E}_{K2} + \mathcal{E}_{P2}. \tag{8.26}$$

8.10 Dimension and Units of Work and Energy

Mechanical work and energy have the same dimension and units. By definition, work done is force times displacement. Therefore, work has the dimension of force times the dimension of length.

$$[\text{Work}] = [\text{Force}][\text{Displacement}] = M\frac{L^2}{T^2}.$$

The units of work and energy in different systems of units are provided in Table 8.1.

Table 8.1 Units of work and energy

System	Units of work and energy	Special name
SI	Newton meter (N-m)	Joule (J)
CGS	Dyne centimeter (dyn-cm)	erg
British	Pound foot (lb-ft)	

8.11 Power

Power, $\mathcal{P}$, is defined as the time rate of work done:

$$\mathcal{P} = \frac{\mathrm{d}W}{\mathrm{d}t}. \qquad (8.27)$$

The work done by a constant force on an object can be determined by considering the dot product of the force and displacement vectors $(W = \underline{F} \cdot \underline{s})$:

$$\mathcal{P} = \frac{\mathrm{d}}{\mathrm{d}t}(\underline{F} \cdot \underline{s}).$$

If the force vector $\underline{F}$ is constant, then:

$$\mathcal{P} = \underline{F} \cdot \frac{\mathrm{d}\underline{s}}{\mathrm{d}t} = (\underline{F} \cdot \underline{v}). \qquad (8.28)$$

In Eq. (8.28), $\underline{v}$ is the velocity vector of the object. If the applied force is collinear with the velocity, then $\mathcal{P} = Fv$. Power is a scalar quantity and has the dimension of force times velocity. The units of power are given in Table 8.2.

Table 8.2 Units of power (1 hp = 550 lb-ft/s = 746 W)

System	Units of work and energy	Special name
SI	N-m/s = J/s	Watt (W)
CGS	dyn-cm/s = erg/s	
British	lb-ft/s	Horsepower (hp)

8.12 Applications of Energy Methods

The work–energy theorem stated in Eq. (8.23) and the principle of conservation of energy stated by Eq. (8.24) provide alternative methods of problem solving in dynamics. The work–energy theorem can be used to analyze problems involving nonconservative forces. On the other hand, the principle of conservation of energy is useful only when the forces involved are conservative. As compared to the applications of the equations of motion, these methods are easier to apply and are particularly useful when the information provided or to be determined is in terms of velocities rather than accelerations. Definitions of important concepts introduced in this chapter and various methods of analyses in kinetics are summarized in Table 8.3. The following examples will demonstrate some of the applications of these methods.

Table 8.3 Summary of equations and formulas

Work done by a varying force	$W = \int_{x_1}^{x_2} F_x \mathrm{d}x$
Work done by a constant force	$W = F_x(x_2 - x_1) = F_x s$
Potential energy	$\mathcal{E}_\mathrm{P} = mgh$
Kinetic energy	$\mathcal{E}_\mathrm{K} = \frac{1}{2}mv^2$
Conservation of energy principle	$\mathcal{E}_{\mathrm{K}1} + \mathcal{E}_{\mathrm{P}1} = \mathcal{E}_{\mathrm{K}2} + \mathcal{E}_{\mathrm{P}2}$
Work–energy theorem	$W_{12} = \mathcal{E}_{\mathrm{K}2} - \mathcal{E}_{\mathrm{K}1}$
Equation of motion	$\sum F_x = ma_x$

Example 8.2 A 20 kg block is pushed up a rough, inclined surface by a constant force of $P = 150$ N that is applied parallel to the incline (Fig. 8.11). The incline makes an angle $\theta = 30°$ with the horizontal and the coefficient of friction between the incline and the block is $\mu = 0.2$.

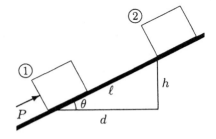

Fig. 8.11 A block is pushed from position 1 to 2

If the block is displaced by $l = 10$ m, determine the work done on the block by force $\underline{P}$, by the force of friction, and by the force of gravity. What is the net work done on the block?

Solution

The free-body diagram of the block is shown in Fig. 8.12. W is the weight of the block, f is the frictional force at the bottom surface of the block, and N is the reaction force applied by the incline on the block. The x and y directions are chosen in such a manner that the motion occurs in the positive x direction, and there is no displacement of the block in the y direction. Force $\underline{P}$ is applied in the same direction as the displacement of the block. Therefore, the

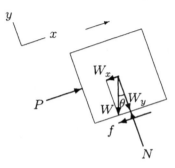

Fig. 8.12 Free-body diagram of the block

work done by $\underline{P}$ to displace the block by a distance of l along the incline is:

$$W_P = Pl. \tag{i}$$

The weight $\underline{W}$ of the block has components along the x and y directions, such that $W_x = W \sin\theta$ and $W_y = W \cos\theta$. Since there is no motion in the y direction, the block is in equilibrium in the y direction. The equilibrium in the y direction requires that $N = W_y$, or since $W = mg$, $N = mg \cos\theta$. The relationship between the frictional force and the normal force at the surfaces of contact is such that $f = \mu N = \mu \cos\theta$. The frictional force acts in a direction parallel to the incline but opposite to that of the displacement of the block. Therefore, the work done by f on the block as the block is displaced by a distance l is:

$$W_f = -fl = -\mu \, mgl \cos\theta. \tag{ii}$$

The force of gravity (weight) always acts downward. In this case, it has a component parallel to the incline in the negative x direction with magnitude $W_x = mg \sin\theta$. The work done by W_x on the block as it moves from position 1 to position 2 is:

$$W_g = -mgl \sin\theta. \tag{iii}$$

Knowing the work done by individual forces acting on the block, we can determine the net work done on the block:

$$W = W_P + W_f + W_g = Pl - mgl(\sin\theta + \mu\cos\theta). \tag{iv}$$

Substituting the numerical values of the parameters involved into Eqs. (i) through (iv) and carrying out the calculations will yield:

$$W_P = (150)(10) = 1{,}500 \text{ J},$$
$$W_f = -(0.2)(20)(9.8)(10)(\cos 30°) = -340 \text{ J},$$
$$W_g = -(20)(9.8)(10)(\sin 30°) = -980 \text{ J},$$
$$W = 1{,}500 - 340 - 980 = 180 \text{ J}.$$

Example 8.3 Figure 8.13 illustrates a pendulum with mass m and length l. The mass is pulled to position 1 that makes an angle θ with the vertical and is released to swing.

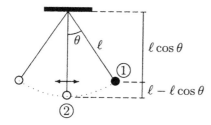

Fig. 8.13 The pendulum

Assuming that frictional effects and air resistance are negligible, determine the speed v_2 of the mass when it is at position 2.

Solution
The mass has zero speed and no kinetic energy at the instant of release (position 1). If we choose position 2 to be the datum from which heights are measured, then the mass is located at a height $h_1 = l(1 - \cos\theta)$ at position 1, and the height of the mass at position 2 is zero. Applying the conservation of energy principle between positions 1 and 2:

$$\mathcal{E}_{K1} + \mathcal{E}_{P1} = \mathcal{E}_{K2} + \mathcal{E}_{P2},$$

$$\frac{1}{2}mv_1{}^2 + mgh_1 = \frac{1}{2}mv_2{}^2 + mgh_2.$$

Substituting $v_1 = 0$, $h_1 = l(1 - \cos\theta)$, and $h_2 = 0$ into this equation and solving it for the speed of the mass at position 2 will yield:

$$v_2 = \sqrt{2gl(1 - \cos\theta)}.$$

Example 8.4 As illustrated in Fig. 8.14, consider a ski jumper moving down a track to acquire sufficient speed to accomplish the ski jumping task. The length of the track is $l = 25$ m and the track makes an angle $\theta = 45°$ with the horizontal.

Fig. 8.14 A ski jumper

If the skier starts at the top of the track with zero initial speed, determine the takeoff speed of the skier at the bottom of the track using (a) the work–energy theorem, (b) the conservation of energy principle, and (c) the equation of motion along with the kinematic relationships. Assume that the effects of friction and air resistance are negligible.

Solution (a): *Work–energy method*
The free-body diagram of the ski jumper is shown in Fig. 8.15. The forces acting on the ski jumper are the

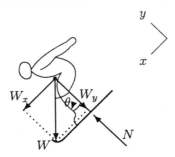

Fig. 8.15 The free-body diagram of the ski jumper

$$\frac{1}{2}mv_1^2 + mgh_1 = \frac{1}{2}mv_2^2 + mgh_2. \qquad (iv)$$

In Eq. (*iv*), the first term on the left-hand side is zero since $v_1 = 0$. If we measure heights relative to the bottom of the track (or selecting 2 to be the datum as shown in Fig. 8.16), then height $h_2 = 0$ and the height of the top of the track is $h_1 = l \sin \theta$. Therefore, the second term on the right-hand side of Eq. (*iv*) is zero as well. Substituting $h_1 = l \sin \theta$ into Eq. (*iv*), eliminating the repeated parameter m, and solving Eq. (*iv*) for v_2 will again yield Eq. (*iii*).

gravitational force $\underline{W}$ and the reaction force applied by the track on the skis is in a direction perpendicular to the track. The x direction is chosen to coincide with the direction of motion and y is perpendicular to the track. Therefore, the weight of the ski jumper has components along the x and y directions, such that $W_x = W \sin \theta = mg \sin \theta$ and $W_y = W \cos \theta = mg \cos \theta$. On the other hand, $\underline{N}$ acts in the y direction. Note that W_x is the driving force for the skier.

Since there is only one force component in the x direction, the work done by that force component is also the net work done on the ski jumper. Labeling the top and the bottom of the track as positions 1 and 2, the work done by W_x to move the skier from position 1 to 2 that are l distance apart is:

$$W_{12} = W_x l = mgl \sin \theta. \qquad (i)$$

According to the work–energy theorem, W_{12} must be equal to the change in kinetic energy of the skier between positions 1 and 2:

$$W_{12} = \mathcal{E}_{K2} - \mathcal{E}_{K1} = \frac{1}{2}mv_2^2 - \frac{1}{2}mv_1^2. \qquad (ii)$$

The second term on the right-hand side of Eq. (*ii*) is zero because the initial speed of the skier is $v_1 = 0$. Substituting Eq. (*i*) into Eq. (*ii*), eliminating the repeated parameter m (the mass of the ski jumper), and solving Eq. (*ii*) for the takeoff speed v_2 of the ski jumper will yield:

$$v_2 = \sqrt{2gl \sin \theta}. \qquad (iii)$$

Solution (b): *Conservation of energy method*
Since the effects of nonconservative forces due to friction and air resistance are assumed to be negligible, this problem can also be analyzed by utilizing the principle of conservation of energy. Between positions 1 and 2 of the ski jumper:

$$\mathcal{E}_{K1} + \mathcal{E}_{P1} = \mathcal{E}_{K2} + \mathcal{E}_{P2}, \qquad$$

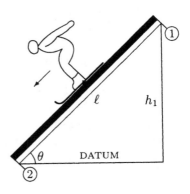

Fig. 8.16 $h_1 = \ell \sin \theta$ and $h_2 = 0$

Solution (c): *Using the equation of motion*
The equation of motion in the direction of motion (x) is:

$$\sum F_x = ma_x : W_x = ma_x. \qquad (v)$$

Substituting $W_x = mg \sin \theta$ in Eq. (*v*), eliminating m, and solving Eq. (*v*) for the acceleration of the ski jumper in the x direction will yield:

$$a_x = g \sin \theta. \qquad (vi)$$

Since the acceleration of the ski jumper is due to gravity only, what we have is a one-dimensional motion with constant acceleration. By definition, acceleration is the time rate of change of velocity, or velocity is the integral of acceleration with respect to time. Since a_x is constant and the initial velocity of the ski jumper at position 1 is zero, we can write:

$$v_x = a_x t. \qquad (vii)$$

The kinematic relationship between the velocity and displacement is such that displacement is equal to the integral of velocity. If we measure the displacement relative to the initial position of the ski jumper, then the initial displacement is zero. Therefore, the equation relating displacement, acceleration, and time is:

$$x = \frac{1}{2} a_x t^2. \tag{viii}$$

Equation (*vii*) can be solved for time $t = v_x/a_x$, which can then be substituted into Eq. (*viii*) so as to eliminate t. This will yield:

$$x = \frac{1}{2} \frac{v_x^2}{a_x}.$$

Solving this equation for v_x will give:

$$v_x = \sqrt{2 x a_x}. \tag{ix}$$

This is a general solution relating the acceleration, speed, and displacement of the ski jumper when the ski jumper is anywhere along the track. $x = l$ and $v_x = v_2$ when the ski jumper reaches the bottom of the track, and the acceleration of the ski jumper is always $a_x = g \sin \theta$. Substituting these parameters into Eq. (*ix*) will again yield Eq. (*iii*).

Finally, substituting the numerical values of $g = 9.8$ m/s^2, $l = 25$ m, and $\theta = 45°$ into Eq. (*iii*) and carrying out the calculations will yield $v_2 = 18.6$ m/s.

Remarks

- It is clear that for problems involving displacement, speed, and force, applications of the methods based on the work–energy theorem and the conservation of energy principle are more straightforward as compared to the application of equations of motion. In general, one should try work–energy or conservation of energy methods first before resorting to the equations of motion.
- Since the effects of nonconservative forces due to friction and air resistance are neglected, the solution of the problem is independent of the shape of the track or how the skier covers the distance between the top and bottom of the track. The most important parameter in this problem affecting the takeoff speed of the skier is the total vertical distance between 1 and 2. This implies that the problem could be simplified by noting that the skier undergoes a "free fall" between 1 and 2 which are $h_1 = l \sin \theta$ distance apart. This is illustrated in Fig. 8.17. Applying the principle of conservation of energy between 1 and 2 will again yield Eq. (*iii*).

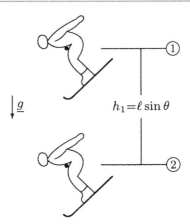

Fig. 8.17 The solution of the problem is independent of the path of motion

8.13 Exercise Problems

Problem 8.1 Figure 8.18 shows a person pushing a block of mass m on a surface that makes an angle θ with the horizontal. The coefficient of kinetic friction between the block and the inclined surface is μ.

Fig. 8.18 Problem 8.1

If the person is applying a force with constant magnitude P and in a direction parallel to the incline, show that the acceleration of the block in the direction of motion can be expressed as:

$$a_x = \frac{P}{m} - g(\mu \cos \theta + \sin \theta).$$

Problem 8.2 Consider that a force with magnitude F_x that varies with displacement along the x direction is applied on an object. Assume that the variation of the force is as shown in Fig. 8.19 where force is measured in Newtons and displacement is measured in meters.

Fig. 8.19 Problem 8.2

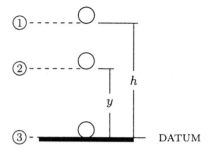

Fig. 8.21 Problem 8.4

Determine the work done by F_x on the object as the object moved from $x = 0$ to $x = 15$ m.

Answer: 100 J.

Problem 8.3 A force with varying magnitude F_x is applied on an object and the displacement of the object is recorded in terms of x. The applied force is then plotted as a function of displacement and the curve shown in Fig. 8.20 is obtained. It is observed that between $x = 0$ and $x = 9$ m, the force is proportional to the square root of displacement:

$$F_x = c\sqrt{x}.$$

Here, F_x is measured in Newtons and x in meters, and the constant of proportionality between F_x and x is estimated to be $c = 6$.

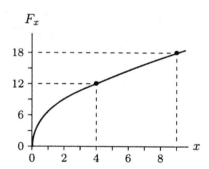

Fig. 8.20 Problem 8.3

Determine the work done by F_x on the object to move the object from:
(a) $x = 0$ to $x = 4$ m
(b) $x = 0$ to $x = 9$ m
(c) $x = 4$ to $x = 9$ m.

Answers: (a) 32 J, (b) 108 J, and (c) 76 J.

Problem 8.4 As illustrated in Fig. 8.21, a ball is dropped from a height h measured from the ground level. If the air resistance is neglected, show that the speed of the ball as a

function of height y measured from the ground level can be expressed as:

$$v = \sqrt{2g(h - y)}.$$

Here, g is the magnitude of the gravitational acceleration.

Problem 8.5 As illustrated in Fig. 8.22, consider a ski jumper moving down a track to acquire sufficient speed to accomplish the ski jumping task. The length of the track is l, the track makes an angle θ with the horizontal, and the coefficient of friction between the track and the skis is μ.

Fig. 8.22 Problem 8.5

If the ski jumper starts at the top of the track with zero initial speed, determine expressions for:
(a) The takeoff speed v_2 of the ski jumper at the bottom of the track using the work–energy theorem.
(b) The acceleration a_x of the ski jumper using the equation of motion.
Assume that the effects of air resistance are negligible.

Answers:
(a) $v_2 = \sqrt{2lg(\sin\theta - \mu\cos\theta)}$.
(b) $a_x = g(\sin\theta - \mu\cos\theta)$.

Angular Kinematics

9

9.1 Polar Coordinates

Two-dimensional angular motions of bodies are commonly described in terms of a pair of parameters, r and θ (theta), which are called the *polar coordinates*. Polar coordinates are particularly well suited for analyzing motions restricted to circular paths. As illustrated in Fig. 9.1, let O and P be two points on a two-dimensional surface. The location of P with respect to O can be specified in many different ways. For example, in terms of rectangular coordinates, P is a point with coordinates x and y. Point P is also located at a distance r from point O with r making an angle θ with the horizontal. Both x and y, and r and θ specify the position of P with respect to O uniquely, and O forms the origin of both the rectangular and polar coordinate systems. Note that these pairs of coordinates are not mutually independent. If one pair is known, then the other pair can be calculated because they are associated with a right-triangle: r is the hypotenuse, θ is one of the two acute angles, and x and y are the lengths of the adjacent and opposite sides of the right-triangle with respect to angle θ. Therefore:

$$x = r \cos \theta,$$
$$y = r \sin \theta. \tag{9.1}$$

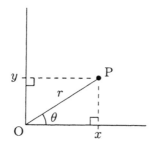

Fig. 9.1 Rectangular and polar coordinates of point P

Expressing r and θ in terms of x and y:

$$r = \sqrt{x^2 + y^2},$$
$$\theta = \arctan\left(\frac{y}{x}\right). \tag{9.2}$$

9.2 Angular Position and Displacement

Consider an object undergoing a rotational motion in the xy-plane about a fixed axis. Let O be a point in the xy-plane along the axis of rotation of the object, and P be a fixed point on the object located at a distance r from O (Fig. 9.2). Point P will move in a circular path of radius r and center located at O. Assume that at some time t_1, the point is located at P_1 with OP_1 making an angle θ_1 with the horizontal. At a later time t_2, the point is at P_2, with OP_2 making an angle θ_2 with the horizontal. Angles θ_1 and θ_2 define the *angular positions* of the point at times t_1 and t_2, respectively. If θ denotes the change in angular position of the point in the time interval

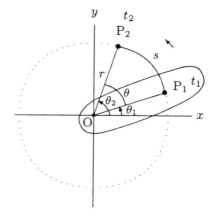

Fig. 9.2 $\theta = \theta_2 - \theta_1$ is the angular displacement in the time interval between t_1 and t_2

N. Özkaya et al., *Fundamentals of Biomechanics: Equilibrium, Motion, and Deformation*,
DOI 10.1007/978-1-4614-1150-5_9, © Springer Science+Business Media, LLC 2012

between t_1 and t_2, then $\theta = \theta_2 - \theta_1$ is called the *angular displacement* of the point in the same time interval between. In the same time interval, the point travels a distance s measured along the circular path. The equation relating the radius r of the circle, angle θ, and arc length s is:

$$s = r\theta \quad \text{or} \quad \theta = \frac{s}{r}. \tag{9.3}$$

In Eq. (9.3), angle θ must be measured in radians, rather than in degrees. As reviewed in Appendix C, radians and degrees are related in that there are $360°$ in a complete circle that must correspond to an arc length equal to the circumference, $s = 2\pi r$, of the circle, with $\pi = 3.14$ approximately. Therefore, $\theta = s/r = 2\pi r/r = 2\pi$ for a complete circle or $360° = 2\pi$. One radian is then equal to $360°/2\pi = 57.3°$. The following formula can be used to convert angles given in degrees to corresponding angles in radians:

$$\theta(\text{radians}) = \frac{\pi}{180}\theta(\text{degrees}).$$

Selected angles and their equivalents in radians are listed in Table 9.1.

Table 9.1 Selected angles in degrees and radians

Degrees (°)	Radians (rad)
30	$\pi/6 = 0.524$
45	$\pi/4 = 0.785$
60	$\pi/3 = 1.047$
90	$\pi/2 = 1.571$
180	$\pi = 3.142$
270	$3\pi/2 = 4.712$
360	$2\pi = 6.283$

9.3   Angular Velocity

The time rate of change of angular position is called *angular velocity*, and it is commonly denoted by the symbol ω (omega). If the angular position of an object is known as a function of time, its angular velocity can be determined by taking the derivative of the angular position with respect to time:

$$\omega = \frac{d\theta}{dt} = \dot{\theta}. \tag{9.4}$$

The *average angular velocity* ($\bar{\omega}$) of an object in the time interval between t_1 and t_2 is defined by the ratio of change in angular position of the object divided by the time interval:

$$\bar{\omega} = \frac{\Delta\theta}{\Delta t} = \frac{\theta_2 - \theta_1}{t_2 - t_1}. \tag{9.5}$$

In Eq. (9.5), θ_1 and θ_2 are the angular positions of the object at times t_1 and t_2, respectively.

9.4   Angular Acceleration

The angular velocity of an object may vary during motion. The time rate of change of angular velocity is called *angular acceleration*, usually denoted by the symbol α (alpha). If the angular velocity of a body is given as a function of time, then its angular acceleration can be determined by considering the derivative of the angular velocity with respect to time:

$$\alpha = \frac{d\omega}{dt}. \tag{9.6}$$

The *average angular acceleration*, $\bar{\alpha}$, is equal to the change in angular velocity over the time interval in which the change occurs. If ω_1 and ω_2 are the instantaneous angular velocities of a body measured at times t_1 and t_2, respectively, then the average angular acceleration of the body in the time interval between t_1 and t_2 is:

$$\bar{\alpha} = \frac{\Delta\omega}{\Delta t} = \frac{\omega_2 - \omega_1}{t_1 - t_2}. \tag{9.7}$$

Note that using the definition of angular velocity in Eq. (9.4), angular acceleration can alternatively be expressed in the following forms:

$$\alpha = \frac{d\omega}{dt} = \frac{d}{dt}\left(\frac{d\theta}{dt}\right) = \frac{d^2\theta}{dt^2} = \ddot{\theta}. \tag{9.8}$$

Also note that Eqs. (9.4) and (9.6) are the kinematic equations relating angular quantities θ, ω, and α.

Angular displacement, velocity, and acceleration are vector quantities. Therefore, their directions must be stated as well as their magnitudes. For two-dimensional problems, the motion is either in the clockwise or counterclockwise direction. Angular displacement and velocity are positive in the direction of motion. Angular acceleration is positive when angular velocity is increasing over time, and it is negative when angular velocity is decreasing over time.

9.5 Dimensions and Units

From Eq. (9.3), the angular displacement θ of an object undergoing circular motion is equal to the ratio of the arc length s and radius r of the circular path. Both arc length and radius have the dimension of length. Therefore, the dimension of angular displacement is 1 or it is a *dimensionless* quantity:

$$[\text{Angular displacement}] = \frac{L}{L} = 1.$$

By definition, angular velocity is the time rate of change of angular position, and angular acceleration is the time rate of change of angular velocity. Therefore, angular velocity has the dimension of 1 over time, and angular acceleration has the dimension of angular velocity divided by time or 1 over time squared.

Note that angular quantities θ, ω, and α differ dimensionally from their linear counterparts x, v, and a by a length factor.

The units of angular quantities in different unit systems are the same. Angular displacement is measured in radians (rad), angular velocity is measured in radians per second (rad/s) or s^{-1}, and angular acceleration is measured in radians per second squared (rad/s^2) or s^{-2}.

9.6 Definitions of Basic Concepts

To be able to define concepts common in angular motions, consider the simple pendulum illustrated in Fig. 9.3. The pendulum consists of a mass attached to a string. The string is fixed to the ceiling at one end and the mass is free to swing. Assume that l is the length of the string and it is attached to the ceiling at O. If the mass is simply released, it would stretch the string and come to a rest at B that represents the *neutral* or *equilibrium position* of the mass. If the mass is pulled to the side, to position A, so that the string makes an angle θ with the vertical and is then released, the mass will oscillate or swing back and forth about its neutral position in a circular arc path of radius l. Due to

internal friction and air resistance, the oscillations will die out over time and eventually the pendulum will come to a stop at its neutral position. An analysis of the motion characteristics of this relatively simple system may give us considerable insight into the nature of other more complex dynamic systems.

For the sake of simplicity, we ignore the air resistance and frictional effects, and assume that once the pendulum is excited, it will oscillate forever. Also, assume that there is a roll of paper behind the pendulum that moves in a prescribed manner. (For example, 10 mm of paper rolls up in each second.) Furthermore, the mass has a dye on it that marks the position of the mass on the paper. In other words, as the mass swings back and forth, it draws its motion path on the paper that would look like the one illustrated in Fig. 9.4. In Fig. 9.4, θ represents the angle that the pendulum makes with the vertical and t is time. Angle θ is a measure of the instantaneous angular position of the pendulum.

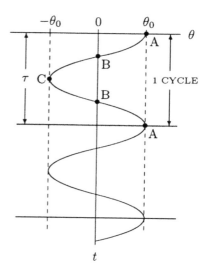

Fig. 9.4 Simple harmonic motion

The motion described in Fig. 9.4 is known as the *simple harmonic motion*. At time $t = 0$ that corresponds to the instant when the mass is first released, the mass is located at A which makes an angle θ_0 with the vertical. The mass swings, passes through B where $\theta = 0$, and reaches C where $\theta = -\theta_0$. Here, it is assumed that θ is positive between A and B, zero at B, and negative between B and C. At C, the mass momentarily stops and then reverses its direction of motion from clockwise to counterclockwise. It passes through B again and returns to A, thus completing one full *cycle* in a time interval of τ (tau) seconds, which is called the *period* of harmonic motion. The total angle covered by the pendulum between A and C is called the *range of motion* (ROM) and, in this case, it is equal to $2\theta_0$. Also, the entire

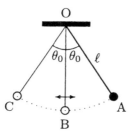

Fig. 9.3 Pendulum

motion of the pendulum is confined between $+\theta$ and $-\theta$ that set the limits of the ROM. Half of the ROM is called the *amplitude* of the oscillations measured in radians and here is equal to θ_0. Note that in this case, both the amplitude and period of the harmonic motion are constants. Also note that since the effects of friction and air resistance are neglected, the series of events between A, B, C, B, and A are repeated forever in τ time intervals.

From Fig. 9.4, it is clear that angular position θ is a function of time t. Furthermore, θ is a harmonic, cyclic function of t that must remind us trigonometric functions. As discussed in Appendix C, the θ versus t graph in Fig. 9.4 can be compared to the graphs of known functions to establish the functions that relate θ and t. It can be shown that:

$$\theta = \theta_0 \cos(\varphi t).$$

In this equation, the parameter θ_0 multiplied with the cosine function is the amplitude of the harmonic motion and φ (phi) is called the *angular frequency* measured in radians per second (rad/s). The period and angular frequency are related:

$$\varphi = \frac{2\pi}{\tau} \quad (\pi = 3.1416).$$

For oscillatory motions, the reciprocal of the period is called the *frequency*, f, measured in Hertz (Hz) that represents the total number of cycles occurring per second:

$$f = \frac{1}{\tau} = \frac{\varphi}{2\pi}.$$

Note that for the simple harmonic motion discussed herein, the parameters involved (ROM, amplitude, period, and frequency) are constants. Also note that the validity of the function relating angular position and time can be checked by assigning values to t and calculating corresponding θ values. For example, at A: $t = 0$, $\varphi t = 0$, $\cos(0) = 1$, and $\theta = \theta_0$. At B: $t = \tau/4$, $\varphi t = \pi/2 = 90°$, $\cos(90) = 0$, and $\theta = 0$. At C: $t = \tau/2$, $\varphi t = \pi = 180°$, $\cos(180) = -1$, and $\theta = -\theta_0$. All of these are consistent with the observations in Fig. 9.4.

Now that we have defined most of the important parameters involved, we can also determine the angular velocity and angular acceleration of the pendulum. Utilizing Eqs. (9.4) and (9.6):

$$\omega = \frac{d\theta}{dt} = -\theta_0\varphi \sin(\varphi t),$$

$$\alpha = \frac{d\omega}{dt} = -\theta_0\varphi^2 \cos(\varphi t).$$

These functions are plotted in Figs. 9.5 and 9.6. Notice that the amplitude of the angular velocity of the pendulum

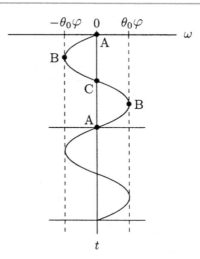

Fig. 9.5 Angular velocity ω versus time t

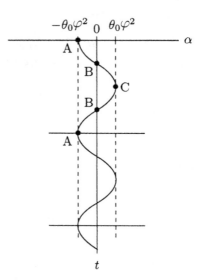

Fig. 9.6 Angular acceleration α versus time t

is $\theta_0\varphi$, and the amplitude of its angular acceleration is $\theta_0\varphi^2$. They are the terms multiplied by the sine and cosine functions. At A, the angular velocity is zero. Between A and B, the mass accelerates and the magnitude of its angular velocity increases in the clockwise direction. The angular velocity reaches a peak value of $\theta_0\varphi$ at B. The angular velocity is negative and the angular acceleration is positive between B and C. Therefore, the mass decelerates (its angular velocity decreases in the clockwise direction) between B and C. The angular velocity reduces to zero at C. In the meantime, the magnitude of the angular acceleration reaches its peak value of $\theta_0\varphi^2$. Between C and B, the mass accelerates in the counterclockwise direction, the magnitude of its angular velocity returns to a peak at B, slows down between B and A, and momentarily comes to rest at A. This series of events is repeated over time.

Next, consider that the mass is again pulled to A so that the pendulum makes an angle θ_0 with the vertical and is released (Fig. 9.7). The mass will oscillate about its neutral position in a circular arc path of radius l. Due to internal friction and air resistance, the oscillations will die out over time and eventually the pendulum will come to a rest at its neutral position, B. This type of motion is called *damped oscillations*. To help understand some aspects of damped oscillations, consider the angular position θ versus time t graph shown in Fig. 9.8. The pendulum completes four full cycles in t_f seconds before coming to a stop. The period of each cycle is equal, but the amplitude of the harmonic oscillations decreases linearly with time and to zero at time t_f. That is, we have a harmonic motion with a constant period but varying amplitude. For measured θ_0, τ, and t_f, the θ versus time graph shown in Fig. 9.8 can be represented as:

$$\theta = \theta_0 \left(1 - \frac{t}{t_f}\right) \cos(\varphi t).$$

Here, φ is again the angular frequency of harmonic oscillations and is equal to $2\pi/\tau$. What is different in this case is that the harmonic oscillations of the pendulum are confined between two converging straight lines that can be represented by the functions $\theta = \theta_0(1 - t/t_f)$ and $\theta = -\theta_0(1 - t/t_f)$, and that the oscillations of the pendulum are "damped-out" by friction and air resistance. Knowing the angular position of the pendulum as a function of time enables us to determine the angular velocity and acceleration of the pendulum. Using Eqs. (9.4) and (9.6), and applying the product and chain rules of differentiation (see Appendix C):

$$\omega = \frac{d\theta}{dt} = -\frac{\theta_0}{t_f} \cos(\varphi t) - \theta_0 \varphi \left(1 - \frac{t}{t_f}\right) \sin(\varphi t),$$

$$\alpha = \frac{d\omega}{dt} = \frac{2\theta_0 \varphi}{t_f} \sin(\varphi t) - \theta_0 \varphi^2 \left(1 - \frac{t}{t_f}\right) \cos(\varphi t).$$

These functions are relatively complex. Their graphs are shown in Fig. 9.9, which are obtained simply by assuming a value for τ, assigning values to t, calculating corresponding ω and α, and plotting them.

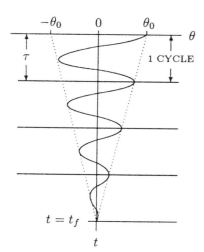

Fig. 9.7 Pendulum under the effect of air resistance

Fig. 9.8 Damped oscillations

Fig. 9.9 Angular velocity ω (*open circles*) and angular acceleration α (*open diamonds*) versus time t

Example 9.1 Shoulder abduction Figure 9.10 shows a person doing shoulder abduction in the frontal plane. O represents the axis of rotation of the shoulder joint in the frontal plane, line OA represents the position of the arm when it is stretched out parallel to the ground (horizontal), line OB represents the position of the arm when the hand is at its highest elevation, and line OC represents the position of the arm when the hand is closest to the body. In other words, for this activity, OB and OC are the arm's limits of ROM. Assume that the angle between OA and OB is equal to the angle between OA and OC, which are represented by

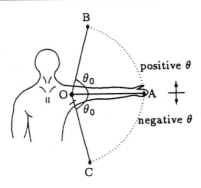

Fig. 9.10 Shoulder abduction

angle θ_0. The motion of the arm is symmetric with respect to line OA. Also assume that the time it takes for the arm to cover the angles between OA and OB, OB and OA, OA and OC, and OC and OA are approximately equal.

Derive expressions for the angular displacement, velocity, and acceleration of the arm. Take the period of angular motion of the arm to be 3s and the angle θ_0 to be 80°.

Solution
Notice the similarities between the motion of the arm in this example and the simple harmonic motion of the pendulum discussed previously. In this case, angle θ_0 represents the amplitude of the angular displacement of the arm while undergoing a harmonic motion about line OA. The ROM of the arm is equal to twice that of angle θ_0. The period of the angular motion is given as $\tau = 3$ s, and the angular frequency of harmonic oscillations of the arm about line OA (the horizontal) can be calculated as $\varphi = 2\pi/\tau = 2.09$ rad/s . If we let θ represent the angular displacement of the arm measured relative to the position defined by line OA, then θ can be written as a sine function of time:

$$\theta = \theta_0 \sin(\varphi t). \qquad (i)$$

The angular displacement of the arm as given in Eq. (i) is plotted as a function of time in Fig. 9.11. Notice that θ is zero when the arm is at position A. θ assumes positive values between A and B, and it is negative while the arm is between A and C. θ reaches its peak at B and C, and θ_0 is the

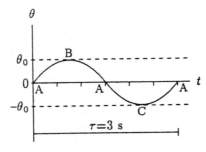

Fig. 9.11 Graph of function $\theta = \theta_0 \sin(\phi t)$ with $\phi = 2\pi/\tau$

amplitude of angular displacement of the arm. Since all of these are consistent with the information provided in the statement of the problem, Eq. (i) does represent the angular displacement of the arm.

To derive expressions for the angular velocity and acceleration of the arm, we have to consider time derivatives of the function given in Eq. (i). The time rate of change of angular displacement is defined as angular velocity:

$$\omega = \frac{d\theta}{dt} = \theta_0 \varphi \cos(\varphi t). \qquad (ii)$$

The time rate of change of angular velocity is angular acceleration:

$$\alpha = \frac{d\omega}{dt} = -\theta_0 \varphi^2 \sin(\varphi t). \qquad (iii)$$

Equations (ii) and (iii) can alternatively be written as:

$$\omega = \omega_0 \cos(\varphi t), \qquad (iv)$$

$$\alpha = -\alpha_0 \sin(\varphi t). \qquad (v)$$

Here, ω_0 is the amplitude of the angular velocity and α_0 is the amplitude of the angular acceleration of the arm, such that:

$$\omega_0 = \theta_0 \varphi = \theta_0 \frac{2\pi}{\tau},$$

$$\alpha_0 = \theta_0 \varphi^2 = \theta_0 \frac{4\pi^2}{\tau^2}.$$

Notice that the amplitude of the angular velocity is a linear function of the angular frequency, and the amplitude of angular acceleration is a quadratic function of angular frequency. Angular frequency, on the other hand, is inversely proportional with the period of harmonic oscillations. Therefore, low period indicates high frequency, which indicates high angular velocity and acceleration amplitudes.

We can use the numerical values of $\theta_0 = 80° = 1.40$ rad and $\varphi = 2.09$ rad/s to calculate ω_0 and α_0 as 2.93 rad/s and 6.12 rad/s², respectively. Equations (i), (iv), and (v) can now be expressed as:

$$\theta = 1.40 \sin(2.09t), \qquad (vi)$$

$$\omega = 2.93 \cos(2.09t), \qquad (vii)$$

$$\alpha = -6.12 \sin(2.09t) \qquad (viii)$$

Equations (vi) through (viii) can be used to calculate the instantaneous angular position, velocity, and acceleration of the arm at any time t. These equations can also be used to plot θ, ω, and α versus t graphs for the arm by assigning values to time and calculating corresponding θ, ω, and α values. A set of sample graphs are shown in Fig. 9.12 for a single cycle.

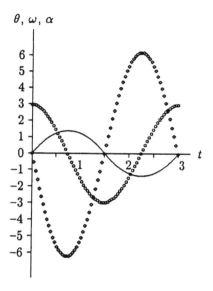

Fig. 9.12 Angular position, velocity (*open circles*), and acceleration (*open diamonds*) versus time (θ in rad, ω in rad/s, α in rad/s², and t in seconds)

Example 9.2 Flexion-extension test Figure 9.13 illustrates a computer-controlled dynamometer that can be used to measure angular displacement, angular velocity, and torque output of the trunk. During a repetitive flexion-extension test in the sagittal plane (plane that passes through the chest and divides the body into right-hand and left-hand parts), a subject is placed in the dynamometer, positioned in the machine so that the subject's fifth lumbar vertebra (L5/S1) is aligned with the flexion-extension axis (indicated as O) of

Fig. 9.13 Dynamometer

the machine, tied to the equipment firmly, and asked to perform trunk flexion and extension as long as possible, exerting as much effort as possible. The angular position of the subject's trunk relative to the upright position is measured and recorded. The data collected is then plotted to obtain an angular displacement θ versus time t graph. The curves obtained for this particular subject in different cycles are observed to be qualitatively and quantitatively similar except for the first and the last few cycles. A couple of sample cycles are provided in Fig. 9.14, in which the angular displacement of the trunk measured in degrees is plotted as a function of time measured in seconds.

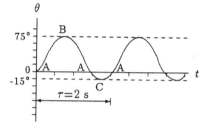

Fig. 9.14 Angular position versus time graph (θ is in degrees and t is in seconds)

The angular position measurements are made relative to the upright position in which the angular displacement of the trunk is zero. The subject flexes between A and B, and reaches a peak flexion at B. The extension phase is identified with the motion of the trunk from B toward A. The angular displacement of the trunk is positive between A and B. Between A and C, the trunk undergoes hyperextension and reaches a peak extension at C. In this range, the angular displacement of the trunk assumes negative values.

The purpose of this example is to demonstrate the means of analyzing experimentally collected data. The specific task is to find a function that can express the angular displacement of the subject's trunk as a function of time, from which we can derive expressions for the angular velocity and acceleration of the trunk.

Solution

The problem may be easier to visualize if we form an analogy between the upper body and a mechanical system called the *inverted pendulum*, shown in Fig. 9.15. An inverted pendulum consists of a concentrated mass m attached to a very light rod of length l that is hinged to the ground through an axis about which it is allowed to rotate. In this case, the concentrated mass represents the total mass of the upper body. The hinge corresponds to the disk between the fifth lumbar vertebra and the sacrum, about which the upper body rotation occurs in the sagittal plane. Length l is the distance between the fifth lumbar vertebra and the center of gravity of the upper body.

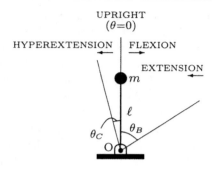

UPRIGHT
($\theta=0$)

HYPEREXTENSION | FLEXION

EXTENSION

Fig. 9.15 Inverted pendulum

It is clear from Fig. 9.14 that θ is a harmonic (sine or cosine) function of time. From Fig. 9.14, it is possible to read the peak angles the trunk makes with the upright position during flexion and extension phases, and the period of harmonic motions. However, it is not easy to determine exactly how θ varies with time. To obtain a function relating θ and t, we must work through several steps.

Let τ be the period of harmonic oscillations, and θ_B and θ_C be the peak angular displacements of the trunk in the flexion and extension phases, respectively. From Fig. 9.14 or using the experimentally obtained raw data, $\tau = 2$ s, $\theta_B = 75°$, and $\theta_C = -15°$. Knowing the period, the angular frequency of the harmonic oscillations can be determined:

$$\varphi = \frac{2\pi}{\tau} = \frac{2\pi}{2} = \pi \text{ rad/s}.$$

Using θ_B and θ_C, we can also calculate the ROM of the trunk. By definition, ROM is the total angle covered by the rotating object. Therefore:

$$\text{ROM} = \theta_B + \theta_C = 75° + 15° = 90°.$$

Figure 9.14 is redrawn in Fig. 9.16 in which two sets of coordinates are used. In addition to θ and t, we have a second set of coordinates Θ (capital theta) and T that is obtained by

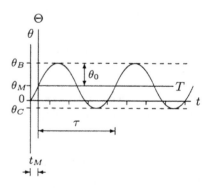

Fig. 9.16 Translating the θ versus t coordinate frame to Θ versus T coordinate frame

translating the origin of the θ versus t coordinate system to a point with coordinates $t = t_M$ and $\theta = \theta_M$. Here, θ_M designates the mean angular displacement that can be calculated as:

$$\theta_M = \frac{\theta_B - \theta_C}{2} = \frac{75° - 15°}{2} = 30°.$$

Time t_M corresponds to the time when $\theta = \theta_M$. t_M can be determined from the experimentally collected data. In this case, $t_M = 0.232$ s.

We define a second set of coordinates so that, with respect to Θ and T, the function representing the angular displacement versus time curve is simply a sine function:

$$\Theta = \theta_0 \sin(\varphi T). \tag{i}$$

In Eq. (*i*), θ_0 is the amplitude of the harmonic oscillations and is equal to one-half of the ROM:

$$\theta_0 = \frac{\text{ROM}}{2} = \frac{90°}{2} = 45° \left(\frac{\pi}{4} \text{rad}\right).$$

We now have a function representing the experimentally obtained curve in terms of Θ and T. If we can relate Θ to θ and T to t, then we can derive a function in terms of θ and t. This can be achieved by employing *coordinate transformation*. Notice that $\Theta = 0$ when $\theta = \theta_M$. Therefore:

$$\Theta = \theta - \theta_M. \tag{ii}$$

Also notice that $T = 0$ when $t = t_M$. Hence:

$$T = t - t_M. \tag{iii}$$

Substituting Eqs. (*ii*) and (*iii*) into Eq. (*i*) will yield:

$$\theta = \theta_M + \theta_0 \sin[\varphi(t - t_M)]. \tag{iv}$$

In Eq. (*iv*), the angular displacement of the trunk is defined as a function of time, representing the experimentally obtained curve shown in Fig. 9.14. We can also obtain expressions for the angular velocity and acceleration of the trunk by considering the time derivatives of θ in Eq. (*iv*):

$$\omega = \frac{d\theta}{dt} = \theta_0 \varphi \cos[\varphi(t - t_M)], \tag{v}$$

$$\alpha = \frac{d\omega}{dt} = -\theta_0 \varphi^2 \sin[\varphi(t - t_M)]. \tag{vi}$$

The numerical values of θ_M, θ_0, t_M, and φ can be substituted into the above equations to obtain:

$$\theta = \frac{\pi}{6} + \frac{\pi}{4}\sin[\pi(t - 0.232)], \qquad (vii)$$

$$\omega = \frac{\pi^2}{4}\cos[\pi(t - 0.232)], \qquad (viii)$$

$$\alpha = -\frac{\pi^3}{4}\sin[\pi(t - 0.232)]. \qquad (ix)$$

These functions are plotted in Fig. 9.17 to obtain angular displacement, velocity, and acceleration versus time graphs for the trunk.

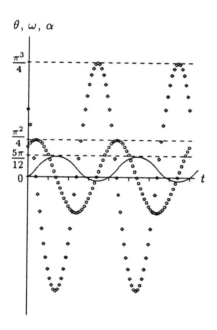

Fig. 9.17 Angular position, velocity (*open circles*), and acceleration (*open diamonds*) versus time

Note that the validity of Eq. (*vii*) can be checked by assigning values to t and calculating corresponding θ values using Eq. (*vii*). For example, $\theta = 0$ when $t = 0$ and $t = \tau = 2s$, and $\theta = \pi/6 = 0.52$ rad or $30°$ when $t = t_M = 0.232s$. These are consistent with the initial data presented in Fig. 9.14.

Also note that angular velocity is a cosine function of time. The amplitude of the ω versus t curve shown in Fig. 9.17 is equal to the coefficient $\pi^2/4 = 2.47$ rad/s in front of the cosine function in Eq. (*viii*). Similarly, the amplitude of the angular acceleration is $\pi^3/4 = 7.75$ rad/s².

9.7 Rotational Motion About a Fixed Axis

Consider the arbitrarily shaped object in Fig. 9.18. Assume that the object is undergoing a rotational motion in the *xy*-plane about a fixed axis that is perpendicular to the *xy*-plane. Let O and P be two points in the *xy*-plane, such that O is along the axis of rotation of the object and P is a fixed point on the rotating object located at a distance r from point O. Due to the rotation of the object, the point P will experience a circular motion with r being the radius of its circular path.

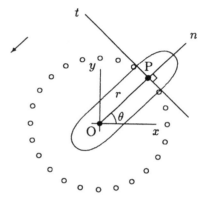

Fig. 9.18 n and t are the normal (radial) and tangential directions at point P

To describe circular motions, it is usually convenient to define velocity and acceleration vectors with respect to two mutually perpendicular directions normal (radial) and tangential to the circular path of motion. These directions are indicated as n and t in Fig. 9.18, and are also known as *local coordinates*. By definition, the velocity vector $\underline{v}$ is always tangent to the path of motion. Therefore, for a circular motion, the velocity vector can have only one component tangent to the circular path of motion (Fig. 9.19). $\underline{v}$ is called the *tangential* or *linear velocity*. The magnitude, v of the velocity vector

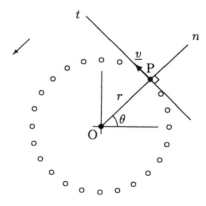

Fig. 9.19 Velocity vector $\underline{v}$ is always tangent to the path of motion

can be determined by considering the time rate of change of relative position of point P along the circular path:

$$v = \frac{ds}{dt}. \qquad (9.9)$$

For a circular motion, the acceleration vector can have both tangential and normal components (Fig. 9.20). The *tangential acceleration* a_t is related to the change in magnitude of the velocity vector and has a magnitude:

$$a_t = \frac{dv}{dt}. \qquad (9.10)$$

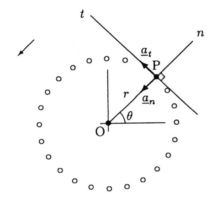

Fig. 9.20 a_t and a_n are the tangential and normal components of the acceleration vector

The *normal acceleration* a_n is related to the change in direction of the velocity vector and has a magnitude:

$$a_n = \frac{v^2}{r}. \qquad (9.11)$$

For an object undergoing a rotational motion, a_t is zero if the object is rotating with constant v. On the other hand, a_n is always present because it is associated with the direction of $\underline{v}$ that changes continuously throughout the motion.

The direction of $\underline{a_t}$ is the same as the direction of $\underline{v}$ if v is increasing, or opposite to that of $\underline{v}$ if v is decreasing over time. The normal component of the acceleration vector is also known as *radial* or *centripetal* (center-seeking), and it is always directed toward the center of rotation of the body.

If the tangential and normal acceleration components are known, then the net or resultant acceleration of a point on a body rotating about a fixed axis can also be determined (Fig. 9.21). If $\underline{t}$ and $\underline{n}$ are unit vectors indicating positive tangential and normal directions, respectively, then the resultant acceleration vector can be expressed as:

$$\underline{a} = \underline{a_t} + \underline{a_n} = a_t\underline{t} + a_n\underline{n}. \qquad (9.12)$$

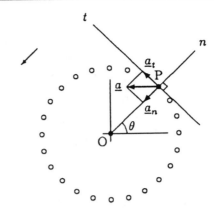

Fig. 9.21 $\underline{a}$ is the resultant linear acceleration vector

The magnitude of the resultant acceleration vector can be determined as:

$$a = \sqrt{a_t{}^2 + a_n{}^2}. \qquad (9.13)$$

On the other hand, the velocity vector can be expressed as:

$$\underline{v} = v\underline{t}. \qquad (9.14)$$

Note that v and a are linear quantities. v has the dimension of length divided by time, and both a_t and a_n have the dimension of length divided by time squared. Also note that it is customary to take the positive normal direction (the direction of $\underline{n}$) to be outward (from the center of rotation toward the rim), and the positive tangential direction (the direction of $\underline{t}$) to be counterclockwise.

9.8 Relationships Between Linear and Angular Quantities

Recall from Eq. (9.3) that $s = r\theta$. For a circular motion, radius r is constant and Eq. (9.9) can be evaluated as follows:

$$v = \frac{d}{dt}(r\theta) = r\frac{d\theta}{dt}.$$

By definition, time rate of change of angular displacement is angular velocity. Therefore:

$$v = r\omega. \qquad (9.15)$$

Equation (9.15) states that the magnitude of the linear velocity of a point in a body that is undergoing a rotational motion about a fixed axis is equal to the distance of that point from the center of rotation multiplied by the angular velocity of the body. Notice that at a given instant, every point on the body has the same angular velocity but may have different linear velocities. The magnitude of the linear velocity increases with increasing radial distance, or as one moves outward from the center of rotation toward the rim.

Using the relationship given in Eq. (9.15), Eq. (9.10) can be evaluated for a motion in a circular path as follows:

$$a_t = \frac{d}{dt}(r\omega) = r\frac{d\omega}{dt}.$$

By definition, time rate of change of angular velocity is angular acceleration. Therefore:

$$a_t = r\alpha. \tag{9.16}$$

Similarly, substituting Eq. (9.11) into Eq. (9.11) will yield:

$$a_n = r\omega^2. \tag{9.17}$$

Equations (9.15)–(9.17) relate linear quantities v, a_t, and a_n to angular quantities r, ω, and α. Equation (9.16) states that the tangential component of linear acceleration of a point on a body rotating about a fixed axis is equal to the distance of that point from the axis of rotation times the angular acceleration of the body.

9.9   Uniform Circular Motion

Uniform circular motion occurs when the angular velocity of an object undergoing a rotational motion about a fixed axis is constant. When angular velocity is constant, angular acceleration is zero. Therefore, for a point located at a radial distance r from the center of rotation of an object undergoing uniform circular motion:

$$v = r\omega \quad \text{(constant)},$$

$$a_t = 0,$$

$$a_n = r\omega^2 \quad \text{(constant)}.$$

9.10   Rotational Motion with Constant Acceleration

In Chap. 7, a set of kinematic equations (Eqs. 7.11 through 7.14) were derived to analyze the motion characteristics of bodies undergoing translational motion with constant acceleration. Similar equations can also be derived for rotational motion about a fixed axis with constant angular acceleration:

$$\omega = \omega_0 + \alpha_0 t, \tag{9.18}$$

$$\theta = \theta_0 + \omega_0 t + \frac{1}{2}\alpha_0 t^2, \tag{9.19}$$

$$\theta = \theta_0 + \frac{1}{2}(\omega + \omega_0)t, \tag{9.20}$$

$$\omega^2 = \omega_0^2 + 2\alpha_0(\theta - \theta_0). \tag{9.21}$$

In Eqs. (9.18) through (9.21), α_0 is the constant angular acceleration, and θ_0 and ω_0 are the initial angular position and velocity of the object at time $t_0 = 0$, respectively.

9.11   Relative Motion

A motion observed in different frames of reference may be different. For example, the motion of a train observed by a stationary person would be different than the motion of the same train observed by a passenger in a moving car. The motion of a ball thrown up into the air by a person riding in a moving vehicle would have a vertical path as observed by the person riding in the same vehicle (Fig. 9.22a), but a curved path for a second, stationary person watching the ball (Fig. 9.22b). The general approach in analyzing such physical situations requires defining the motion of the moving body with respect to a convenient moving coordinate frame, defining the motion of this frame with respect to a fixed coordinate frame, and combining the two.

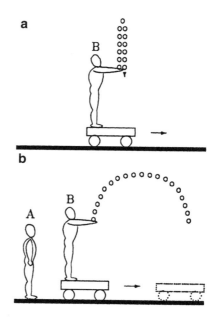

Fig. 9.22 A motion observed by different observers in different reference frames may be different

Assume that the motion of a point P in a moving body is to be analyzed. Let *XYZ* and *xyz* refer to two coordinate frames with origins at A and B, respectively (Fig. 9.23). Assume that the *XYZ* frame is fixed (stationary) and the *xyz* frame is moving, such that the respective coordinate directions (e.g., *x* and *X*) remain parallel throughout the motion. This implies that the *xyz* coordinate frame is undergoing a translational motion only, and that the same set of unit vectors $\underline{i}$, $\underline{j}$, and $\underline{k}$ can be used in both reference frames. The motion of the moving *xyz* frame can be identified by specifying the motion of its origin B. If $\underline{r}_B$ denotes the position vector of B with respect to the fixed coordinate frame, then the velocity and acceleration vectors of B with respect to the *XYZ* coordinate frame are:

$$\underline{v}_B = \frac{d}{dt}(\underline{r}_B) = \dot{\underline{r}}_B; \quad \underline{a}_B = \frac{d}{dt}(\underline{v}_B) = \ddot{\underline{r}}_B.$$

Fig. 9.23 *XYZ* is a fixed and *xyz* is a moving coordinate frame

Similarly, the motion of point P with respect to the moving coordinate frame *xyz* can be defined by the position vector $\underline{r}_{P/B}$ of point P relative to the origin B of the *xyz* frame. The first and second time derivatives of $\underline{r}_{P/B}$ will yield the velocity and acceleration vectors of point P relative to the *xyz* frame:

$$\underline{v}_{P/B} = \frac{d}{dt}\left(\underline{r}_{P/B}\right) = \dot{\underline{r}}_{P/B}; \quad \underline{a}_{P/B} = \frac{d}{dt}\left(\underline{v}_{P/B}\right) = \ddot{\underline{r}}_{P/B}.$$

Finally, the position vector $\underline{r}_P$, velocity vector $\underline{v}_P$, and acceleration vector $\underline{a}_P$ of point P with respect to the fixed coordinate frame *XYZ* can be obtained by superposition:

$$\underline{r}_P = \underline{r}_B + \underline{r}_{P/B}, \tag{9.22}$$

$$\underline{v}_P = \underline{v}_P + \underline{v}_{P/B}, \tag{9.23}$$

$$\underline{a}_P = \underline{a}_P + \underline{a}_{P/B}. \tag{9.24}$$

The motion of point B (which happens to be the origin of the moving coordinate frame *xyz*) with respect to the fixed *XYZ* coordinate frame is called the *absolute motion* of B and is denoted by the subscript B. Similarly, the motion of point P observed relative to the *XYZ* frame is the absolute motion of P. The motion of point P with respect to the moving coordinate frame is called the *relative motion* of P and is denoted by the subscript P/B. Note here that the position vector $\underline{r}_{P/B}$ refers to a vector drawn from point B to point P. Also note that the position vector of point P relative to the *XYZ* coordinate frame could also be expressed as $\underline{r}_{P/A}$. However, by convention, $\underline{r}_P$ implies that the position vector is defined relative to the fixed coordinate frame.

Example 9.3 Consider the motion described in Fig. 9.22. A person (B) riding on a vehicle that is moving toward the right by a constant speed of 2 m/s throws a ball straight up into the air with an initial speed of 10 m/s.

Describe the motion of the ball as observed by a stationary person (A) in the time interval between when the ball is first released and when it reaches its maximum elevation.

Solution

This is a two-dimensional problem and can be analyzed in three steps. First, let *x* and *y* represent a coordinate frame moving with the vehicle. With respect to the *xy* frame, the ball thrown up into the air will undergo one-dimensional linear motion (translation) in the *y* direction (Fig. 9.24). Because of the constant downward gravitational acceleration, the ball will decelerate in the positive *y* direction, reach its maximum elevation, change its direction of motion, and begin to descend. With respect to the *xy* coordinate frame moving with the vehicle, or as observed by person B moving with the vehicle, the speed of the ball in the *y* direction between the instant of release and when the ball reaches its peak elevation can be determined from (see Chap. 7):

$$v_y = v_{y_0} - gt.$$

Fig. 9.24 Relative to the *xy* frame, the ball is undergoing a translational motion in the *y* direction

Here, $v_{y_0} = 10$ m/s is the initial speed of the ball and $g \approx 10$ m/s^2 is the magnitude of gravitational acceleration. This equation is valid in the time interval between $t = 0$ (the instant of release) and $t = v_{y_0}/g = 10/10 = 1$s (the time it takes for the ball to reach its maximum elevation where $v_y = 0$). As observed by person B, the ball has no motion in the x direction. Therefore, the velocity $\underline{v}_{P/B}$ of the ball relative to person B can be expressed as:

$$\underline{v}_{P/B} = v_y \underline{j}.$$

Next, let X and Y represent a coordinate frame fixed to the ground. With respect to the XY frame, or with respect to the stationary person A, the vehicle is moving in the positive X direction with a constant speed of $v_{x_0} = 2$ m/s (Fig. 9.25). Therefore:

$$\underline{v}_B = v_{x_0} \underline{i}.$$

Fig. 9.25 Relative to the XY frame, the vehicle is undergoing a translational motion in the X direction with constant velocity

Finally, to determine the velocity of the ball relative to person A, we have to add velocity vectors $\underline{v}_B$ and $\underline{v}_{P/B}$ together:

$$\underline{v}_P = \underline{v}_B + \underline{v}_{P/B} = v_{x_0} \underline{i} + v_y \underline{j}.$$

Or, by substituting the known parameters:

$$\underline{v}_P = 2\underline{i} + (10 - 10t)\underline{j}.$$

For example, half a second after the ball is released, the ball has a velocity:

$$\underline{v}_P = 2\underline{i} + 5\underline{j}.$$

That is, according to person A or relative to the XY coordinate frame, the ball is moving to the right with a speed of 2 m/s and upward with a speed of 5 m/s (Fig. 9.26). At this instant, the magnitude of the net velocity of the ball is $v_P = \sqrt{(2)^2 + (5)^2} = 5.4$ m/s.

Fig. 9.26 Relative to the XY frame, the ball moves both in the X and Y directions

9.12 Linkage Systems

A *linkage system* is composed of several parts connected to each other and/or to the ground by means of hinges or joints, such that each part constituting the system can undergo motion relative to the other segments. An example of such a system is the *double pendulum* shown in Fig. 9.27. A double pendulum consists of two bars hinged together and to the ground. Linkage systems are also known as *multilink systems*.

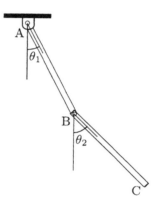

Fig. 9.27 Double pendulum

If the angular velocity and acceleration of individual parts are known, then the principles of relative motion can be applied to analyze the motion characteristics of each part constituting the multilink system. The following example will illustrate the procedure of analyzing the motion of a double pendulum. However, the procedure to be introduced can be generalized to analyze any multilink system.

An important concept associated with linkage systems is the number of independent coordinates necessary to describe the motion characteristics of the parts constituting the system. The number of independent parameters required defines the *degrees of freedom* of the system. For example, the two-dimensional motion characteristics of the simple pendulum shown in Fig. 9.28 can be fully described by θ that defines the location of the pendulum uniquely. Therefore, a simple

Fig. 9.28 Pendulum

pendulum has one degree of freedom. On the other hand, parameters θ_1 and θ_2 are necessary to analyze the coplanar motion of bar BC of the double pendulum shown in Fig. 9.27, and therefore, a double pendulum has two degrees of freedom.

Example 9.4 Double pendulum Assume that arms AB and BC of the double pendulum shown in Fig. 9.29 are undergoing coplanar motion. Let $l_1 = 0.3$ m and $l_2 = 0.3$ m be the lengths of arms AB and BC, and θ_1 and θ_2 be the angles of arms AB and BC make with the vertical. The angular velocity and acceleration of arm AB are measured as $\omega_1 = 2$ rad/s (counterclockwise) and $\alpha_1 = 0$ relative to point A. The angular velocity and acceleration of arm BC are measured as $\omega_2 = 4$ rad/s (counterclockwise) and $\alpha_2 = 0$ relative to point B.

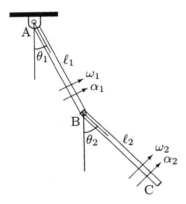

Fig. 9.29 Double pendulum

Determine the linear velocity and acceleration of point B on arm AB and of point C on arm BC at an instant when $\theta_1 = 30°$ and $\theta_2 = 45°$.

Solution

Let X and Y refer to a set of rectangular coordinates with origin located at A, and x and y be a second set of rectangular coordinates with origin at B. The XY coordinate frame is stationary, while the xy frame can move as point B moves. Since the angular velocity and acceleration of arm AB are given relative to point A, the motion characteristics of any point on arm AB can be determined with respect to the XY coordinate frame. Similarly, the motion of any point on arm BC can easily be analyzed relative to the xy coordinate frame.

Motion of point B as observed from point A:

Every point on arm AB undergoes a rotational motion about a fixed axis passing through point A with constant angular velocity of $\omega_1 = 2$ rad/s. Every point on arm AB experiences a uniform circular motion in the counterclockwise direction. As illustrated in Fig. 9.30, point B moves in a circular path of radius l_1. Magnitudes of linear velocity in the tangential direction and linear acceleration in the normal direction of point B can be determined using:

$$v_B = l_1\omega_1,$$
$$a_B = l_1\omega_1^2.$$

Fig. 9.30 Circular motion of B as observed from point A

The magnitude of the tangential component of the acceleration vector is zero since ω_1 is constant or since $\alpha_1 = 0$. Therefore, v_B and a_B are essentially the magnitudes of the resultant linear velocity and acceleration vectors. To express these quantities in vector forms, let n_1 and t_1 represent the normal and tangential directions to the circular path of point B when arm AB makes an angle θ_1 with the horizontal (Fig. 9.31). Also let $\underline{n}_1$ and $\underline{t}_1$ be unit vectors in the positive n_1 and t_1 directions, such that the positive $\underline{n}_1$ direction is

Fig. 9.31 Tangential velocity and normal acceleration of B

outward (i.e., from A toward B) and positive t_1 direction is pointing in the direction of motion (i.e., counterclockwise). The normal (centripetal) acceleration is always directed toward the center of motion, and is acting in the negative n_1 direction:

$$\underline{v}_B = v_B \underline{t}_1 = l_1 \omega_1 \underline{t}_1,$$

$$\underline{a}_B = -a_B \underline{n}_1 = -l_1 \omega_1^2 \underline{n}_1.$$

Notice that directions defined by unit vectors n_1 and t_1 change continuously as point B moves along its circular path. That is, $\underline{n}_1$ and $\underline{t}_1$ define a set of *local coordinate* directions that vary in time. By employing proper coordinate transformations, we can express these unit vectors in terms of Cartesian unit vectors $\underline{i}$ and $\underline{j}$. Cartesian coordinate directions, which are *global* as opposed to local, are not influenced by the motion of point B. The coordinate transformation can be done by expressing unit vectors $\underline{n}_1$ and t_1 in terms of Cartesian unit vectors $\underline{i}$ and $\underline{j}$. It can be observed from the geometry of the problem that (Fig. 9.32):

$$\underline{n}_1 = \sin\theta_1 \underline{i} - \cos\theta_1 \underline{j},$$

$$\underline{t}_1 = \cos\theta_1 \underline{i} + \sin\theta_1 \underline{j}.$$

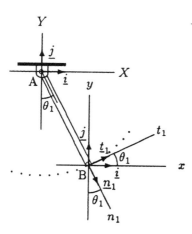

Fig. 9.32 Expressing unit vectors $\underline{n}_1$ and $\underline{t}_1$ in terms of Cartesian unit vectors $\underline{i}$ and $\underline{j}$

Therefore, the velocity and acceleration vectors of point B with respect to the XY coordinate frame and in terms of Cartesian unit vectors are:

$$\underline{v}_B = l_1 \omega_1 (\cos\theta_1 \underline{i} + \sin\theta_1 \underline{j}),$$

$$\underline{a}_B = -l_1 \omega_1^2 (\sin\theta_1 \underline{i} - \cos\theta_1 \underline{j}).$$

If we substitute the numerical values of $l_1 = 0.3$ m, $\theta_1 = 30°$, and $\omega_1 = 2$ rad/s, and carry out the necessary calculations we obtain:

$$\underline{v}_B = 0.52\underline{i} + 0.30\underline{j}, \tag{i}$$

$$\underline{a}_B = -0.60\underline{i} + 1.04\underline{j}. \tag{ii}$$

Motion of point C as observed from point B:
The motion of point C as observed from point B is similar to the motion of point B as observed from point A. Point C rotates with a constant angular velocity of ω_2 in a circular path of radius l_2 about point B (Fig. 9.33). Therefore, the derivation of velocity and acceleration vectors for point C relative to the xy coordinate frame follows the same procedure outlined for the derivation of velocity and acceleration vectors for point B relative to the XY coordinate frame. The magnitudes of the tangential velocity and normal acceleration vectors of point C relative to B are:

$$v_{C/B} = l_2 \omega_2,$$

$$a_{C/B} = l_2 \omega_2^2.$$

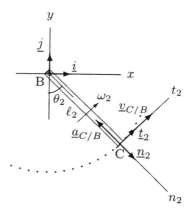

Fig. 9.33 Circular motion of C as observed from point B

If $\underline{n}_2$ and $\underline{t}_2$ are unit vectors in the normal and tangential directions to the circular path of C when arm BC makes an angle θ_2 with the vertical, then:

$$\underline{v}_{C/B} = v_{C/B} \underline{t}_2 = l_2 \omega_2 \underline{t}_2,$$

$$\underline{a}_{C/B} = -a_{C/B} \underline{n}_2 = -l_2 \omega_2^2 \underline{n}_2.$$

From Fig. 9.34, unit vectors $\underline{n}_2$ and $\underline{t}_2$ can be expressed in terms of Cartesian unit vectors $\underline{i}$ and $\underline{j}$ as:

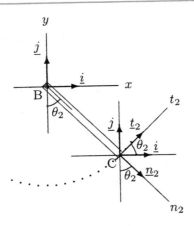

Fig. 9.34 Expressing unit vectors $\underline{n}_2$ and $\underline{t}_2$ in terms of Cartesian unit vectors $\underline{i}$ and $\underline{j}$

$$\underline{n}_2 = \sin\theta_2 \underline{i} - \cos\theta_2 \underline{j},$$

$$\underline{t}_2 = \cos\theta_2 \underline{i} + \sin\theta_2 \underline{j}.$$

Therefore, the velocity and acceleration vectors of point C relative to the xy coordinate frame can be written as:

$$\underline{v}_{C/B} = l_2 \omega_2 (\cos\theta_2 \underline{i} + \sin\theta_2 \underline{j}),$$

$$\underline{a}_{C/B} = -l_2 \omega_2{}^2 (\sin\theta_2 \underline{i} - \cos\theta_2 \underline{j}).$$

Substituting the numerical values of $l_2 = 0.3$ m, $\theta_2 = 45°$, and $\omega_2 = 4$ rad/s, and carrying out the necessary calculations we obtain:

$$\underline{v}_{C/B} = 0.85\underline{i} + 0.85\underline{j}, \tag{iii}$$

$$\underline{a}_{C/B} = -3.39\underline{i} + 3.39\underline{j}. \tag{iv}$$

Motion of point C as observed from point A:
We determined the velocity and acceleration of point C relative to B, and velocity and acceleration of point B with respect to A. Now, we can apply the principles of relative motion to determine the velocity and acceleration of point C as observed from point A or with respect to the XY coordinate frame:

$$\underline{v}_C = \underline{v}_B + \underline{v}_{C/B}, \tag{v}$$

$$\underline{a}_C = \underline{a}_B + \underline{a}_{C/B}. \tag{vi}$$

Since we have already expressed $\underline{v}_B$, $\underline{v}_{C/B}$, $\underline{a}_B$, and $\underline{a}_{C/B}$ in terms of Cartesian unit vectors, we can simply substitute Eqs. (i) and (iii) into Eq. (v), and Eqs. (ii) and (iv) into Eq. (vi):

$$\underline{v}_C = \left(0.52\underline{i} + 0.30\underline{j}\right) + \left(0.85\underline{i} + 0.85\underline{j}\right),$$

$$\underline{a}_C = \left(-0.60\underline{i} + 1.04\underline{j}\right) + \left(-3.39\underline{i} + 3.39\underline{j}\right).$$

Collecting the horizontal and vertical components together:

$$\underline{v}_C = 1.37\underline{i} + 1.15\underline{j},$$

$$\underline{a}_C = -3.39\underline{i} + 4.43\underline{j}.$$

The magnitudes of the velocity and acceleration vectors are:

$$v_C = \sqrt{(1.37)^2 + (1.15)^2} = 1.79 \text{ m/s},$$

$$a_C = \sqrt{(3.99)^2 + (4.43)^2} = 5.96 \text{ m/s}^2.$$

9.13 Exercise Problems

Problem 9.1 Consider an object undergoing a rotational motion in the xy-plane about a fixed axis perpendicular to the plane of motion (Fig. 9.35). Let O be a point in the xy-plane along the axis of rotation, and P is a fixed point on the object. Due to the rotation of the object, point P will move in a circular path with a radius $r = 0.8$ m. The relative position of point P along its circular path is given as a function of time $S = 0.45\, t^{4/3}$. Determine the distance (S) traveled by point P along its path, and the magnitude of its linear velocity (v), tangential (a_t), normal (a_n), and net accelerations 3s after the motion began.

Answers: $S = 1.95$ m; $v = 0.86$ m/s; $a_t = 0.1$ m/s^2; $a_n = 0.92$ m/s^2; $a = 0.93$ m/s^2.

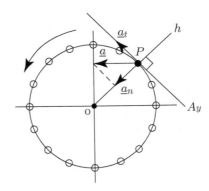

Fig. 9.35 Problem 9.1 and 9.2

Problem 9.2 Consider an object undergoing a rotational motion in the *xy*-plane about a fixed axis perpendicular to the plane of motion (Fig. 9.35). Let O be a point in the *xy*-plane along the axis of rotation and P is a fixed point on the object. Due to the rotation of the object, point P will experience a circular motion with the radius of its circular path $r = 0.6$ m.

Assume that at some point in time, the angular acceleration of the point P is $\alpha = 5$ rad/s and an angle between the vectors of its tangential and net acceleration is $\beta = 30°$.

Determine the magnitude of linear velocity (v) of point P and the magnitude of its tangential (a_t), normal (a_n), and net (a) acceleration vectors.

Answers: $v = 1.02$ m/s; $a_t = 3$ m/s^2; $a_n = 1.72$ m/s^2; $a = 3.46$ m/s^2.

Problem 9.3 Consider an arbitrarily shaped object undergoing a rotational motion in the *xy*-plane (Fig. 9.36). Let O be a point in the *xy*-plane along the axis of rotation, and P, N, and M are fixed points on the object located at distances r, $0.5\,r$, and $0.25\,r$, from the center of rotation, respectively. Due to the rotational motion of the object, points P, N, and M will move along their circular path.

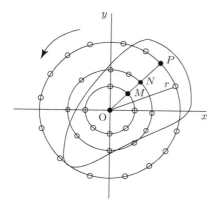

Fig. 9.36 Problem 9.3

If the linear velocity of point P is $v_p = 4.5$ m/s and it is located at distance $r = OP = 1.2$ m from the axis of rotation, determine the magnitude of angular velocity of points P, N, and M, and the magnitude of the linear velocity of points N and M.

Answers: $\omega_P = \omega_N = \omega_M = 3.75$ rad/s; $v_N = 2.25$ m/s; $v_M = 1.125$ m/s.

Problem 9.4 Consider a gymnast doing giant circles around a high bar (Fig. 9.37). Assume that the center of gravity of the gymnast is located at distance r from the bar and is undergoing a uniform circular motion with a linear velocity of $v = 5.0$ m/s. After completing several cycles, the gymnast releases the bar at the instant when his center of gravity is directly beneath the bar, and then he undergoes a projectile motion and lands on the floor at point P. The distance between point P and the projection of the point of release on the floor (point O') is $l = 4.0$ m.

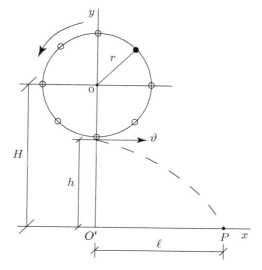

Fig. 9.37 Problem 9.4

Determine the time elapsed between the instant of release and landing (t), the height of the gymnast's center of gravity above the floor at the point of release (h), and the height of the bar above the floor (H).

Answers: $t = 0.8$ s; $h = 3.14$ m; $H = 4.39$ m.

10.1 Kinetics of Angular Motion

The kinetic characteristics of objects undergoing translational motion are discussed in Chap. 8. Kinetic analyses utilize Newton's second law of motion that can be formulated in terms of the equations of motion and work and energy methods. Similar methods can be employed to analyze the kinetic characteristics of objects undergoing rotational motion.

Consider an object undergoing a rotational motion in the xy-plane about a fixed point O (Fig. 10.1). Let P be a point on the object located at a distance r from point O. As the object rotates, point P moves in a circular path of radius r and center located at point O. To be able to analyze the kinetic characteristics of point P using the equations of motion, forces acting on the object and the acceleration of point P can be expressed in terms of their components normal and tangential to the circular path of motion. If n and t designate the normal (radial) and tangential directions at point P, then the equations of motion can be expressed as:

$$\sum F_n = ma_n,$$ (10.1)

$$\sum F_t = ma_t.$$ (10.2)

Here, $\sum F_n$ is the net force acting in the normal direction, $\sum F_t$ is the net force acting in the tangential direction, a_n is the magnitude of the centripetal acceleration (always directed toward the center of rotation), and a_t is the magnitude of tangential acceleration. While applying Eq. (10.1), the forces acting toward the center of rotation (centripetal forces) must be taken to be positive, and the forces directed outward (centrifugal forces) must be negative. For rotational motion about a fixed axis, the motion characteristics are completely known if the linear velocity (with magnitude v and direction tangent to the circular path) and the radius r of

the circular path are known. Since $a_n = v^2/r$ and $a_t = dv/dt$, Eqs. (10.1) and (10.2) can alternatively be written as:

$$\sum F_n = m\frac{v^2}{r},$$ (10.3)

$$\sum F_t = m\frac{dv}{dt}.$$ (10.4)

Note that $v = r\omega$ and $dv/dt = r\alpha$. Therefore, if the angular velocity and angular acceleration are known, then the kinetic characteristics of the problem can be analyzed using:

$$\sum F_n = mr\omega^2,$$ (10.5)

$$\sum F_t = mr\alpha.$$ (10.6)

Note that for rotational motion there is always a normal component of the acceleration vector, and therefore, a force acting in the normal direction, but the tangential components of the force and acceleration vectors may or may not exist.

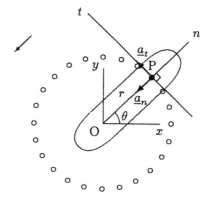

Fig. 10.1 Rotational motion

N. Özkaya et al., *Fundamentals of Biomechanics: Equilibrium, Motion, and Deformation*,
DOI 10.1007/978-1-4614-1150-5_10, © Springer Science+Business Media, LLC 2012

Example 10.1 Figure 10.2 illustrates a 60-kg gymnast swinging on a high bar. The rotational motion of the gymnast may be simplified by modeling the gymnast as a particle attached to a string such that the mass of the particle is equal to the mass of the gymnast and the length of the string is equal to the distance between the high bar and the center of gravity of the gymnast. As the gymnast moves, the center of gravity undergoes a circular motion.

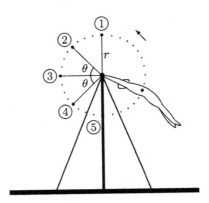

Fig. 10.2 A gymnast on the high bar

Assume that the center of gravity of the gymnast is located at a distance $r = 1$ m from the high bar, the speed of the center of gravity at position 1 is almost zero, and that the effects of air resistance are negligible. Position 1 is directly above the high bar and it represents the highest elevation reached by the center of gravity of the gymnast.

(a) By using the conservation of energy principle, calculate the speeds of the gymnast's center of gravity at positions 2, 3, 4, and 5. As shown in Fig. 10.2, positions 2 and 4 make an angle $\theta = 45°$ with the horizontal, position 3 is along the same horizontal line as the high bar, and position 5 is directly under the high bar.

(b) Calculate the angular velocities of the gymnast at positions 1, 2, 3, 4, and 5.

(c) Calculate the normal component of the linear accelerations of the gymnast's center of gravity at positions 1, 2, 3, 4, and 5.

(d) Calculate the forces applied on gymnast's arms at positions 1, 2, 3, 4, and 5.

(e) Calculate the tangential component of the linear accelerations of the gymnast's center of gravity at positions 1, 2, 3, 4, and 5.

(f) Calculate the angular accelerations of the gymnast at positions 1, 2, 3, 4, and 5.

Solution

(a) Different positions of the gymnast's center of gravity are illustrated in Fig. 10.3. Point O in Fig. 10.3 represents the high bar. The conservation of energy principle states that if an object is moving under the effect of conservative forces, then the total energy (sum of potential and kinetic energies) will remain constant throughout the motion. Between any two positions 1 and 2:

$$\mathcal{E}_{P1} + \mathcal{E}_{K1} = \mathcal{E}_{P2} + \mathcal{E}_{K2}$$

$$mgh_1 + \frac{1}{2}mv_1^2 = mgh_2 + \frac{1}{2}mv_2^2.$$

Here, $m = 60$ kg is the total mass of the gymnast, $g = 9.8$ m/s^2 is the magnitude of gravitational acceleration, $v_1 = 0$ is the speed of the gymnast at position 1, v_2 is the unknown speed of the gymnast at position 2, and h_1 and h_2 are the heights of positions 1 and 2 relative to a datum. Once a datum is chosen, h_1 and h_2 can be calculated and the above equation can be solved for the unknown parameter v_2. To calculate v_3, v_4, and v_5, the above procedure must be repeated.

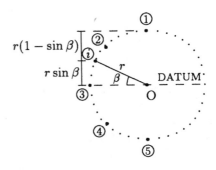

Fig. 10.3 For different β values, i represents positions 1, 2, 3, 4, and 5

Instead of carrying out the same procedure four times, consider the position of the gymnast labeled as "i" in Fig. 10.3. Assume that the line connecting point O and position i makes an angle β with the horizontal that passes through O such that $\beta = 90°$ at position 1, $\beta = 45°$ at position 2, $\beta = 0°$ at position 3, $\beta = -45°$ at position 4, and $\beta = -90°$ at position 5. We can apply the conservation of energy principle between positions 1 and i:

$$mgh_1 + \frac{1}{2}mv_1^2 = mgh_i + \frac{1}{2}mv_i^2.$$

Since $v_1 = 0$, the second term on the left-hand side of this equation is zero. Also, m is a common parameter in all of the terms and can be eliminated. To calculate heights h_1 and h_i, we need to chose a datum. If we chose the datum to coincide with the level of high bar, then from the geometry of the problem $h_1 = r$ and $h_i = r \sin \beta$. We can substitute h_1 and h_i into the above equation and solve it for v_i:

$$v_i = \sqrt{2gr(1 - \sin \beta)}.$$

This is a general solution valid for any position on the circular path of motion of the gymnast's center of gravity. For example, when angle $\beta = 45°$, $i = 2$ and $v_2 = \sqrt{2(9.8)(1)(1 - \sin 45)} = 2.39 \, \text{m/s}$. When $\beta = 0°$, $i = 3$ and $v_3 = \sqrt{2(9.8)(1)(1 - \sin 0)} = 4.43 \, \text{m/s}$. Similarly, $v_4 = 5.77 \, \text{m/s}$ and $v_5 = 6.26 \, \text{m/s}$. These results are used to plot a speed versus angular position (measured in terms of angle β) graph in Fig. 10.4.

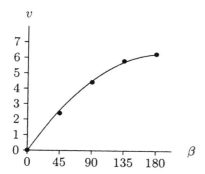

Fig. 10.4 Speed measured in m/s versus angle β in degrees

(b) The relationship between the angular velocity and linear velocity of a point on an object undergoing a rotational motion about a fixed axis is

$$\omega = \frac{v}{r}.$$

For example, for position 2, $\omega_2 = v_2/r = 2.39/1 = 2.39 \, \text{rad/s}$. Similarly, $\omega_1 = 0$, $\omega_3 = 4.43 \, \text{rad/s}$, $\omega_4 = 5.57 \, \text{rad/s}$, and $\omega_5 = 6.26 \, \text{rad/s}$.

(c) The normal (radial) component of the acceleration of the gymnast's center of gravity can be calculated using

$$a_n = (v^2/r) = r\omega^2.$$

For example, $a_{n2} = r\omega_2{}^2 = (1)(2.39)^2 = 5.74 \, \text{m/s}^2$ for position 2. On the other hand, $a_{n1} = 0$, $a_{n3} = 19.62 \, \text{m/s}^2$, $a_{n4} = 33.41 \, \text{m/s}^2$, and $a_{n5} = 39.19 \, \text{m/s}^2$.

(d) To calculate the forces applied on the gymnast's arms, consider Fig. 10.5 that shows the free-body diagrams of the gymnast's center of gravity. The forces acting on the gymnast are those applied by gravity and on the gymnast's arms by the high bar. The force applied by gravity is always directed vertically downward. The force applied by the high bar is in the radial direction. The equation of motion in the normal direction is

$$\sum F_n = ma_n.$$

At position 1, $a_{n1} = 0$ and the gymnast is in equilibrium. Therefore, since the weight of the gymnast is acting

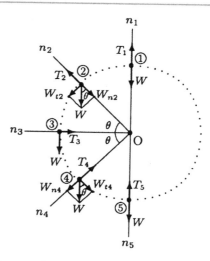

Fig. 10.5 Forces acting on the gymnast's center of gravity

downward, the force, T_1, applied by the high bar on the gymnast's arms must be upward. Applying the equation of equilibrium,

$$\sum F_{n1} = 0 : W - T_1 = 0$$
$$T_1 = W = mg$$
$$T_1 = (60)(9.8) = 588 \, \text{N}$$

At position 2, forces acting in the normal direction are the radial component, W_{n2}, of the gymnast's weight and T_2 applied by the high bar on the gymnast's arms. From the geometry of the problem, $W_{n2} = W \cos(45)$ and the direction of $\underline{W}_{n2}$ is toward the center of rotation. At this point, we do not know the direction of $\underline{T}_2$. We can assume that it is centrifugal. Now, we can apply the equation of motion in the normal direction:

$$\sum F_{n2} = ma_{n2} : W_{n2} - T_2 = ma_{n2}$$
$$T_2 - W_{n2} = ma_{n2}$$
$$T_2 = mg \sin(45) = ma_{n2}$$
$$T_2 = (60)(9.8)(\sin 45) = (60)(5.74) = 71 \, \text{N}$$

Since we calculated a positive value for T_2, the direction we assumed for $\underline{T}_2$ was correct. In other words, the effect of the high bar is such that it is "pushing" the arms of the gymnast at position 2.

At position 3, the gymnast's weight has no component in the radial direction. The force acting in the normal direction is T_3 applied by the high bar. In this case, assume that $\underline{T}_3$ is centripetal (toward O). Writing the equation of motion for position 3:

$$\sum F_{n3} = ma_{n3} : T_3 = ma_{n3}$$
$$T_3 = (60)(19.62) = 1,177 \, \text{N}$$

Again, since we calculated a positive value for T_3, the direction we assumed for $\underline{T}_3$ was correct. At position 3, the tendency of the gymnast is to move away from the center of rotation and what is holding the gymnast in the circular path of motion is the "pulling" effect of the high bar.

At position 4, forces acting in the radial direction are T_4 applied by the high bar and $W_{n4} = W \sin(45)$ component of the gymnast's weight. From the geometry of the problem, $\underline{W}_{n4}$ is centrifugal. Assuming that $\underline{T}_4$ is centripetal and applying the equation of motion:

$$\sum F_{n4} = ma_{n4} : \quad T_4 - W_{n4} = ma_{n4}$$
$$T_4 = W_{n4} + ma_{n4}$$
$$T_4 = mg \sin(45) + ma_{n4}$$
$$T_4 = (60)(9.8)(\sin 45) + (60)(33.41)$$
$$T_4 = 2,420 \,\text{N}.$$

Similarly at position 5:

$$\sum F_{n5} = ma_{n5} : \quad T_5 - W = ma_{n5}$$
$$T_5 = mg + ma_{n5}$$
$$T_5 = (60)(9.8) + (60)(39.19) = 2,939 \,\text{N}.$$

In Fig. 10.6, these results are used to plot a force applied by the high bar on the arms of the gymnast versus angular position (measured in terms of angle β in Fig. 10.3) graph. Note that between positions 1 and 2, the high bar has a pushing effect on the arms. In other words, the force applied by the high bar on the arms is compressive. Just after position 2, the force applied by the high bar is zero, and thereafter it has a pulling or tensile effect on the arms.

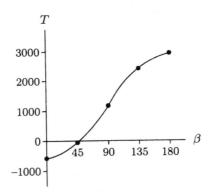

Fig. 10.6 Force applied by the high bar on the gymnast's arms (measured in Newtons) versus angle β in degrees

(e) The equation of motion in the tangential direction is

$$\sum F_t = ma_t \quad \text{or} \quad a_t = \frac{\sum F_t}{m}.$$

Forces acting in the tangential direction for different positions of the gymnast's center of gravity are shown in Fig. 10.5. At position 1, there is no force in the tangential direction, and therefore, $a_{t1} = 0$. At position 2, $W_{t2} = W \cos(45)$ is the only tangential force. Therefore,

$$a_{t1} = \frac{W_{t2}}{m} = \frac{mg \cos(45)}{m} = g \cos(45) = 6.93 \,\text{m/s}^2.$$

At position 3, $W = mg$ is the only tangential force:

$$a_{t3} = \frac{W_{t3}}{m} = \frac{mg}{m} = g = 9.80 \,\text{m/s}^2.$$

At position 4, $W_{t4} = W \cos(45)$ is the only tangential force:

$$a_{t4} = \frac{W_{t4}}{m} = \frac{mg \cos(45)}{m} = g \cos(45) = 6.93 \,\text{m/s}^2.$$

There is no tangential force at position 5, and therefore, $a_{t5} = 0$.

(f) Now that we calculated the tangential components of the acceleration vector, we can also calculate the angular acceleration of the gymnast using:

$$\alpha = \frac{a_t}{r}.$$

Note that $r = 1$ m. Therefore, $\alpha_1 = 0$, $\alpha_2 = 6.93 \,\text{rad/s}^2$, $\alpha_3 = 9.80 \,\text{rad/s}^2$, $\alpha_4 = 6.93 \,\text{rad/s}^2$, and $\alpha_5 = 0$.

Example 10.2 Figure 10.7 illustrates the circular end region of a ski jump track. The radius of curvature of the track at this region is $r = 50$ m. At the end of the track, the direction normal to the track coincides with the vertical and the direction tangential to the track coincides with the horizontal.

Fig. 10.7 Circular end region of a ski jump track

Consider a 70-kg ski jumper who is decelerating at a rate of 1.5 m/s² due to air resistance. If the friction on the track is negligible and the ski jumper reaches the end of the track with a horizontal velocity of $v = 20$ m/s, determine the forces applied on the skier by the air resistance and the track.

Solution

The free-body diagram of the ski jumper at the very end of the track (just before takeoff) is shown in Fig. 10.8. The forces acting on the ski jumper are $\underline{W}$ due to gravity, $\underline{R_1}$ due to air resistance, and the reaction force $\underline{R_2}$ applied by the track on the skis. Note that $\underline{R_2}$ is applied in the vertical direction or direction normal to the track. It is assumed that $\underline{R_1}$ due to air resistance is applied in the horizontal direction or direction tangent to the track.

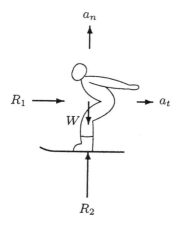

Fig. 10.8 Free-body diagram of the ski jumper before takeoff

At the circular end region of the track, the ski jumper undergoes a motion in a circular path with radius $r = 50$ m. The speed of the ski jumper at the very end of the track is $v = 20$ m/s. Therefore, the magnitude of the ski jumper's acceleration in the normal direction is

$$a_n = \frac{v^2}{r} = \frac{(20)^2}{50} = 8.0 \text{ m/s}^2.$$

Due to air resistance, the ski jumper is decelerating at a rate of 1.5 m/s² in the direction of motion (toward the left). Or, the ski jumper is accelerating at a rate of 1.5 m/s² in the direction opposite to the direction of motion (toward the right). Therefore, the tangential acceleration of the ski jumper toward the right is

$$a_t = 1.5 \text{ m/s}^2.$$

Now we can utilize the equations of motion. In the tangential direction:

$$\sum F_t = ma_t: \quad R_1 = ma_t$$
$$R_1 = (70)(1.5) = 105 \text{ N}.$$

Equation of motion in the normal direction:

$$\sum F_n = ma_n: \quad R_2 = W = ma_n$$
$$R_2 = W + ma_n = mg + ma_n$$
$$R_2 = (70)(9.8) + (70)(8.0) = 1{,}246 \text{ N}.$$

Therefore, at the very end of the ski jump track, air resistance is applying a horizontal force of $R_1 = 105$ N to retard the motion of the skier and the track is applying a vertical force of $R_2 = 1{,}246$ N on the skis. Note that R_2 includes the effects of the weight W and rotational inertia ma_n of the ski jumper.

10.2 Torque and Angular Acceleration

Torque is the quantitative measure of the ability of a force to rotate an object. The mathematical definition of torque is the same as that of moment, studied in detail in Chap. 3. Consider the bolt and wrench arrangement illustrated in Fig. 10.9. Force $\underline{F}$ applied on the wrench rotates the wrench, which advances the bolt into the wall by rotating it in the clockwise direction. The magnitude of torque $\underline{M}$ due to force $\underline{F}$ about point O is

$$M = rF_t = rF \sin \varphi. \tag{10.7}$$

The line of action of $\underline{M}$ is perpendicular to the plane of rotation and its direction can be determined by using the right-hand rule (in this case, clockwise).

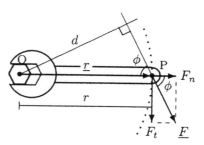

Fig. 10.9 Force $\underline{F}$ applied on the wrench produces a clockwise torque about the centerline of the bolt

An object would rotate about an axis if the rotational motion of the object is not constrained and if there is a net torque acting on the object about that axis. The angular acceleration of an object undergoing a rotational motion is directly proportional to the resultant torque acting on it. To derive the relationship between torque and angular acceleration, consider a particle of mass m undergoing a rotational motion about a fixed axis. Let O be a point on this axis, r be the radius of the circular path of motion, and $\underline{F}_t$ be the tangential force causing the rotational motion (Fig. 10.10). The equation of motion in the tangential direction can be written as:

$$F_t = ma_t. \tag{10.8}$$

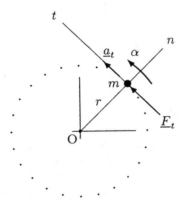

Fig. 10.10 $F_t = ma_t$ and $M_O = I_O\alpha$

In Eq. (10.8), a_t is the magnitude of the tangential acceleration of the particle. If the angular acceleration, α, of the particle is known, then $a_t = r\alpha$. Replacing a_t by $r\alpha$ and multiplying both sides of Eq. (10.8) by r will yield:

$$rF_t = (mr^2)\alpha. \tag{10.9}$$

Note that the left-hand side of Eq. (10.9) is the magnitude M_O of the torque generated by force $\underline{F}_t$ about O. The term mr^2 on the right-hand side is known as the *mass moment of inertia* of the particle about O. Denoting the mass moment of inertia with I_O, Eq. (10.9) can also be written as:

$$M_O = I_O\alpha. \tag{10.10}$$

If there is more than one torque-generating force applied to the particle, then M_O in Eq. (10.10) represents the net torque acting on the particle about O. The general form of Eq. (10.10) can be obtained by representing the torque and angular acceleration as vector quantities:

$$\underline{M} = I\underline{\alpha}. \tag{10.11}$$

This is the rotational analogue of Newton's second law of motion, which states that angular acceleration is directly proportional to the net torque and inversely proportional to the mass moment of inertia.

10.3 Mass Moment of Inertia

In general, the term inertia implies resistance to change. When a rotation-causing force is applied to a pivoted body, its tendency to resist angular acceleration depends on its mass moment of inertia. The larger the mass moment of inertia of a body, the more difficult it is to accelerate it in rotation. For a particle of mass m, the mass moment of inertia about an axis is defined as the mass times the square of the shortest distance, r, between the particle and the axis about which the mass moment of inertia is to be determined:

$$I = mr^2. \tag{10.12}$$

A rigid body that is not a particle can be assumed to consist of many particles, the sum of masses of which is equal to the total mass of the body itself. The mass moment of inertia of the entire body can be determined by considering the sum of the mass of each particle of the body multiplied by the square of its distance from the axis of rotation.

Notice that the mass moment of inertia of a rigid body is proportional to its mass, which is a function of its density and volume. Therefore, the mass moment of inertia of a body depends upon its material and geometric properties as well as the location and orientation of the axis about which it is to be determined. The ability of a body to resist changes in its angular velocity is dependent not only upon the mass of the body, but also upon the distribution of the mass. The greater the concentration of mass at the periphery, the greater the mass moment of inertia and the more difficult it is to change the angular velocity. For a rigid body with a simple, symmetrical geometry and homogeneous composition, the mass moment of inertia about an axis coinciding with an axis of symmetry, called a *centroidal axis*, can be calculated relatively easily. In Table 10.1, moments of inertia for some geometric shapes are provided about their centroidal axes.

The mass moment of inertia is a scalar quantity. It has a dimension $[M][L^2]$, and is measured in terms of $\text{kg} - \text{m}^2$ in SI.

10.4 Parallel-Axis Theorem

If the mass moment of inertia of a body about a centroidal axis is known, then the mass moment of inertia of the same body about any other axis parallel to that centroidal axis can

Table 10.1 Moments of inertia of homogeneous rigid bodies with different geometries about their centroidal axes. Their volumes are also provided.

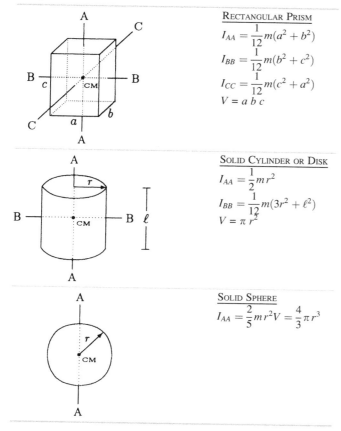

RECTANGULAR PRISM

$$I_{AA} = \frac{1}{12}m(a^2 + b^2)$$

$$I_{BB} = \frac{1}{12}m(b^2 + c^2)$$

$$I_{CC} = \frac{1}{12}m(c^2 + a^2)$$

$$V = a\,b\,c$$

SOLID CYLINDER OR DISK

$$I_{AA} = \frac{1}{2}mr^2$$

$$I_{BB} = \frac{1}{12}m(3r^2 + \ell^2)$$

$$V = \pi r^2$$

SOLID SPHERE

$$I_{AA} = \frac{2}{5}mr^2 \qquad V = \frac{4}{3}\pi r^3$$

Fig. 10.11 According to the parallel-axis theorem, $I_{DD} = I_{AA} + mr_c^2$

moving segment rather than about its mass center or centroidal axes. Furthermore, mass moment of inertia measurements can only be made about a joint center. If needed, the parallel-axis theorem can be utilized to determine the mass moment of inertia of a segment about its mass center.

10.5 Radius of Gyration

Consider a rigid body of mass m. Let I be the mass moment of inertia of the rigid body about a given axis AA. Also consider a point mass m located at a distance ρ (rho) from the same axis such that its mass moment of inertia $m\rho^2$ about AA is equal to the mass moment of inertia I of the rigid body about AA. That is,

$$\rho = \sqrt{\frac{I}{m}}. \qquad (10.14)$$

ρ is called the *radius of gyration*, and for rotational motion analysis, the rigid body can be treated as a point with the mass equal to the total mass of the body and located at a distance ρ from the axis of rotation.

be determined using the *parallel-axis theorem*. This theorem can be stated as:

$$I = I_c + mr_c^2. \qquad (10.13)$$

In Eq. (10.13), m is the total mass of the body, I_c is the mass moment of inertia of the body about one of its centroidal axes, I is the required mass moment of inertia about an axis parallel to the centroidal axis, and r_c is the shortest distance between the two axes. For example, consider the solid cylinder shown in Fig. 10.11. From Table 10.1, the mass moment of inertia of the cylinder about AA is $I_{AA} = (1/2)mr^2$. The mass moment of inertia of the same cylinder about DD, which is parallel to AA and located at a distance r_c from AA, is

$$I_{DD} = I_{AA} + mr_c^2 = \frac{1}{2}mr^2 + mr_c^2.$$

Note that in the case of human body segments, each segment or limb rotates about the joints at either end of the

10.6 Segmental Motion Analysis

The information provided in the previous sections can be utilized to develop mathematical models for analyzing the motion characteristics of human body segments. Here, the general procedure for developing a dynamic model of a body segment will be outlined, and then applied to analyze the rotational motion of the lower leg about the knee joint.

The first step of a dynamic model analysis involves defining the forces acting on the body segment. These may include the gravitational (weight), external, inertial, muscle, and joint reaction forces. The weight of the body segment can be assumed to act at its center of gravity, and therefore, the center of gravity of the segment must be known. The magnitude, point of application, and direction of any external force present must be specified. Inertial forces are those present due to the dynamics of the problem under consideration. One way of incorporating inertial effects into the model is through the use of the radius of gyration. Muscle and joint reaction forces are the unknowns to be determined as a result of these analyses. It is important to draw the free-body diagram of the segment to be analyzed, and to identify all the known and unknown forces acting on it.

The next step of dynamic analysis is the identification of measurable quantities. In general, both the angular displacement of the moving body segment and the net torque generated about its axis of rotation can be measured as functions of time over the range of segmental motion. The angular displacement measurement techniques include goniometric, dynamometric, and photogrammetric methods. Through the use of kinematic relationships, the angular displacement data can be used to calculate the angular velocity and angular acceleration of the moving segment. If the angular displacement θ is known as a function of time t, then the angular velocity ω and angular acceleration α can be determined by considering the first and second derivatives of θ with respect to t:

$$\omega = \frac{d\theta}{dt}, \qquad \alpha = \frac{d\omega}{dt} = \frac{d^2\theta}{dt^2}.$$

If it is not possible to find a function representing the relationship between the angular displacement and time, then numerical differentiation techniques can be employed. Once the angular acceleration is determined, the relationship between torque, mass moment of inertia, and angular acceleration can be used to calculate the net torque produced about the joint center, provided that the mass moment of inertia (or the radius of gyration) of the segment about the joint center is known.

For a two-dimensional (planar) motion analysis of a segment about its joint center, if the net torque M produced about the joint axis is measured as a function of time and the mass moment of inertia I of the moving segment about the same axis is known, then the angular acceleration of the segment can be determined from:

$$\alpha = \frac{M}{I}.$$

If needed, the angular velocity and displacement of the moving segment can also be determined by considering the integral of the function representing angular acceleration with respect to time.

It is clear from this discussion that anthropometric information about the moving segment must be available. For this purpose, anthropometric data tables listing average segmental weights, lengths, and radius of gyration can be utilized. Another important consideration is that the instantaneous center of rotation of the moving segment must also be known. Note, however, that the instantaneous center of rotation about a given joint may vary.

The final step of dynamic model analyses involves the computation of muscle and joint reaction forces. Note that the net torque measured or calculated includes the effects of all forces acting on the moving segment. The torque generated by the muscles crossing the joint can be determined by subtracting the effects of external and gravitational forces from the net torque measured or calculated about the joint center. If the forces generated by individual muscles are required, then additional factors must be considered. For example, the locations of muscle attachments and the lines of action (lines of pull) of the muscle forces must be known. The distribution of forces among different muscles must also be specified. If the segmental motion is achieved primarily by a single muscle group, then the rotational component of the muscle force can be determined by applying Newton's second law of motion. Since the line of action of the muscle force is assumed to be known, the magnitude of the muscle force can also be determined.

Example 10.3. Knee Extension The angular motion of the lower leg about the knee joint, and the forces and torques produced by the muscles crossing the knee joint during knee flexions and extensions have been investigated by a number of researchers utilizing different experimental techniques. One of these techniques is discussed here.

Consider the person illustrated in Fig. 10.12. The test subject is sitting on a table, with the back placed against a back rest and the lower legs free to rotate about the knee joint. The subject's torso is strapped to the back rest and the

Fig. 10.12 Knee extension

right thigh is strapped firmly to the table. A well-padded sawhorse is placed in front of the subject to prevent hyper-extension at the knee joint (not shown in Fig. 10.12). An electrogoniometer is attached to the subject's right leg. The arms of the goniometer are aligned with the estimated long axes of the thigh and shank, and the axis of rotation of the goniometer is aligned with the estimated axis of rotation of the knee joint. The subject is then asked to extend the lower leg as rapidly as possible. The signals received from the electrogoniometer's potentiometer are stored in a computer, and are used to calculate the angular displacement θ of the lower leg as measured from its initial vertical position. Using a finite difference (numerical differentiation) technique, the angular velocity ω and angular acceleration α of the lower leg are also computed.

Some of the forces acting on the lower leg are shown in Fig. 10.13, along with the geometric parameters of the model under consideration. This model is based on the assumption that the quadriceps muscle is the primary muscle group responsible for knee extension. Point O represents the instantaneous center of rotation of the knee joint. The patellar tendon is attached to the tibia at A. For the position of the lower leg relative to the upper leg shown in Fig. 10.13, it is estimated that the line of pull of the patellar tendon force $\underline{F}_m$ makes an angle β with the long axis of the tibia. The lever arm of $\underline{F}_m$ relative to O can be represented by a distance a that changes as the lower leg moves up through the range of motion. The total weight of the lower leg is W and its center of gravity is located at B, which is at a distance b from O

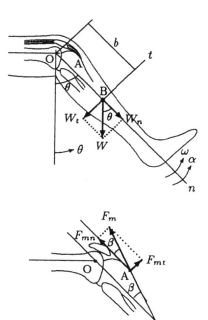

Fig. 10.13 Some of the forces acting on the lower leg

measured along the long axis of the tibia. The intended direction of motion is counterclockwise (extension).

At an instant when $\theta = 60°$, $\omega = 5$ rad/s, and $\alpha = 200$ rad/s^2, and assuming that $W = 50$ N, $a = 4$ cm, $b = 22$ cm, $\beta = 24°$, and the mass moment of inertia of the lower leg about the knee joint is $I_o = 0.25$ kg–m^2, determine:

(a) The net torque produced about the knee joint
(b) The tension in the patellar tendon
(c) The reaction force at the knee joint

Solution

(a) From Newton's second law of motion, the net torque M_o generated about the knee joint is

$$M_o = I_o\alpha = (0.25)(200) = 50 \text{ N–m} \quad \text{(ccw)}.$$

(b) Note that M_o is the magnitude of the net torque about the knee joint and it includes the rotational effects of all of the external forces acting on the lower leg. Relative to the knee joint, there are two external forces with rotational effects: patellar tendon force $\underline{F}_m$ and weight of the lower leg $\underline{W}$. For the position of the lower leg that makes an angle $\theta = 60°$ with the vertical, the lever arm of the patellar tendon force relative to O is estimated to be $a = 0.04$ m. Therefore, the torque generated by $\underline{F}_m$ relative to O is $M_m = aF_m$ (counterclockwise). On the other hand, $\underline{W}$ acts downward. The torque generated by $\underline{W}$ about O is $M_w = W_t b = \sin \theta b$ (clockwise). Therefore, the magnitude of the net torque, M_O, about the knee joint due to $\underline{F}_m$ and $\underline{W}$ is

$$M_o = M_m - M_w = aF_m - bW \sin \theta.$$

This equation can be solved for the unknown force, F_m,

$$F_m = \frac{M_o + bW \sin\theta}{a}.$$

Substituting the known parameters and carrying out the calculations will yield:

$$F_m = \frac{50 + (0.22)(50)(\sin 60°)}{0.04} = 1,488 \text{ N}.$$

(c) The free-body diagram of the lower leg is shown in Fig. 10.14. W is the weight of the lower leg, F_m is the magnitude of the patellar tendon force applied on the tibia at A, F_{jn} is the component of the tibiofemoral joint reaction force along the long axis of the tibia, and F_{jt} is the component of the tibiofemoral joint reaction force in a direction perpendicular to the long axis of the tibia. The components of $\underline{W}$ and $\underline{F}_m$ along the long axis of the tibia and in the perpendicular direction are

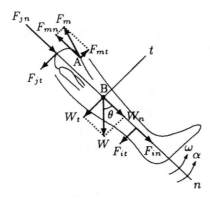

Fig. 10.14 Free-body diagram of the lower leg

also shown in Fig. 10.14, as well as the components of inertial force F_i.

One way of taking into account the inertial effects of a moving body, in this case a rotating body segment, is by means of the d'Alembert's principle. If the mass m, distance r between the mass center and the axis of rotation, angular velocity ω, and angular acceleration α, of the rotating body are known, then the magnitudes of inertial forces F_{in} and F_{it} which are normal and tangent to the path of motion can be calculated using Eqs. (10.5) and (10.6) provided in Sect. 10.1:

$$F_{in} = ma_n = mr\omega^2,$$

$$F_{it} = ma_t = mr\alpha.$$

In this case, we have $m = W/g = 50/9.8 = 5.1\,\mathrm{kg}$, $r = b = 0.22\,\mathrm{cm}$, $\omega = 12\,\mathrm{rad/s}$, and $\alpha = 200\,\mathrm{rad/s^2}$. a_n and a_t are the components of the acceleration vector of the lower leg in the directions normal and tangential to its path of motion. a_n is always toward the center of rotation, and since the motion is counterclockwise, a_t is counterclockwise. Therefore, the inertial forces $\underline{F}_{in}$ and $\underline{F}_{it}$ are such that $\underline{F}_{in}$ is centripetal (toward the center of rotation) and $\underline{F}_{it}$ is trying to rotate the leg in the counterclockwise direction. As illustrated in Fig. 10.14, the d'Alembert's principle can be applied by assuming that $\underline{F}_{in}$ is a centrifugal force (rather than centripetal) trying to pull the leg outward, $\underline{F}_{it}$ is trying to rotate the lower leg in the clockwise direction (rather than in the counterclockwise direction), and the system is in static equilibrium. The conditions of the equilibrium of the system can be represented by the following equations valid along the directions normal and tangential to the path of motion:

$$\sum F_n = 0 : \quad F_{jn} - F_{mn} + F_{in} + W_n = 0,$$

$$\sum F_t = 0 : \quad F_{jt} - F_{mt} + F_{it} + W_t = 0.$$

Solving these equations for the components of the joint reaction force will yield:

$$F_{jn} = F_{mn} - F_{in} - W_n = 0,$$

$$F_{jt} = F_{mt} - F_{it} - W_t = 0.$$

Substituting the known parameters by their mathematical expressions will yield:

$$F_{jn} = F_m \cos \beta - mb\omega^2 - W \cos \theta,$$

$$F_{jt} = F_m \sin \beta - mb\alpha - W \sin \theta.$$

Now, substituting the numerical values and carrying out the calculations will yield:

$$F_{jn} = (1{,}488)(\cos 24) - (5.1)(0.22)(5)^2 - (50)(\cos 60)$$
$$= 1{,}306\,\mathrm{N},$$

$$F_{jt} = (1{,}488)(\sin 24) - (5.1)(0.22)(200) - (50)(\sin 60)$$
$$= 338\,\mathrm{N}.$$

Therefore, the magnitude of the resultant force applied by the femur on the tibia is $F_j = \sqrt{(F_{jn})^2 + (F_{jt})^2} = 1{,}349\,\mathrm{N}$.

10.7 Rotational Kinetic Energy

Assume that the rigid body shown in Fig. 10.15 is composed of many small particles and that the body rotates about a fixed axis with an angular velocity ω. If m_i and v_i are the mass and the speed of the ith particle in the body, respectively, then the kinetic energy of the particle is

$$\mathcal{E}_{Ki} = \frac{1}{2} m_i v_i^2.$$

At any instant, every particle in the body has the same angular velocity ω, but the linear velocity of each particle depends on its distance measured from the axis of rotation.

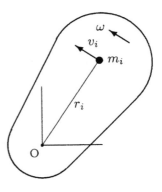

Fig. 10.15 Rotational motion of a body about a fixed axis

If r_i is the perpendicular distance between the ith particle and the axis of rotation (i.e., radius of the circular motion path of the ith particle), then $v_i = r_i \omega$ and its kinetic energy is $\mathcal{E}_{Ki} = \frac{1}{2} m_i r_i^2 \omega^2$. Each particle in the body has a kinetic energy, and the total kinetic energy, $\mathcal{E}_K$, of the rotating body is the sum of the kinetic energies of the individual particles in the body. That is,

$$\mathcal{E}_K = \sum_{i=1}^{n} \mathcal{E}_{Ki} = \frac{1}{2} \left(\sum_{i=1}^{n} m_i r_i^2 \right) \omega^2.$$

The quantity in parentheses is the mass moment of inertia I of the body. Therefore,

$$\mathcal{E}_K = \frac{1}{2} I \omega^2. \tag{10.15}$$

Equation (10.15) defines the rotational kinetic energy of a body in terms of the mass moment of inertia and angular velocity of the body, and it is analogous to the kinetic energy $\mathcal{E}_K = \frac{1}{2} m v^2$ associated with linear motion.

10.8 Angular Work and Power

By definition, the work done by a force is equal to the magnitude of the force times the corresponding displacement. The *angular work done* by a force applied on a rotating body is related to the angular displacement of the body. Consider a body rotating about a fixed axis at O due to an applied force $\underline{F}$. As illustrated in Fig. 10.16, let P_1 and P_2 represent the positions of a point in the body at times t_1 and t_2, respectively. In the time interval between t_1 and t_2, the body rotates through an arc of length s or angle θ. The work done by $\underline{F}$ on the body is equal to the magnitude of the component of the force vector in the direction of motion (tangential component, F_t), times the displacement s:

$$W = F_t s.$$

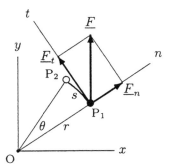

Fig. 10.16 A partial located at P_1 is displaced by an angle θ or arc length s to position P_2

The arc length is related to the angular displacement through the radius of the circular path of motion as $s = r\theta$. Therefore,

$$W = F_t r \theta.$$

By definition, $F_t r$ is the magnitude M of the torque generated by force $\underline{F}$ about O. Hence,

$$W = M \theta. \tag{10.16}$$

In other words, the work done by a rotation-producing force is equal to the torque generated by the force times the angular displacement of the body. Notice that the normal (radial) component of the force vector does not work on a body undergoing rotational motion because there is no motion in the normal direction.

It must be pointed out here that the relationship between angular work done, torque, and angular displacement given in Eq. (10.16) is valid when the torque is constant. The work done by a torque, which is a function of angular displacement, on a body to rotate the body from position 1 to 2 is

$$W = \int_{\theta_1}^{\theta_2} M \, d\theta. \tag{10.17}$$

Here, θ_1 and θ_2 are the angular displacements of the body at positions 1 and 2, respectively. Equation (10.17) can also be written in terms of the change in angular velocity by noting that $M = I\alpha$ and $\alpha = d\omega/dt$. Using the chain rule of differentiation,

$$M = I\alpha = I \frac{d\omega}{dt} = I \frac{d\omega}{d\theta} \frac{d\theta}{dt} = I \omega \frac{d\omega}{d\theta}.$$

Substituting this into Eq. (10.17),

$$W = \int_{\omega_1}^{\omega_2} I\omega \, d\omega = \frac{1}{2} I \omega_2^2 - \frac{1}{2} I \omega_1^2. \tag{10.18}$$

In Eq. (10.18), ω_1 and ω_2 are the angular velocities of the body at positions 1 and 2, respectively. Equation (10.18), known as the *work–energy theorem in rotational motion*, states that the net angular work done on a rigid body in rotating the body about a fixed axis is equal to the change in the body's rotational kinetic energy.

The rate at which work is done is known as power. The *angular power* describes the rate at which angular work is done. For a constant torque,

$$P = \frac{dW}{dt} = M \frac{d\theta}{dt} = M\omega. \tag{10.19}$$

That is, the angular power is equal to the product of the applied torque and the angular velocity of the body.

Example 10.4 Consider the knee extension problem analyzed in Example 14.3. As illustrated in Fig. 10.17, the person is seated on a table. The upper body is strapped to a back rest and the right thigh is strapped firmly on the table with the lower leg hanging vertically downward. The person is then asked to extend the right lower leg. The angular displacement of the lower leg during knee extension is determined via a goniometer attached to the leg. After a series of computations, it is determined that the lower leg was extended from $\theta = 0°$ to $90°$ in a time period of 0.5 s with an average angular velocity of 3 rad/s by producing an average extensor muscle torque of 90 N−m.

Fig. 10.17 Knee extension

Assuming that the mass moment of inertia of the lower leg about the center of rotation of the knee joint is 92 kg–m², calculate the average angular kinetic energy produced, angular work done, and angular power generated by the knee extensor muscles to extend the lower leg from $\theta = 0°$ to $90°$.

Solution The range of motion of the lower leg is $\Delta\theta = 90°$, which is covered in a time period of $\Delta t = 0.5$ s (Fig. 10.18).

Fig. 10.18 Movement of the lower leg

The mass moment of inertia of the lower leg about the knee joint is given as $I_o = 92$ kg–m². The average angular velocity of the lower leg is calculated to be $\overline{\omega} = 3$ rad/s and the average torque produced by the knee extensors is $\overline{M} = 90$ N−m. Therefore, the average angular kinetic energy produced by the knee extensor muscles is

$$\overline{\varepsilon_K} = \frac{1}{2}I_o\overline{\omega}^2 = \frac{1}{2}(92)(3)^2 = 414 \text{ J}.$$

The average work done by the muscles to extend the lower leg at an angle of $90°$ or $90 \times \pi/180 = 1.57$ rad is

$$\overline{W} = \overline{M}\Delta\theta = (90)(1.57) = 141.3 \text{ J}.$$

The average power generated by the extensors is

$$\overline{P} = \overline{M}\overline{\omega} = (90)(3) = 270W.$$

11.1   Introduction

In Chap. 8, Newton's second law of motion is presented in the form of "equations of motion." In Chap. 10, the concepts of work and energy are introduced. Based on the same law, "work–energy" and "conservation of energy" methods are devised to facilitate the solutions of specific problems in kinetics. In this chapter, the concepts of linear momentum and impulse are defined. Newton's second law of motion is reformulated to introduce other methods for kinetic analyses based on the "impulse–momentum theorem" and the principle of "conservation of linear momentum." These methods are then applied to analyze the impact and collision of bodies.

11.2   Linear Momentum and Impulse

Consider an object with mass m acted upon by an external force $\underline{F}$. Let $\underline{a}$ be the acceleration of the object under the action of the applied force. The relationship between $\underline{F}$, m, and $\underline{a}$ is described by the equation of motion in the following vector form:

$$\underline{F} = m\underline{a}. \tag{11.1}$$

By definition, acceleration is the time rate of change of velocity. If the mass of the object is constant, then Eq. (11.1) can be written as:

$$\underline{F} = \frac{d}{dt}(m\underline{v}). \tag{11.2}$$

The vector $m\underline{v}$ is called the *linear momentum* (or simply the momentum) of the object, and is denoted by $\underline{p}$:

$$\underline{p} = m\underline{v}. \tag{11.3}$$

Momentum is a vector quantity. The line of action and direction of the momentum vector is the same as the velocity vector of the object. The magnitude of the momentum is equal to the product of the mass and the speed of the object.

The equation of motion can now be expressed in terms of momentum by substituting Eq. (11.3) into Eq. (11.2):

$$\underline{F} = \frac{d\underline{p}}{dt}. \tag{11.4}$$

If there is more than one force acting on the object, then $\underline{F}$ in Eq. (11.4) must be replaced by the vector sum of all forces (the resultant force) acting on the object. Equation (11.4) states that the time rate of change of momentum of an object is equal to the resultant force on the object.

The concept of momentum is particularly useful for analyzing the effects of forces applied in very short time intervals. Such forces are called *impulsive forces*, and the motions associated with them are called *impulsive motions*. Consider an object moving with velocity $\underline{v}_1$ at time t_1 (Fig. 11.1). Assume that a force $\underline{F}$ is applied on the object in the time interval between t_1 and t_2, and the velocity of the object is changed to $\underline{v}_2$ at time t_2. Multiplying Eq. (11.4) by dt and integrating it between t_1 and t_2 will yield

$$\int_{t_1}^{t_2} \underline{F}\, dt = \underline{p}_2 - \underline{p}_1 = m\underline{v}_2 - m\underline{v}_1. \tag{11.5}$$

Fig. 11.1 Impulsive force $\underline{F}$ changes the momentum of the object from $\underline{p}_1$ to $\underline{p}_2$

N. Özkaya et al., *Fundamentals of Biomechanics: Equilibrium, Motion, and Deformation*,
DOI 10.1007/978-1-4614-1150-5_11, © Springer Science+Business Media, LLC 2012

The integral on the left-hand side of Eq. (11.5) represents the *linear impulse* of force F on the object in the time interval $\Delta t = t_2 - t_1$. The right-hand side of Eq. (11.5) is equal to the change in momentum $\Delta \underline{p} = \underline{p}_2 - \underline{p}_1$ of the object in the time interval Δt. Therefore,

$$\int_{t_1}^{t_2} \underline{F}\, \mathrm{d}t = \Delta \underline{p}. \qquad (11.6)$$

This equation is the mathematical representation of the *impulse–momentum theorem*.

Impulse is a vector quantity. It has the same line of action and direction as the impulsive force. In general, the magnitude and direction of the impulsive force may vary in the time interval Δt. If the force is known as a function of time, the impulse can be determined by integrating $\underline{F}$ with respect to time which will essentially yield the area under the force versus time curve. If the impulsive force is constant (Fig. 11.2), then the impulse is simply

$$\underline{F} = \Delta t = \Delta p \quad (\underline{F} \text{ constant}). \qquad (11.7)$$

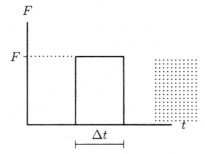

Fig. 11.2 Impulse is the area under the force versus time curve

Equation (11.7) can also be used to determine the impulse of a varying force by approximating the force with an average force value (Fig. 11.3).

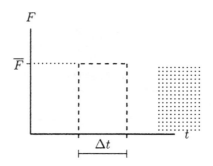

Fig. 11.3 Impulse of a varying force can be determined by considering an average force and an average time

In the equations derived so far, all of the parameters involved (except for the mass m) are vector quantities. These parameters can be represented in terms of their rectangular components along the Cartesian coordinate directions x, y, and z. This approach will yield three independent scalar equations valid along the x, y, and z directions. For example, Eq. (11.5) will yield

$$\int_{t_1}^{t_2} F_x\, \mathrm{d}t = \Delta p_x = p_{x2} - p_{x1} = mv_{x2} - mv_{x1}$$

$$\int_{t_1}^{t_2} F_y\, \mathrm{d}t = \Delta p_y = p_{y2} - p_{y1} = mv_{y2} - mv_{y1} \qquad (11.8)$$

$$\int_{t_1}^{t_2} F_z\, \mathrm{d}t = \Delta p_z = p_{z2} - p_{z1} = mv_{z2} - mv_{z1}.$$

For one-dimensional problems, any one of the above equations is sufficient to analyze the problem. For plane problems, two equations may be needed. While using Eqs. (11.8), it should be kept in mind that force, velocity, momentum, and impulse are vector quantities. For example, consider an object moving in the positive x direction. Assume that between times t_1 and t_2, an impulsive force $\underline{F}_x$ is applied on the object in the direction of motion (Fig. 11.4a). The momentum of the object will become

$$p_{x2} = p_{x1} + \int_{t_1}^{t_2} F_x\, \mathrm{d}t.$$

That is, the final momentum is equal to the initial momentum plus the impulse. If the force is applied in the direction opposite to that of the motion of the object (Fig. 11.4b), then the final momentum of the object is equal to the initial momentum minus the impulse:

$$p_{x2} = p_{x1} - \int_{t_1}^{t_2} F_x\, \mathrm{d}t.$$

Fig. 11.4 Scalar and vector additions of impulse and momentum

If the motion is in one direction and the impulsive force is applied in a different direction (Fig. 11.4c), then it is easier to treat impulse and momentum as vector quantities.

Impulse and momentum have the same dimension (ML/T) and units. Their units in different systems are listed in Table 11.1.

Table 11.1 Units of impulse and momentum

System	Units of impulse and momentum
SI	kg–m/s = N–s
CGS	g–cm/s = dyn–s
British	slug–ft/s = lb–s

11.3 Applications of the Impulse–Momentum Method

The concepts of impulse and momentum have numerous applications. These concepts may facilitate the analysis of situations where impact and impulsive forces play a significant role. The impulse–momentum method can be utilized to determine the forces involved in various athletic, sports, and daily activities. This method enables us to investigate the forces exerted by the foot on a football, or by the foot or the forehead on a soccer ball, which is in fact the same force exerted by the ball on the foot or the forehead. Forces involved while walking and running and the forces applied on various joints of the human body, which are transmitted through the feet during landing after a long jump or after a routine in gymnastics, may also be analyzed with the same technique. Other applications of the method include the analyses of forces exerted by a club on a golf ball, a bat on a baseball, and a racket on a tennis ball.

Example 11.1 As illustrated in Fig. 11.5, consider a ball of mass $m = 0.25$ kg dropped from a height $h_0 = 1$ m. The ball hits the floor, bounces, and reaches a height $h_3 = 0.75$ m.

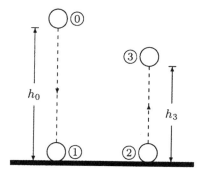

Fig. 11.5 The ball hits the floor and bounces

Determine the momentum of the ball immediately before and after it collides with the floor. Assuming that the duration of collision (duration of contact) is $\Delta t = 0.01$ s, determine an average force exerted by the floor on the ball during impact.

Solution

Let 0, 1, 2, and 3 correspond to each stage of the action: release, impact, rebound, and reaching the height 0.75 m for the ball. The speed v_1 of the ball at the instant of impact can be determined by utilizing the conservation of energy principle between 0 and 1. Since $v_0 = 0$ and $h_1 = 0$

$$mgh_0 = \frac{1}{2}mv_1^2.$$

Solving this equation for v_1 will yield

$$v_1 = \sqrt{2gh_0} = \sqrt{2(9.8)(1)} = 4.43 \text{ m/s}.$$

Similarly, the speed v_2 of the ball immediately after collision can be determined by considering the motion of the ball between 2 and 3. Since $h_2 = 0$ and $v_3 = 0$

$$\frac{1}{2}mv_2^2 = mgh_3.$$

Solving this equation for v_2 will yield

$$v_2 = \sqrt{2gh_3} = \sqrt{2(9.8)(0.75)} = 3.83 \text{ m/s}.$$

Momenta of the ball immediately before and after collision are

$$p_1 = mv_1 = (0.25)(4.43) = 1.11 \text{ kg–m/s} \quad (\downarrow),$$

$$p_2 = mv_2 = (0.25)(3.83) = 0.96 \text{ kg–m/s} \quad (\uparrow).$$

In vector form,

$$\underline{P}_1 = -1.11\underline{j}(\text{kg–m/s}),$$

$$\underline{P}_2 = 0.96\underline{j}(\text{kg–m/s}).$$

The change in momentum of the ball during the course of collision is

$$\Delta\underline{p} = \underline{p}_2 - \underline{p}_1 = (0.96\underline{j}) - (-1.11\underline{j}) = 2.07\underline{j}(\text{kg–m/s}).$$

The force exerted by the floor on the ball during the course of collision can be determined by assuming that the impulsive force is approximately constant during the collision. Using the impulse–momentum equation given in Eq. (11.7):

$$\underline{F}\Delta t = \Delta\underline{p}.$$

Solving this equation for the unknown force $\underline{F}$:

$$\underline{F} = \frac{\Delta p}{\Delta t} = \frac{2.07\underline{j}}{0.01} = 207\underline{j}\,(\text{N}).$$

Remarks

- The magnitude of the average impulsive force calculated is much greater than the force of gravity on the ball which is about 2.5 N. This result suggests that for problems involving collisions, the impulsive force between the colliding bodies (in this case the ball and the floor) is the most dominant, and the effects of all other external forces which may be present can be ignored.
- The average force calculated depends heavily on the duration of contact Δt.
- The fact that $v_2 = 3.83$ m/s is less than $v_1 = 4.43$ m/s suggests that some of the kinetic energy of the ball just before collision is lost as heat during the course of collision. The amount of energy lost can be calculated by considering the difference between the kinetic energies of the ball before and after collision.

Example 11.2 Consider a soccer player kicking a stationary ball (Fig. 11.6). If the effect of air resistance is negligible, the ball will undergo a projectile motion (Fig. 11.7).

Fig. 11.6 The player applies an impulsive force on the ball

Fig. 11.7 The soccer ball undergoes a projectile motion

Assuming that the mass of the ball is $m = 0.5$ kg, the horizontal range of motion of the ball is $l = 40$ m, the maximum height the ball reaches in the air is $h = 4$ m, and the time at which the foot of the soccer player remains in contact with the ball is $\Delta t = 0.1$ s, determine the momentum of the ball at the instant of takeoff and the impulsive force applied by the player on the ball.

Solution

To determine the momentum of the ball at the instant of takeoff, we need to calculate the takeoff speed of the ball and its angle of takeoff first. For this purpose, we can utilize the formulas given in Sect. 7.10. The angle of takeoff is calculated (8.39) in terms of the maximum height and horizontal range of motion of the projectile motion in the following form:

$$\theta = \arctan\left(\frac{4h}{l}\right). \qquad (\text{v})$$

Substituting the numerical values of $h = 4$ m and $l = 40$ m into the above equation, and carrying out the calculations will yield

$$\theta = \arctan\left[\frac{4(4)}{40}\right] = 21.8°.$$

Using the equations in Sect. (7.10), the speed of takeoff can also be determined:

$$v = \frac{\sqrt{2gh}}{\sin\theta} = \frac{\sqrt{2(9.8)(4)}}{\sin 21.8°} = 23.8\,\text{m/s.} \qquad (\text{vi})$$

The velocity of takeoff $\underline{v}$ has components both in the x and y directions with magnitudes:

$$v_x = v\cos\theta = (23.8)(\cos 21.8°) = 22.1\,\text{m/s} \quad (\rightarrow),$$

$$v_y = v\sin\theta = (23.8)(\sin 21.8°) = 8.8\,\text{m/s} \quad (\uparrow).$$

Therefore, the takeoff velocity of the ball in vector form is

$$\underline{v} = 22.1\underline{i} + 8.8\underline{j} \quad (\text{m/s}).$$

The momentum of the soccer ball at the instant of takeoff (immediately after impact) can be determined as:

$$\underline{p} = m\underline{v} = 0.5(22.1\underline{i} + 8.8\underline{j}) = 11.1\underline{i} + 4.4\underline{j} \quad (\text{kg–m/s}).$$

The ball is stationary and its momentum is zero immediately before the soccer player kicks the ball. Therefore, the change in momentum of the ball during the course of contact with the player's foot is equal to its takeoff momentum. The impulse–momentum equation (11.7) can now be utilized to determine an average impulsive force exerted by the soccer player on the ball:

$$\underline{F} = \frac{\Delta\underline{p}}{\Delta t} = \frac{11.0\underline{i} + 4.4\underline{j}}{0.1} = 110\underline{i} + 44\underline{j}\,(\text{N}).$$

Therefore, the net or the resultant force applied on the ball by the player is

$$F = \sqrt{(110)^2 + (44)^2} = 118.5\,\text{N}.$$

Note that since action and reaction must have equal magnitudes, F is also the magnitude of the force applied by the ball on the player's foot.

Example 11.3 A force platform, as illustrated in Fig. 11.8, is a flat, rectangular, force-sensitive device that electronically records forces exerted against its upper surface. This device can be used to measure the impulsive forces involved during walking, running, jumping, and other activities.

Fig. 11.8 A force platform

Consider the force versus time recording shown in Fig. 11.9 for an athlete making vertical jumps on a force platform. The force scale is normalized with the weight of the athlete so that the force reading is zero when the athlete is stationary (standing still or crouching). The reason for normalizing the force measurement is to be able to disregard the effect of gravitational acceleration on the impulse calculations. In this case, a positive force means a force exerted on the platform due to factors other than the weight of the person. The force versus time graph has three distinct regions: An initial "takeoff push" during which the athlete exerts a positive force on the platform, an "airborne" region during which the athlete is not in contact with the platform, and a "landing" period in which the athlete again exerts impulsive forces on the platform. These regions are approximated with rectangular areas A_1 and A_2, and a triangular area A_3. Boundaries of these regions are shown with dashed lines in Fig. 11.9. The approximate force applied and the duration of takeoff push are about $F_1 = 600\,\text{N}$ and $\Delta t_1 = 0.3\,\text{s}$. The force reading is about $F_2 = -800\,\text{N}$ while the athlete is airborne, and the athlete remains in the air for about $\Delta t_2 = 0.4\,\text{s}$. This suggests that the weight of the athlete is about 800 N. Therefore, the

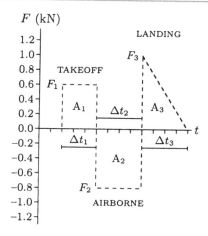

Fig. 11.9 Normalized force versus time plot for the athlete (1 kN = 1,000 N)

athlete has a mass of about $m = (800\,\text{N})/(9.8\,\text{m/s}^2) = 82\,\text{kg}$. The maximum impulsive force the athlete exerts on the platform during landing is about $F_3 = 1,000\,\text{N}$, which reduces to zero (in the normalized scale) in a time interval of about $\Delta t_3 = 0.4\,\text{s}$.

Determine an approximate takeoff velocity of the center of gravity of the athlete, calculate the height of jump, and determine the impulse and momentum of the athlete during landing by using the approximated areas under the force versus time curve.

Solution
Under the suggested assumptions, the analysis of the problem is quite straightforward. All we need to do is to utilize the impulse–momentum equation. The rectangular area designated by A_1 under the force versus time curve is equal to the impulse during takeoff:

$$A_1 = F_1 \Delta t_1 = (600)(0.3) = 180\,\text{N}-\text{s}.$$

The velocity of the athlete before takeoff is zero, and therefore, the impulse is converted into momentum at the instant of takeoff. If v_T is the speed of takeoff, then the impulse–momentum equation takes the following simplified form:

$$F_1 \Delta t_1 = \Delta p = m v_T.$$

Solving this equation for the takeoff speed will yield:

$$v_T = \frac{F_1 \Delta t_1}{m} = \frac{180}{82} = 2.2\,\text{m/s}.$$

To determine the height of the jump, we can utilize conservation of energy principle. At the instant of takeoff, the athlete has a kinetic energy of $m v_t^2/2$ and zero potential energy. As the athlete ascends, the kinetic energy is converted to potential energy. At the instant when the athlete reaches the maximum height h, the kinetic energy of the athlete becomes zero and the potential energy is mgh. Therefore,

$$\frac{1}{2}mv_t^2 = mgh.$$

Solving this equation for the maximum elevation of athlete's center of gravity will yield

$$h = \frac{v_2^2}{2g} = \frac{(2.2)^2}{2(9.8)} = 0.25\,\text{m}.$$

The impulse applied by the athlete on the force platform during landing is equal to the triangular area, A_3, under the force versus time curve:

$$A_3 = \frac{F_3 \Delta t_3}{2} = \frac{(1,000)(0.4)}{2} = 200\,\text{N}-\text{s}.$$

During the course of landing, the downward velocity of the athlete is reduced to zero and the impulse calculated during landing is essentially equal to the momentum, p_L, of the athlete at the instant of landing:

$$p_L = A_3 = 200\,\text{N}-\text{s}.$$

Notice that we can also calculate the landing speed of the athlete as:

$$v_L = \frac{p_L}{m} = \frac{200}{82} = 2.4\,\text{m/s}.$$

Example 11.4 A laboratory crash test is set up to measure the endurance of seat belts for automobile passengers (Fig. 11.10). The initial horizontal speed of the test vehicle carrying a 70-kg dummy is set to 100 km/h. The speed of the vehicle and the dummy is brought to zero in a time interval of 0.1 s.

Fig. 11.10 A crash test

Assuming that the frictional effects are negligibly small, determine an average horizontal force applied by the dummy on the seat belt.

Solution

In the absence of frictional forces, the dummy would have continued moving along the positive x direction (toward the right) if the seat belt were not constraining its motion. The seat belt is applying a force on the dummy in the negative x direction, which brings the dummy to rest in a time interval of $\Delta t = 0.1$ s. At the instant when the brakes are applied, the speed of the vehicle and the dummy is $v_1 = 100$ km/h or 27.8 m/s. Therefore, the momentum of the dummy at the same instant is

$$p_1 = mv_1 = (70)(27.8) = 1,946\,\text{kg}-\text{m/s}.$$

In vector form, $\underline{p}_1 = 1,946\underline{i}\,\text{kg}-\text{m/s}$. In a time interval of $\Delta t = 0.1$ s, the speed of the dummy is brought to zero. Therefore, the momentum of the dummy at the end of 0.1 s is zero:

$$\underline{p}_2 = 0.$$

According to the impulse–momentum theorem, the change in momentum of the dummy must be equal to the impulse applied by the seat belt on the dummy:

$$\underline{F}\Delta t = \underline{p}_2 - \underline{p}_1$$

or

$$\underline{F} = \frac{\underline{p}_1}{\Delta t} = \frac{1,946\underline{i}}{0.1} = -19,460\underline{i}\,(\text{N}).$$

Therefore, the seat belt applied a force of $F = 19,460$ N on the dummy in the negative x direction. Since action and reaction must have equal magnitudes, F is also the magnitude of the force applied by the dummy on the seat belt. If the objective was to design a proper seat belt, this result suggests that the seat belt material must be chosen to withstand a force of more than about 19.5 kN.

11.4 Conservation of Linear Momentum

The equation of motion (Eq. 11.4) relates the resultant of forces applied on an object and the time rate of change of momentum of the object. When the resultant force on an object is zero (i.e., when the object is in equilibrium), then the time rate of change of momentum is also zero, and the momentum of the object is constant. This condition is known as the *principle of conservation of linear momentum*. Conservation of momentum is particularly useful for impact and collision analyses.

Consider two objects that interact with each other. Assume that these objects are isolated from their surroundings so that there are no external forces present except for the forces they exert onto each other. In other words, the effects of external forces are negligibly small as compared to the forces they exert onto each other, which is particularly true in the case of impact and collision. Suppose that at some time t, the two objects have momenta $\underline{p}_1$ and $\underline{p}_2$, respectively. Let $\underline{F}_{12}$ be the force on object 1 applied by object 2, and $\underline{F}_{21}$ the force on object 2 applied by object 1. Using Eq. (11.4)

$$\underline{F}_{12} = \frac{\mathrm{d}\underline{p}_1}{\mathrm{d}t}, \quad \underline{F}_{21} = \frac{\mathrm{d}\underline{p}_2}{\mathrm{d}t}.$$

According to Newton's third law, the action and reaction must be equal in magnitude and opposite in direction ($\underline{F}_{12} = -\underline{F}_{21}$), or

$$\underline{F}_{12} + \underline{F}_{21} = \frac{\mathrm{d}\underline{p}_1}{\mathrm{d}t} + \frac{\mathrm{d}\underline{p}_2}{\mathrm{d}t} = \frac{\mathrm{d}}{\mathrm{d}t}(\underline{p}_1 + \underline{p}_2) = 0.$$

This condition of equilibrium is equivalent to

$$\underline{p}_1 + \underline{p}_2 = \text{constant}. \tag{11.9}$$

Equation (11.9) represents the principle of conservation of linear momentum for two interacting bodies that form an isolated system. It states that whenever two objects collide their total momentum will remain constant, regardless of the nature of the forces involved between the two.

11.5 Impact and Collisions

When two bodies collide, they deform to a certain extent because of the forces involved. We know that the amount of deformation depends on the objects' material properties, the extent and duration of applied forces, and other conditions such as temperature. In general, an object may undergo an *elastic* deformation, a *plastic* deformation, or both. The elastic deformation is recoverable upon release of the force causing the deformation, whereas plastic deformations are permanent.

There are several types of collisions, which may be distinguished either with respect to the orientations of the impact velocities of the two objects, or according to the nature of deformations they undergo. As illustrated in Fig. 11.11, two objects may collide "head-on" which is known as the *direct central impact*. In this case, the velocities of the objects are collinear with the line of impact.

The *line of impact* (AA) is a fictitious line passing through the mass centers of the colliding objects, and it is perpendicular to the *plane of contact* (BB) which is tangent to the contacting surfaces. Figure 11.12 illustrates the *oblique central impact* of two objects. In this case, the impact velocities of the objects are not collinear with the line of impact.

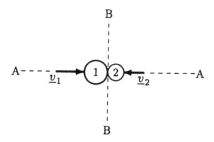

Fig. 11.11 Direct central impact

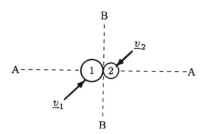

Fig. 11.12 Oblique central impact

Collisions can also be distinguished based on the nature of the deformations occurring during the course of a collision. An *elastic* or *perfectly elastic collision* is defined as a collision in which both the total momentum and total kinetic energy of the system (i.e., the two objects) are conserved. For an *inelastic* or *plastic collision*, on the other hand, only the total momentum of the system is conserved. During an inelastic collision, some of the kinetic energies of the colliding objects are dissipated as heat. The extreme case of inelastic collision is called the *perfectly inelastic collision*. This is a collision in which the objects stick together after the collision and move with a common velocity.

Whether a collision is an elastic or an inelastic one, the total momentum of the system is conserved. Therefore, the analyses of impact and collision problems are based on the conservation of momentum principle. Consider the collision of two objects with masses m_1 and m_2. Let $\underline{v}_1$ and $\underline{v}_2$ refer to the velocities of objects 1 and 2, respectively. Also let subscripts "i" and "f" denote the instants immediately before and after the collision. In the time interval between t_i and t_f, the principle of conservation of momentum Eq. (11.9) can be expressed as:

$$m_1\underline{v}_{1i} + m_2\underline{v}_{2i} = m_1\underline{v}_{1f} + m_2\underline{v}_{2f}. \tag{11.10}$$

Characteristics of different types of collisions will be provided by considering one-dimensional cases first. These concepts will then be expanded to analyze two-dimensional collisions by utilizing the vectorial properties of the parameters involved.

11.6   One-Dimensional Collisions

For a collision along a straight line (i.e., direct central impact), Eq. (11.10) can be simplified in the following scalar form:

$$m_1 v_{1i} + m_2 v_{2i} = m_1 v_{1f} + m_2 v_{2f}. \tag{11.11}$$

This equation is valid both for elastic and inelastic collisions.

11.6.1   Perfectly Inelastic Collision

A perfectly inelastic collision is one in which the objects stick together and move with a common velocity v_f after the collision (Fig. 11.13). To determine v_f, it is sufficient to consider the conservation of momentum principle during the collision. Substituting $v_{1f} = v_{2f} = v_f$ into Eq. (11.11), and solving it for v_f will yield

$$v_f = \frac{m_1 v_{1i} + m_2 v_{2i}}{m_1 + m_2}. \tag{11.12}$$

Before collision:

After collision:

Fig. 11.13 Perfectly inelastic collision

Example 11.5 Figure 11.14 illustrates a *ballistic pendulum* which may be a block of wood suspended by light wires. This simple device can be used to measure the velocity of a bullet. A bullet of mass m_1 fired at a stationary block of mass m_2 will penetrate the block, and the bullet–block system with mass $m_1 + m_2$ will swing to a height h.

Assuming that the bullet remains in the block, determine an expression for the initial speed v_{1i} of the bullet immediately before impact in terms of m_1, m_2, and h.

Fig. 11.14 Ballistic pendulum

Solution

This is a typical example of perfectly inelastic collision. The velocity of the block before impact is zero. Immediately after the collision, the block gains a kinetic energy. Designating the speed immediately after impact by v_f, then the bullet–block system has a kinetic energy of $(m_1 + m_2)v_f^2/2$ which is converted to a potential energy of $(m_1 + m_2)gh$ as the block swings to a height h. Applying the conservation of energy principle,

$$\frac{1}{2}(m_1 + m_2)v_f^2 = (m_1 + m_2)gh.$$

Solving this equation for the common speed v_f of the bullet and the block right after impact:

$$v_f = \sqrt{2gh}. \tag{i}$$

During the course of impact the momentum is conserved. Since the velocity of the block before impact is zero, Eq. (11.12) takes the following simpler form:

$$v_f = \frac{m_1 v_{1i}}{m_1 + m_2}.$$

Solving this equation for v_{1i} and using Eq. (*i*) will yield

$$v_{1i} = \left(\frac{m_1 + m_2}{m_1}\right)\sqrt{2gh}.$$

11.6.2   Perfectly Elastic Collision

A perfectly elastic collision is one in which the total kinetic energy is conserved as well as the total momentum of the objects involved (Fig. 11.15). The condition of conservation of total kinetic energy during the course of collision can be written as:

$$\frac{1}{2}m_1 v_{1i}^2 + \frac{1}{2}m_2 v_{2i}^2 = \frac{1}{2}m_1 v_{1f}^2 + \frac{1}{2}m_2 v_{2f}^2. \tag{11.13}$$

Before collision:

After collision:

Fig. 11.15 Perfectly elastic collision

In this case, Eqs. (11.11) and (11.13) must be solved simultaneously for the unknowns v_{1f} and v_{2f}. This will yield

$$v_{1f} = \left(\frac{m_1 - m_2}{m_1 + m_2}\right) v_{1i} + \left(\frac{2m_2}{m_1 + m_2}\right) v_{2i}, \qquad (11.14)$$

$$v_{2f} = \left(\frac{2m_1}{m_1 + m_2}\right) v_{1i} + \left(\frac{m_2 - m_1}{m_1 + m_2}\right) v_{2i}. \qquad (11.15)$$

At this point, it should be remembered that velocity is a vector quantity and that appropriate signs for all velocities involved must be included in Eqs. (11.12), (11.14), and (11.15). For this purpose, a positive direction of motion must be chosen. Velocities acting in the same direction must be considered to be positive, and those acting in the opposite direction must be taken to be negative.

Example 11.6 As illustrated in Fig. 11.16, consider a perfectly elastic collision of two billiard balls with equal masses m. Before the collision, ball 1 has a speed of v_{1i} and ball 2 is stationary.

Before collision:

After collision:

Fig. 11.16 Perfectly elastic collision of two identical billiard balls

Assuming a direct central impact, determine the velocities of the billiard balls immediately after the collision.

Solution
The fact that the masses of the balls are equal (i.e., $m_1 = m_2 = m$) and that one of the balls was stationary before the collision ($v_{2i} = 0$) simplifies the problem considerably. Since the masses are equal, $m_1 - m_2 = m_2 - m_1 = 0$ and $2m_1/(m_1 + m_2) = 1$. Therefore, the first term on the right-hand side of Eq. (11.14) and the second term on the

right-hand side of Eq. (11.15) are zero. The second term on the right-hand side of Eq. (11.14) is zero as well because $v_{2i} = 0$. Hence, choosing the right to be the positive direction of motion, Eqs. (11.14) and (11.15) yield

$$v_{1f} = 0,$$

$$v_{2f} = v_{1i} \quad (\rightarrow).$$

Notice that the kinetic energy and momentum of ball 1 immediately before the collision is totally transferred into kinetic energy and momentum for ball 2 during the collision.

11.6.3 Elasto-Plastic Collision

In reality, a material can undergo both elastic and plastic deformations. To facilitate the analyses of the impact characteristics of objects made up of such materials, a concept called the *coefficient of restitution* is developed which is defined as the ratio of the relative velocity of separation and the relative velocity of approach. If e represents the coefficient of restitution between two materials, then

$$e = \frac{\text{relative velocity of separation}}{\text{relative velocity of approach}}. \qquad (11.16)$$

The coefficient of restitution can take values between 0 and 1, such that $e = 0$ for perfectly inelastic impact and $e = 1$ for perfectly elastic impact. The coefficient of restitution is a positive number, and while calculating e the absolute values of the relative velocities of separation and approach must be considered. Consider the two objects shown in Fig. 11.17. Before the collision, object 1 is moving toward the right with velocity v_{1i} and object 2 is moving toward the left with velocity v_{2i}. Therefore, the relative velocity of approach is $v_{1i} + v_{2i}$. Assume that after the collision, the objects move in opposite directions with velocities v_{1i} and v_{2f}. Then the relative velocity of separation is $v_{1f} + v_{2f}$. If the objects were moving in the same direction before the collision, then the relative velocity of approach is

Velocities of approach:

Velocities of separation:

Fig. 11.17 The coefficient of restitution between two colliding objects is defined as the ratio of relative velocities of separation and approach

$v_{1i} - v_{2i}$. If they move in the same direction after the collision, then the relative velocity of separation is $v_{1f} - v_{2f}$. If one of the objects is stationary before and after the collision, then the coefficient of restitution can be determined by considering the ratio of the final and initial velocities of the object. In fact, this is the easiest to measure the coefficient of restitution between two materials.

The coefficient of restitution depends on the material properties of the objects involved in a collision. Temperature is also a factor that can influence the coefficient of restitution, because temperature can alter the mechanical properties of materials. For example, a ball will bounce better after being heated. If the ball is air-filled, such as a tennis ball, heat can also increase the internal pressure of the ball. A highly inflated ball will bounce better than a flat ball. When a ball is deformed during impact, some of the energy is dissipated as heat. The rise in temperature is important in games like tennis and squash where the ball is impacted at a relatively high rate. Another factor that may affect the coefficient of restitution between two materials is the relative velocity of approach. The higher the velocity of approach is, the lower the coefficient of restitution will be.

Example 11.7 Consider the ball analyzed in Example 11.1, which has a mass $m = 0.25$ kg. As illustrated in Fig. 11.18, the ball is dropped from a height $h_0 = 1$ m. After hitting the floor, the ball bounces and reaches a height $h_3 = 0.75$ m. Using the conservation of energy principle between 0 and 1, and between 2 and 3, we calculated that the ball has speeds $v_1 = 4.43$ m/s and $v_2 = 3.83$ m/s immediately before and after impact.

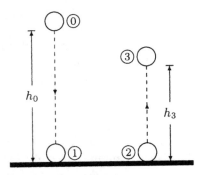

Fig. 11.18 The ball hits the floor and bounces

What is the coefficient of restitution between the ball and the floor, and how much energy is lost during the course of impact?

Solution
By definition, the coefficient of restitution is equal to the ratio of the velocities of separation and approach. In this case, the floor is stationary before and after the impact. Therefore, the speeds v_1 and v_2 of the ball before and after

impact are also the magnitudes of the relative velocities of approach and separation, respectively. Therefore, the coefficient of restitution between the ball and the floor is

$$e = \frac{v_2}{v_1}. \tag{i}$$

Substituting the numerical values and carrying out the calculations

$$e = \frac{3.83}{4.43} = 0.86.$$

The amount of energy lost (converted into heat, sound, or internal potential energy) during the course of impact can be determined by calculating the difference in the kinetic energies of the ball before and after the impact:

$$\mathcal{E}_{K2} - \mathcal{E}_{K1} = \frac{1}{2}m(v_2{}^2 - v_1{}^2) = -0.62J.$$

The coefficient of restitution (represented by Eq. (*i*) for the ball hitting a stationary surface and bouncing back) can alternatively be written in terms of the initial height h_0 from which the ball is dropped and the final height h_3 to which the ball bounced. In Example 11.1, using the conservation of energy principle, we determined that $v_1 = \sqrt{2gh_0}$ and $v_2 = \sqrt{2gh_3}$. Substituting these into Eq. (*i*) will yield

$$e = \sqrt{\frac{h_3}{h_0}} \tag{ii}$$

The US National Collegiate Athletic Association (NCAA) rule requiring a basketball dropped from a height $h_0 = 1.8$ m to bounce a height of about $h_3 = 1.0$ m is equivalent to requiring a coefficient of restitution of about $e = \sqrt{1.0/1.8} = 0.75$.

11.7 Two-Dimensional Collisions

As discussed in Sect. 11.4, if two interacting objects are isolated from their surroundings (if the effects of external forces are negligible as compared to the impulsive forces present), then the total momentum of the system is conserved. This principle of conservation of momentum is applied to analyze direct central impact, or one-dimensional collision, of some simple systems. By considering the vectorial properties of the parameters involved, the concepts developed for one-dimensional collisions can be expanded to analyze the oblique central impact or two-dimensional collision of two objects.

For a two-dimensional collision problem, the conservation of linear momentum principle must be applied in two coordinate directions. In addition, the nature of the collision must be known. For example, if the collision is perfectly elastic, then the total kinetic energy of the system is also conserved. If the collision is perfectly inelastic, then the velocities of the objects after the collision are equal.

The following example illustrates some of the concepts involved in two-dimensional collision problems.

Example 11.8 Figure 11.19 illustrates an instant during a pool game. What the pool player wishes to do is to hit the stationary target ball (ball 2) by the cue ball (ball 1) so as to move the target ball toward and into the corner pocket. Consider that the cue ball is given a velocity of $v_{1i} = 5$ m/s toward the target ball, and that a line connecting the center of mass and the geometric center of the corner pocket make an angle $\theta = 45°$, as shown in Fig. 11.20. The rectangular coordinates x and y are chosen in such a way that they respectively coincide with the line of impact (perpendicular to the contacting surfaces), and the plane of contact (tangential to the contacting surfaces). Assume that the cue ball hits the target ball at a point along the line of impact, and that the balls have equal mass.

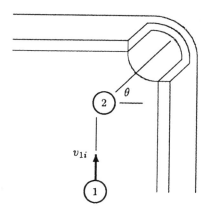

Fig. 11.19 Oblique central impact of two pool balls

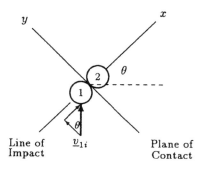

Fig. 11.20 Before collision

Neglecting the effects of friction, rotation, and gravity, determine the velocities of the cue and target balls immediately after collision if the coefficient of restitution between them is $e = 0.8$.

Solution This problem can be analyzed individually along the x and y directions first, and then by combining the results by utilizing the vectorial properties of velocity. Before impact, the target ball is stationary, and the cue ball has velocity components both in the x and y directions. Therefore,

$$(v_{1i})_x = v_{1i} \sin \theta = (5)(\sin 45°) = 3.54 \text{ m/s},$$

$$(v_{1i})_y = v_{1i} \cos \theta = (5)(\cos 45°) = 3.54 \text{ m/s},$$

$$(v_{2i})_x = 0,$$

$$(v_{2i})_y = 0.$$

Since the frictional effects are negligible, the motion along the y direction is equivalent to two balls moving along parallel lines without a chance of collision. The motion along the x direction, on the other hand, is equivalent to the direct central impact of the two balls.

(a) *Motion along the y direction:*

In the y direction, there is no collision. Therefore, the momentum of ball 1 in the y direction is conserved:

$$m(v_{1i})_y = m(v_{1f})_y,$$

$$(v_{1f})_y = (v_{1i})_y = 3.54 \text{ m/s}. \tag{i}$$

Utilizing the conservation of momentum principle for ball 2 in the y direction:

$$m(v_{2i})_y = m(v_{2f})_y,$$

$$(v_{2f})_y = (v_{2i})_y = 0. \tag{ii}$$

(b) *Motion along the x direction:*

In the x direction, the balls undergo a direct central impact, and therefore, the total momentum of the system is conserved. Assume that the positive direction of motion is along the positive x axis (i.e., toward the corner pocket), and that after collision, both balls move along the positive x direction. In other words, both $(v_{1f})_x$ and $(v_{2f})_x$ are positive. Since the balls have equal masses, the conservation of linear momentum along the x direction can be written as:

$$m(v_{1i})_x + m(v_{2i})_x = m(v_{1f})_x + m(v_{2f})_x.$$

Eliminating the masses, and since $(v_{2i})_x = 0$:

$$(v_{1f})_x + (v_{2f})_x = (v_{1i})_x. \qquad \text{(iii)}$$

The coefficient of restitution for the collision is given as $e = 0.8$. By definition, e is equal to the ratio of the relative velocities of separation and approach. Before impact, ball 2 is stationary. Hence, the relative velocity of approach in the x direction is equal to the initial speed of ball 1 in the x direction. Because of the assumed directions of motion after the collision, the relative velocity of separation in the x direction is equal to the difference of the velocity components of the balls along the x direction. Assuming that $(v_{2f})_x$ is larger than $(v_{1f})_x$:

$$e = \frac{(v_{2f})_x - (v_{1f})_x}{(v_{1i})_x}.$$

This equation can also be written as:

$$(v_{2f})_x - (v_{1f})_x = e(v_{1i})_x. \qquad \text{(iv)}$$

We have two equations, Eqs. (iii) and (iv), for two unknowns, $(v_{2f})_x$ and $(v_{1f})_x$. Solving these equations simultaneously will yield:

$$(v_{1f})_x = \frac{(e-1)}{2}(v_{1i})_x, \qquad \text{(v)}$$

$$(v_{2f})_x = \frac{(e+1)}{2}(v_{1i})_x. \qquad \text{(vi)}$$

Substituting $(v_{1f})_x = 3.54$ m/s and $e = 0.8$ into these equations will yield:

$$(v_{1f})_x = -0.35$$

$$(v_{2f})_x = 3.19.$$

Hence, the velocities of the balls after the collision are

$$\underline{v}_{1f} = -0.35\underline{i} + 3.54\underline{j} \quad \text{(m/s)},$$

$$\underline{v}_{2f} = 3.19\underline{i} \quad \text{(m/s)}.$$

Immediately after the collision, the target ball will move with a speed of 3.19 m/s toward the corner pocket (i.e., along the positive x direction). As illustrated in Fig. 11.21, the cue ball will move with a speed of $v_{1f} = \sqrt{(0.35)^2 + (3.54)^2} = 3.56$ m/s along a direction that

makes an angle arctan $(3.54/0.35) = 84°$ with the negative x direction, or at an angle $\beta = 84° - 45° = 39°$ with the horizontal.

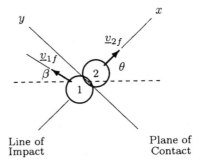

Fig. 11.21 After the collision

While utilizing the equations for the coefficient of restitution and conservation of momentum, we assumed that the velocity components of both balls would be in the positive x direction. As a result of our computations, we determined a negative value for $(v_{1f})_x$ which means that it is acting along the negative x direction.

If the collision were a perfectly elastic one (i.e., $e = 1$), then Eqs. (v) and (vi) would yield $(v_{1f})_x = 0$ and $(v_{1f})_y = (v_{1i})_x$. Therefore, the velocity vectors for the balls after the collision would be:

$$\underline{v}_{1f} = 3.54\underline{j} \quad \text{(m/s)},$$

$$\underline{v}_{2f} = 3.54\underline{i} \quad \text{(m/s)}.$$

Immediately after the collision, the target ball would move with a speed of 3.54 m/s along the positive x direction (toward the corner pocket) and the cue ball would move with the same speed along the positive y direction, as illustrated in Fig. 11.22. The balls would move at right angles to each other after the collision.

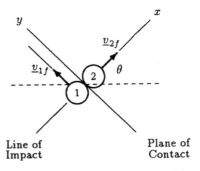

Fig. 11.22 Perfectly elastic collision of two pool balls

11.8   Angular Impulse and Momentum

The rotational analogue of linear momentum is called *angular momentum*. Angular momentum is defined as the product of the mass moment of inertia and the angular velocity of the body undergoing rotational motion and is commonly denoted with L:

$$L = I\omega. \tag{11.17}$$

The *impulse–momentum theorem for rotational motion* relates applied torque and change in angular momentum. If a torque with magnitude M is applied to a rotating body in the time interval between t_1 and t_2 so that the angular momentum of the body is changed from L_1 to L_2, then the impulse–momentum theorem for rotational motion states that:

$$\int_{t_1}^{t_2} M\,dt = L_2 - L_1. \tag{11.18}$$

The left-hand side of Eq. (11.18) is the *angular impulse*, and M is the *impulsive torque*. If the torque is constant, then the integral in Eq. (11.18) can be evaluated to yield:

$$M\Delta t = \Delta L = I\Delta\omega. \tag{11.19}$$

That is, a constant impulsive torque M applied on a body in the time interval $\Delta t = t_2 - t_1$ will change the angular velocity of the body from ω_1 to ω_2. Consequently, the angular momentum of the body will change from L_1 to L_2. The angular velocity and momentum of the body will increase if the torque is applied in the direction of motion.

Notice that rotational kinetic energy, angular work done, and power have the same dimensions and units as their linear counterparts. On the other hand, the dimension of angular momentum is $[M][L^2]/[T]$ and has the unit of $kg-m^2/s$ in SI.

11.9   Summary of Basic Equations

Table 11.2 provides a list of the basic equations necessary for rotational motion analyses about a fixed axis (circular motion) along with the equations for one-dimensional translational motion analyses. Note that the linear and angular quantities are analogous to each other in such pairs as x and θ, v and ω, a and α, m and I, F and M, and p and L.

Table 11.2   Equations of translational and rotational motion

Translational motion	Rotational motion (circular)
Velocity	
$v = \dfrac{dx}{dt}$	$\omega = \dfrac{d\theta}{dt}$
Acceleration	
$a = \dfrac{dv}{dt}$	$\alpha = \dfrac{d\omega}{dt}$
Kinematic relations for constant acceleration	
$x = x_0 + v_0 t + \frac{1}{2}a_0 t^2$	$\theta = \theta_0 + \omega_0 t + \frac{1}{2}\alpha_0 t^2$
$v = v_0 + a_0 t$	$\omega = \omega_0 + \alpha_0 t$
$v^2 = v_0^2 + 2a_0(x - x_0)$	$\omega^2 = \omega_0^2 + 2\alpha_0(x - x_0)$
Equation of motion	
$F = ma$	$M = I\alpha$
Work done	
$W = \int_{x_1}^{x_2} F\,dx$	$W = \int_{\theta_1}^{\theta_2} M\,d\theta$
Kinetic energy	
$\mathcal{E}_K = \frac{1}{2}mv^2$	$\mathcal{E}_K = \frac{1}{2}I\omega^2$
Work–energy	
$W = \frac{1}{2}m(v_2^2 - v_1^2)$	$W = \frac{1}{2}I(\omega_2^2 - \omega_1^2)$
Power	
$P = Fv$	$P = M\omega$
Momentum	
$p = mv$	$L = I\omega$
Impulse–momentum	
$\int_{t_1}^{t_2} F\,dt = p_2 - p_1$	$\int_{t_1}^{t_2} M\,dt = L_2 - L_1$

11.10   Kinetics of Rigid Bodies in Plane Motion

In most situations (e.g., when the effects due to air resistance are neglected), the size and shape of an object do not affect its translational motion characteristics. The size and shape (i.e., inertial effects) must be taken into consideration if the object is undergoing a rotational motion. Now that we have defined the basic concepts behind the rotational motion of rigid bodies, we can integrate our knowledge about translational and rotational motions to investigate their general motion characteristics.

Consider the rigid body illustrated in Fig. 11.23. Let m be the total mass of the body, and C be the location of its mass center. There are three coplanar forces acting on the body. Force F_3 is not producing any torque about C because its line of action is passing through C (i.e., its moment arm is zero). Forces $\underline{F}_1$ and $\underline{F}_2$ are producing clockwise moments about C with magnitudes $M_1 = d_1 F_1$ and $M_2 = d_2 F_2$, respectively. As illustrated in Fig. 11.24, the three-force system can be reduced to a one-force and one-moment system such that $\sum \underline{F} = \underline{F}_1 + \underline{F}_2 + \underline{F}_3$ is the net or the resultant force acting on the body and $\underline{M}_c = \underline{M}_1 + \underline{M}_2$ is the resultant of the couple-moments as measured about C. $\sum \underline{F}$ causes the body to translate and $\underline{M}_c$ causes it to rotate about C.

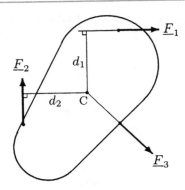

Fig. 11.23 A system of three forces acting on the body

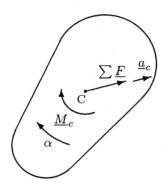

Fig. 11.24 The multiforce system can be reduced to a one-force and one-moment system

Recall that the translational motion of a body depends on its mass and the net force applied on it. Newton's second law of motion states that

$$\sum \underline{F} = m\underline{a}_c \qquad (11.20)$$

where $\underline{a}_c$ is the acceleration of the mass center of the body, and Eq. (11.20) accounts for its translational motion. The rotational motion of the body depends on its mass moment of inertia and the net torque applied on it:

$$\sum \underline{M}_c = I_c\underline{\alpha}. \qquad (11.21)$$

In Eq. (11.21), I_c is the mass moment of inertia of the body about the axis perpendicular to the plane of rotation and passing through the mass center of the body, and α is the angular acceleration. Notice that Eq. (11.20) is valid for any point within the body, but Eq. (11.21) is correct only about the mass center at C. For two-dimensional motion analyses in the xy plane, Eqs. (11.20) and (11.21) will yield three scalar equations:

$$\sum F_x = ma_{cx}, \qquad (11.22)$$

$$\sum F_y = ma_{cy}, \qquad (11.23)$$

$$\sum M_c = I_c\alpha. \qquad (11.24)$$

These three equations are the governing equations of motion for studying the two-dimensional general motion characteristics of bodies.

Notice that if $M_c = 0$ then the body is in pure translation, the body is in pure rotation when $F_x = 0$ and $F_y = 0$, and the body is said to be in equilibrium when $F_x = 0$, $F_y = 0$, and $M_c = 0$.

Introduction to Deformable Body Mechanics

12.1 Overview

Basic concepts of statics were introduced in Chaps. 4 and 5 along with some of their applications. The field of statics is based on Newton's laws (Newtonian mechanics). It constitutes one of the two main branches of the more general field of rigid body mechanics, dynamics being the other branch. The basic assumption in rigid body mechanics is that the bodies involved do not deform under applied loads. This idealization is necessary to simplify the problem under investigation for the sake of analyzing external forces and moments. The field of deformable body mechanics, on the other hand, does not treat the body as rigid, but incorporates the deformability (ability to undergo shape change) and the material properties of the body into the analyses. This field of applied mechanics utilizes the experimentally determined and/or verified relationships between applied forces and corresponding deformations.

Rigid body mechanics has its limitations. One of these limitations was discussed in Sect. 4.6 where the concept of statically indeterminant systems was introduced. A system for which the equations of equilibrium are not sufficient to determine the unknown forces is called *statically indeterminate*. For the analyses of such systems, there is a need for equations in addition to those provided by the conditions of static equilibrium. These additional equations can be derived by considering the material properties of the parts constituting a system and by relating forces to deformations, which is the focus of deformable body mechanics.

The desire to analyze statically determinate systems is only one of the reasons why deformable body mechanics is important. The applications of this field extend to almost all branches of engineering by providing essential design and analysis tools. The task of an engineer—mechanical, civil, electrical, or biomedical—is to determine the safest and most efficient operating condition for a machine, a structure, a piece of equipment, or a prosthetic device. A design engineer can accomplish this task by first assessing the proper operational environment through force analyses, making the correct structural design, and choosing the material that can sustain the forces involved in that environment. The primary concern of a design engineer is to make sure that when loaded, a machine part, a structure, a piece of equipment, or a device will not break or deform excessively.

12.2 Applied Forces and Deformations

Mechanics is concerned with forces and motions. It is possible to distinguish two types of motions. If the resultant of external forces or moments applied on a body is not zero, then the body will undergo gross overall motion (translation and/or rotation). In other words, the position of the body as a whole will change over time. Such movements are studied within the field of dynamics. The second type of motion involves local changes of shape within a body, called *deformations*, which are the primary concern of the field of deformable body mechanics. If a body is subjected to externally applied forces and moments but remains in static equilibrium, then it is most likely that there is some local shape change within the body. The extent of the shape change may depend upon the magnitude, direction, and duration of the applied forces, material properties of the body, and environmental conditions such as heat and humidity.

12.3 Internal Forces and Moments

Consider the arbitrarily shaped object illustrated in Fig. 12.1, which is subjected to a number of externally applied forces. Assume that the resultant of these forces and the net moment acting on the object are equal to zero. That is, the object is in static equilibrium. Also assume that the object is fictitiously separated into two parts by passing an arbitrary plane ABCD through the object. If the object as a whole is in equilibrium, then its individual parts must be in equilibrium as well.

N. Özkaya et al., *Fundamentals of Biomechanics: Equilibrium, Motion, and Deformation*, DOI 10.1007/978-1-4614-1150-5_12, © Springer Science+Business Media, LLC 2012

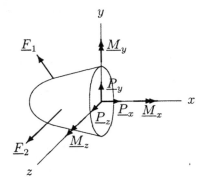

Fig. 12.1 An object subjected to externally applied forces

Fig. 12.3 Internal forces and moments

If one of these two parts is considered, then the equilibrium condition requires that there is a force vector and/or a moment vector acting on the cut section to counterbalance the effects of the external forces and moments applied on that part. These are called the *internal force* and *internal moment* vectors. Of course, the same argument is true for the other part of the object. Furthermore, for the overall equilibrium of the object, the force vectors and moment vectors on either surface of the cut section must have equal magnitudes and opposite directions (Fig. 12.2).

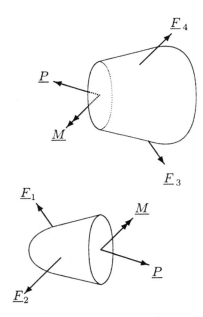

Fig. 12.2 Method of sections

For a three-dimensional object, the internal forces and moments can be resolved into their components along three mutually perpendicular directions, as illustrated in Fig. 12.3. The force and moment vector components measured at the cut sections take special names reflecting their orientation and effects on the cut sections. Assuming that x is the direction normal (perpendicular) to the cut section, the

force component P_x in Fig. 12.3 is called the *axial* or *normal force*, and it is a measure of the pulling or pushing action of the externally applied forces in a direction perpendicular to the cut section. It is called a *tensile force* if it has a pulling action trying to elongate the part, or a *compressive force* if it has a pushing action tending to shorten the part. The force components P_y and P_z are called *shear forces*, and they are measures of resistance to the sliding action of one cut section over the other. Their subscripts indicate their lines of action. The moment component M_x is also called *twisting torque*, and it is a measure of the twisting action of the externally applied forces along an axis normal to the plane of the cut section (in this case, in the x direction). The components M_y and M_z of the moment vector are called the *bending moments*, and they respectively indicate the extent of bending action to which the cut part is subjected in the y and z directions.

Note here that it may be more informative to refer to forces and moments with double subscripted symbols. For example, using P_{xy} instead of P_y would indicate that the force component is acting in the y direction (second subscript) on a section whose normal is in the x direction (first subscript). Similarly, M_{xz} would refer to the component of the moment vector in the z direction measured on the same section.

12.4 Stress and Strain

The purpose of studying the mechanics of deformable bodies or strength of materials is to make sure that the design of a structure is safe against the combined effects of applied forces and moments. The idea is to select the proper material for the structure, or if there is an existing structure, to determine the loading conditions under which the structure can operate safely and efficiently. To make a selection, however, one needs to know the mechanical properties of materials under different loading conditions.

Consider the two bars shown in Fig. 12.4, which are made of the same material, and have the same length but different sizes.

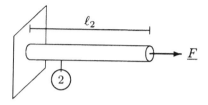

Fig. 12.4 Two bars made of the same material, have the same length, but different cross-sectional areas

Fig. 12.5 Two bars made of the same material, have the same cross-sectional area, but different lengths

The cross-sectional area A_1 of bar 1 is less than the cross-sectional area A_2 of bar 2. Assume that these bars are subjected to successively increasing forces until they break. If the forces F_1 and F_2 at which the bars 1 and 2 break were recorded, it would be observed that the force F_2 required to break bar 2 is greater than the force F_1 required to break bar 1 because bar 2 has a larger cross-sectional area and volume than bar 1. These forces might be an indication of the strength of the bars. However, the fact that the force-to-failure depends on the cross-sectional area of the specimen (in addition to some other factors) makes force an impractical measure of the strength of a material. To eliminate this inconvenience, a concept called *stress* is defined by dividing force with the cross-sectional area:

$$\text{Stress} = \frac{\text{Force}}{\text{Area}}.$$

Although the bars in Fig. 12.4 have different cross-sectional areas and require different forces-to-failure, since they are made of the same material, their stress measurement at failure would be equal.

As stated earlier, the mechanics of deformable bodies is concerned with applied forces and their internal effects on bodies. One of these effects is shape change or deformation. The amount of deformation an object will undergo depends on its size, material properties, and the magnitude and duration of applied forces. Consider the two bars shown in Fig. 12.5, which are made of the same material and have the same cross-sectional area, but have different lengths. The length l_1 of bar 1 is less than the length l_2 of bar 2. Assume that the same force F is applied to both bars, and the elongation of each bar is measured. It would be observed that the increase of length in bar 2 is greater than the increase of length in bar 1, indicating that the amount of elongation

depends on the original length of the specimen. To eliminate the size dependence of deformation measurements, another concept called *strain* is defined by dividing the amount of elongation with the original length of the specimen in the direction of elongation:

$$\text{Strain} = \frac{\text{Amount of elongation}}{\text{Original length}}.$$

Broad definitions of stress and strain are introduced here. More detailed descriptions of these concepts are provided in the following sections.

12.5 General Procedure

A general procedure for analyzing problems in deformable body mechanics, including the purpose of these analyses, is provided below.

- *Static analyses.* At this first stage, the analytical methods of statics are employed to determine the external reaction forces and moments. This stage involves drawing free-body diagrams and applying the conditions of equilibrium to determine the unknown reaction forces and moments by utilizing concepts such as equivalent force systems.

- *Analyses of internal forces and moments.* The internal forces and moments can be determined by the *method of sections.* As discussed briefly in Sect. 12.3, this can be done by separating the body into two sections at the location where the forces and moments need to be calculated. Here, the concern is to determine critical load conditions that correspond to maximum stress levels. These critical loads can be determined by drawing the shear and bending diagrams of the body, which essentially requires the application of the method of sections throughout the body.

- *Stress analyses.* This stage involves the conversion of internal forces and moments, in particular the critical forces and moments, into corresponding stresses by using formulas that also incorporate the material and geometric properties of the problem into the analyses.

- *Material selection.* Materials can be distinguished by their physical and mechanical properties. At this final stage of analysis, a material must be selected for the safe operation of the structure based on the maximum stresses calculated. If the material is already selected and the design is already made, then the maximum stresses calculated are used to set the allowable load conditions.

Note that the prerequisite for analyses in deformable body mechanics is statics. However, the procedure outlined above is not limited to analyzing systems in equilibrium. Under the effect of externally applied forces, a body may deform and undergo overall motion simultaneously. Such a problem can also be analyzed with the procedure outlined above by utilizing the *d'Alembert principle.* This principle is applied by treating the inertial effects due to the acceleration of the body as another external force acting at the center of gravity of the body in a direction opposite to the direction of acceleration.

12.6   Mathematics Involved

The analyses in this part of the text (Chaps. 12–15) use vector algebra and differential and integral calculus as computational tools. Therefore, the reader is advised to review Appendices A–C. Some of the analyses may also require familiarity with ordinary differential equations.

12.7   Topics to Be Covered

At the beginning of Chap. 13, detailed definitions of stress and strain are provided. Based on the stress–strain diagrams, material properties such as ductility, stiffness, and brittleness

are discussed. Elastic and plastic deformations, Hooke's law, the necking phenomenon, and the concepts of work and strain energy are explained. The analyses in Chap. 13 are limited to uniaxial deformations. The concepts introduced are applied to analyze relatively simple systems.

In Chap. 14, more advanced topics in stress–strain analyses are introduced. Two- and three-dimensional stress analyses, techniques of transforming stresses from one plane to another, methods for finding critical stresses, reasons why stress analyses are important in the design of structures, failure theories, concepts such as fatigue, endurance, and stress concentration are also discussed in Chap. 14. Also in Chap. 14, analyses of bodies subjected to torsion, bending, and combined loading are explained.

In Chap. 15, the viscoelastic behavior of materials and empirical models of viscoelasticity are reviewed, and elasticity and viscoelasticity are compared. Also in Chap. 15, the mechanical properties of biological tissues including bone, tendons, ligaments, muscles, and articular cartilage are discussed and the relevance of mechanical concepts introduced earlier to orthopedics are demonstrated.

Suggested Reading

The field of deformable body mechanics has been studied under various titles such as solid mechanics, mechanics of materials, and strength of materials. The subjects covered within deformable body mechanics form the basis for the study of more advanced topics in elasticity, inelasticity, and continuum mechanics. The following books can be reviewed to gain more detailed information on the principles of deformable body mechanics:

Beer, F.P., Johnston, R.E., DeWolf, J.T. and Mazurek, D. 2006. Mechanics of Materials. Boston: McGraw-Hill Higher Education.

Popov, E.P. 1999. Engineering Mechanics of Solids. Upper Saddle River, NJ: Prentice-Hall.

Shames, I.H. 1989. Introduction to Solid Mechanics. Englewood Cliffs, NJ: Prentice-Hall.

13

13.1 Basic Loading Configurations

An object subjected to an external force will move in the direction of the applied force. The object will deform if its motion is constrained in the direction of the applied force. Deformation implies relative displacement of any two points within the object. The extent of deformation will be dependent upon many factors including the magnitude, direction, and duration of the applied force, the material properties of the object, the geometry of the object, and environmental factors such as heat and humidity.

In general, materials respond differently to different loading configurations. For a given material, there may be different mechanical properties that must be considered while analyzing its response to, for example, tensile loading as compared to loading that may cause bending or torsion. Figure 13.1 is drawn to illustrate different loading conditions, in which an L-shaped beam is subjected to forces F_1, F_2, and F_3. The force F_1 subjects the arm AB of the beam to tensile loading. The force F_2 tends to bend the arm AB. The force F_3 has a bending effect on arm BC and a twisting (torsional) effect on arm AB. Furthermore, all of these forces are subjecting different sections of the beam to shear loading.

13.2 Uniaxial Tension Test

The mechanical properties of materials are established by subjecting them to various experiments. The mechanical response of materials under tensile loading is analyzed by the *uniaxial* or *simple tension test* that will be discussed next. The response of materials to forces that cause bending and torsion will be reviewed in the following chapter.

The experimental setup for the uniaxial tension test is illustrated in Fig. 13.2. It consists of one fixed and one moving head with attachments to grip the test specimen. A specimen is placed and firmly fixed in the equipment, a tensile force of known magnitude is applied through the moving head, and the corresponding elongation of the specimen is measured. A general understanding of the response of the material to tensile loading is obtained by repeating this test for a number of specimens made of the same material, but with different lengths, cross-sectional areas, and under tensile forces with different magnitudes.

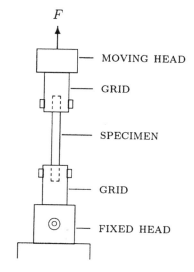

Fig. 13.2 Uniaxial tension test

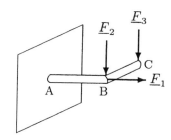

Fig. 13.1 Loading modes

N. Özkaya et al., *Fundamentals of Biomechanics: Equilibrium, Motion, and Deformation*, DOI 10.1007/978-1-4614-1150-5_13, © Springer Science+Business Media, LLC 2012

13.3   Load-Elongation Diagrams

Consider the three bars shown in Fig. 13.3. Assume that these bars are made of the same material. The first and the second bars have the same length but different cross-sectional areas, and the second and third bars have the same cross-sectional area but different lengths. Each of these bars can be subjected to a series of uniaxial tension tests by gradually increasing the applied forces and measuring corresponding increases in their lengths. If F is the magnitude of the applied force and Δl is the increase in length, then the data collected can be plotted to obtain a load versus elongation diagram for each specimen. Effects of geometric parameters (cross-sectional area and length) on the load-bearing ability of the material can be judged by drawing the curves obtained for each specimen on a single graph (Fig. 13.4) and comparing them. At a given force magnitude, the comparison of curves 1 and 2 indicate that the larger the cross-sectional area, the more difficult it is to deform the specimen in a simple tension test, and the comparison of curves 2 and 3 indicates that the longer the specimen, the larger the deformation in tension.

Note that instead of applying a series of tensile forces to a single specimen, it is preferable to have a number of specimens with almost identical geometries and apply one force to one specimen only once. As will be discussed later, a force applied on an object may alter its mechanical properties.

Another method of representing the results obtained in a uniaxial tension test is by first dividing the magnitude of the applied force F with the cross-sectional area A of the specimen, normalizing the amount of deformation by dividing the measured elongation with the original length of the specimen, and then plotting the data on a F/A versus $\Delta l/l$ graph as shown in Fig. 13.5. The three curves in Fig. 13.4, representing three specimens made of the same material, are represented by a single curve in Fig. 13.5. It is obvious that some of the information provided in Fig. 13.4 is lost in Fig. 13.5. That is why the representation in Fig. 13.5 is more advantageous than that in Fig. 13.4. The single curve in Fig. 13.5 is unique for a particular material, independent of the geometries of the specimens used during the experiments. This type of representation eliminates geometry as one of the variables, and makes it possible to focus attention on the mechanical properties of different materials. For example, consider the curves in Fig. 13.6, representing the mechanical

Fig. 13.3   Specimens

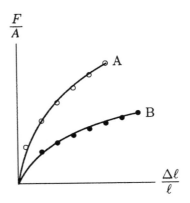

Fig. 13.5   Load over area versus elongation over length diagram

Fig. 13.4   Load-elongation diagrams

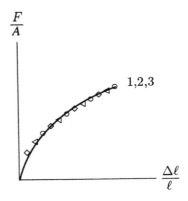

Fig. 13.6   Material A is stiffer than material B

behavior of materials A and B in simple tension. It is clear that material B can be deformed more easily than material A in a uniaxial tension test, or material A is "stiffer" than material B.

13.4   Simple Stress

Consider the cantilever beam shown in Fig. 13.7a. The beam has a circular cross-section, cross-sectional area A, welded to the wall at one end, and is subjected to a tensile force with magnitude F at the other end. The bar does not move, so it is in equilibrium. To analyze the forces induced within the beam, the method of sections can be applied by hypothetically cutting the beam into two pieces through a plane ABCD perpendicular to the centerline of the beam. Since the beam as a whole is in equilibrium, the two pieces must individually be in equilibrium as well. This requires the presence of an internal force collinear with the externally applied force at the cut section of each piece. To satisfy the condition of equilibrium, the internal forces must have the same magnitude as the external force (Fig. 13.7b). The internal force at the cut section represents the resultant of a force system distributed over the cross-sectional area of the beam (Fig. 13.7c). The intensity of the internal force over the cut section (force per unit area) is known as the *stress*. For the case shown in Fig. 13.7, since the force resultant at

the cut section is perpendicular (normal) to the plane of the cut, the corresponding stress is called a *normal stress*. It is customary to use the symbol σ (sigma) to refer to normal stresses. The intensity of this distributed force may or may not be uniform (constant) throughout the cut section. Assuming that the intensity of the distributed force at the cut section is uniform over the cross-sectional area A, the normal stress can be calculated using:

$$\sigma = \frac{F}{A}. \tag{13.1}$$

If the intensity of the stress distribution over the area is not uniform, then Eq. (13.1) will yield an *average normal stress*. It is customary to refer to normal stresses that are associated with tensile loading as *tensile stresses*. On the other hand, *compressive stresses* are those associated with compressive loading. It is also customary to treat tensile stresses as positive and compressive stresses as negative.

The other form of stress is called *shear stress*, which is a measure of the intensity of internal forces acting parallel or tangent to a plane of cut. To get a sense of shear stresses, hold a stack of paper with both hands such that one hand is under the stack while the other hand is above it. First, press the stack of papers together. Then, slowly slide one hand in the direction parallel to the surface of the papers while sliding the other hand in the opposite direction. This will slide individual papers relative to one another and generate frictional forces on the surfaces of individual papers. The shear stress is comparable to the intensity of the frictional force over the surface area upon which it is applied. Now, consider the cantilever beam illustrated in Fig. 13.8a. A downward force with magnitude F is applied to its free end. To analyze internal forces and moments, the method of sections can be applied by fictitiously cutting the beam through a plane ABCD that is perpendicular to the centerline of the beam. Since the beam as a whole is in equilibrium, the two pieces thus obtained must individually be equilibrium as well. The free-body diagram of the right-hand piece of the beam is illustrated in Fig. 13.8b along with the internal force and moment on the left-hand piece. For the equilibrium of the right-hand piece, there has to be an upward force resultant and an internal moment at the cut surface. Again for the equilibrium of this piece, the internal force must have a magnitude F. This is known as the *internal shearing force* and is the resultant of a distributed load over the cut surface (Fig. 13.8c). The intensity of the shearing force over the cut surface is known as the *shear stress*, and is commonly denoted with the symbol τ (tau). If the area of this surface (in this case, the cross-sectional area of the beam) is A, then:

$$\tau = \frac{F}{A}. \tag{13.2}$$

Fig. 13.7 Normal stress

Fig. 13.8 Shear stress

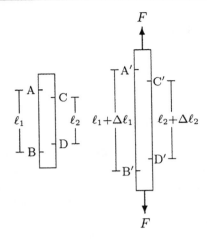

Fig. 13.9 Normal strain

The underlying assumption in Eq. (13.2) is that the shear stress is distributed uniformly over the area. For some cases, this assumption may not be true. In such cases, the shear stress calculated by Eq. (13.2) will represent an average value.

The dimension of stress can be determined by dividing the dimension of force $[F] = [M][L]/[T^2]$ with the dimension of area $[L^2]$. Therefore, stress has the dimension of $[M]/[L][T^2]$. The units of stress in different unit systems are listed in Table 13.1. Note that stress has the same dimension and units as pressure.

Table 13.1 Units of stress

System	Units of stress	Special name
SI	N/m^2	Pascal (Pa)
CGS	dyn/cm^2	
British	lb/ft^2 or lb/in.^{2}	psf or psi

13.5   Simple Strain

Strain, which is also known as unit deformation, is a measure of the degree or intensity of deformation. Consider the bar in Fig. 13.9. Let A and B be two points on the bar located at a distance l_1, and C and D be two other points located at a

distance l_2 from one another, such that $l_1 > l_2$. l_1 and l_2 are called *gage lengths*. The bar will elongate when it is subjected to tensile loading. Let Δl_1 be the amount of elongation measured between A and B, and Δl_2 be the increase in length between C and D. Δl_1 and Δl_2 are certainly some measures of deformation. However, they depend on the respective gage lengths, such that $\Delta l_1 > \Delta l_2$. On the other hand, if the ratio of the amount of elongation to the gage length is calculated for each case and compared, it would be observed that $(\Delta l_1 / \Delta l_1) \simeq (\Delta l_2 / \Delta l_2)$. Elongation per unit gage length is known as *strain* and is a more fundamental means of measuring deformation.

As in the case of stress, two types of strains can be distinguished. The *normal* or *axial strain* is associated with axial forces and defined as the ratio of the change (increase or decrease) in length, Δl, to the original gage length, l, and is denoted with the symbol ε (epsilon):

$$\varepsilon = \frac{\Delta l}{l}. \tag{13.3}$$

When a body is subjected to tension, its length increases, and both Δl and ε are positive. The length of a specimen under compression decreases, and both Δl and ε become negative.

The second form of strain is related to distortions caused by shearing forces. Consider the rectangle (ABCD) shown in Fig. 13.10, which is acted upon by a pair of shearing forces. Shear forces deform the rectangle into a parallelogram (AB'C'D). If the relative horizontal displacement of the top and the bottom of the rectangle is d and the height of the rectangle is l, then the *average shear strain* is defined as the ratio of d and l which is equal to the tangent of angle γ (gamma). This angle is usually very small. For small angles, the tangent of the angle is approximately equal to the angle itself. Hence, the average shear strain is equal

Fig. 13.10 Shear strain

to angle γ (measured in radians), which can be calculated using:

$$\gamma = \frac{d}{l}. \qquad (13.4)$$

Strains are calculated by dividing two quantities having the dimension of length. Therefore, they are dimensionless quantities and there is no unit associated with them. For most applications, the deformations and consequently the strains involved are very small, and the precision of the measurements taken is very important. To indicate the type of measurements taken, it is not unusual to attach units such as cm/cm or mm/mm next to a strain value. Strains can also be given in percent. In engineering applications, the strains involved are of the order of magnitude 0.1% or 0.001.

Figure 13.11 is drawn to compare the effects of tensile, compressive, and shear loading. Figure 13.11a shows a square object (in a two-dimensional sense) under no load.

(a)

(b)

(c)

(d)

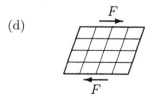

Fig. 13.11 Distortions under tensile (**b**), compressive (**c**), and shear (**d**) loading

The square object is ruled into 16 smaller squares to illustrate different modes of deformation. In Fig. 13.11b, the object is subjected to a pair of tensile forces. Tensile forces distort squares into rectangles such that the dimension of each square in the direction of applied force (axial dimension) increases while its dimension perpendicular to the direction of the applied force (transverse dimension) decreases. In Fig. 13.11c, the object is subjected to a pair of compressive forces that distort squares into rectangles such that the axial dimension of each square decreases while its transverse dimension increases. In Fig. 13.11d, the object is subjected to a pair of shear forces that distort the squares into diamonds.

13.6 Stress–Strain Diagrams

It was demonstrated in Sect. 13.3 that the results of uniaxial tension tests can be used to obtain a unique curve representing the relationship between the applied load and corresponding deformation for a material. This can be achieved by dividing the applied load with the cross-sectional area (F/A) of the specimen, dividing the amount of elongation measured with the gage length ($\Delta l/l$), and plotting a F/A versus $\Delta l/l$ graph. Notice, however, that for a specimen under tension, F/A is the average tensile stress σ and $\Delta l/l$ is the average tensile strain ε. Therefore, the F/A versus $\Delta l/l$ graph of a material is essentially the *stress–strain diagram* of that material.

Different materials demonstrate different stress–strain relationships, and the stress–strain diagrams of two or more materials can be compared to determine which material is relatively stiffer, harder, tougher, more ductile, and/or more brittle. Before explaining these concepts related to the strength of materials, it is appropriate to first analyze a typical stress–strain diagram in detail.

Consider the stress–strain diagram shown in Fig. 13.12. There are six distinct points on the curve that are labeled as O, P, E, Y, U, and R. Point O is the origin of the $\sigma - \varepsilon$ diagram, which corresponds to the initial no load, no deformation stage. Point P represents the *proportionality limit*. Between O and P, stress and strain are linearly proportional, and the $\sigma - \varepsilon$ curve is a straight line. Point E represents the *elastic limit*. The stress corresponding to the elastic limit is the greatest stress that can be applied to the material without causing any permanent deformation within the material. The material will not resume its original size and shape upon unloading if it is subjected to stress levels beyond the elastic limit. Point Y is the *yield point*, and the stress σ_y corresponding to the yield point is called the *yield strength* of the material. At this stress level, considerable elongation (yielding) can occur without a corresponding increase of load. U is the highest stress point on the $\sigma - \varepsilon$ curve. The stress σ_u is the *ultimate strength* of the

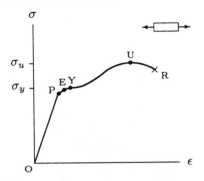

Fig. 13.12 Stress–strain diagram for axial loading

material. For some materials, once the ultimate strength is reached, the applied load can be decreased and continued yielding may be observed. This is due to the phenomena called *necking* that will be discussed later. The last point on the $\sigma - \varepsilon$ curve is R, which represents the *rupture* or *failure point*. The stress at which the rupture occurs is called the *rupture strength* of the material.

For some materials, it may not be easy to determine or distinguish the elastic limit and the yield point. The yield strength of such materials is determined by the *offset method*, illustrated in Fig. 13.13. The offset method is applied by drawing a line parallel to the linear section of the stress–strain diagram and passing through a strain level of about 0.2% (0.002). The intersection of this line with the $\sigma - \varepsilon$ curve is taken to be the yield point, and the stress corresponding to this point is called the *apparent yield strength* of the material.

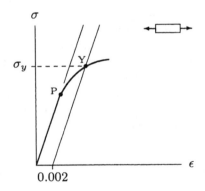

Fig. 13.13 Offset method

Note that a given material may behave differently under different load and environmental conditions. If the curve shown in Fig. 13.12 represents the stress–strain relationship for a material under tensile loading, there may be a similar but different curve representing the stress–strain relationship for the same material under compressive or shear loading. Also, temperature is known to alter the relationship between stress and strain. For a given material and fixed mode of

loading, different stress–strain diagrams may be obtained under different temperatures. Furthermore, the data collected in a particular tension test may depend on the rate at which the tension is applied on the specimen. Some of these factors affecting the relationship between stress and strain will be discussed later.

13.7 Elastic Deformations

Consider the partial stress–strain diagram shown in Fig. 13.14. Y is the yield point, and in this case, it also represents the proportionality and elastic limits. σ_y is the yield strength and ε_y is the corresponding strain. (The $\sigma - \varepsilon$ curve beyond the elastic limit is not shown.) The straight line in Fig. 13.14 represents the stress–strain relationship in the elastic region. *Elasticity* is defined as the ability of a material to resume its original (stress free) size and shape upon removal of applied loads. In other words, if a load is applied on a material such that the stress generated in the material is equal to or less than σ_y, then the deformations that took place in the material will be completely recovered once the applied loads are removed (the material is unloaded).

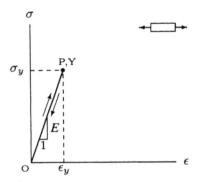

Fig. 13.14 Stress–strain diagram for a linearly elastic material. (↗: loading, ↙: unloading)

An elastic material whose $\sigma - \varepsilon$ diagram is a straight line is called a *linearly elastic* material. For such a material, the stress is linearly proportional to strain, and the constant of proportionality is called the *elastic* or *Young's modulus* of the material. Denoting the elastic modulus with E:

$$\sigma = E\varepsilon. \tag{13.5}$$

The elastic modulus, E, is equal to the slope of the $\sigma - \varepsilon$ diagram in the elastic region, which is constant for a linearly elastic material. E represents the *stiffness* of a material, such that the higher the elastic modulus, the stiffer the material.

The distinguishing factor in linearly elastic materials is their elastic moduli. That is, different linearly elastic materials have different elastic moduli. If the elastic modulus of a material is known, then the mathematical definitions of stress and strain ($\sigma = F/A$ and $\varepsilon = \Delta l/l$) can be substituted into Eq. (13.5) to derive a relationship between the applied load and corresponding deformation:

$$\Delta l = \frac{F\,l}{E\,A}. \tag{13.6}$$

In Eq. (13.6), F is the magnitude of the tensile or compressive force applied to the material, E is the elastic modulus of the material, A is the area of the surface that cuts the line of action of the applied force at right angles, l is the length of the material measured along the line of action of the applied force, and Δl is the amount of elongation or shortening in l due to the applied force. For a given linearly elastic material (or any material for which the deformations are within the linearly elastic region of the $\sigma - \varepsilon$ diagram) and applied load, Eq. (13.6) can be used to calculate the corresponding deformation. This equation can be used when the object is under a tensile or compressive force.

Not all elastic materials demonstrate linear behavior. As illustrated in Fig. 13.15, the stress–strain diagram of a material in the elastic region may be a straight line up to the proportionality limit followed by a curve. A curve implies varying slope and nonlinear behavior. Materials for which the $\sigma - \varepsilon$ curve in the elastic region is not a straight line are known as *nonlinear elastic materials*. For a nonlinear elastic material, there is not a single elastic modulus because the slope of the $\sigma - \varepsilon$ curve is not constant throughout the elastic region. Therefore, the stress–strain relationships for nonlinear materials take more complex forms. Note, however, that even nonlinear materials may have a linear elastic region in their $\sigma - \varepsilon$ diagrams at low stress levels (the region between points O and P in Fig. 13.15).

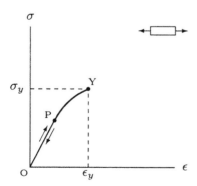

Fig. 13.15 Stress–strain diagram for a nonlinearly elastic material

Some materials may exhibit linearly elastic behavior when they are subjected to shear loading (Fig. 13.16). For such materials, the shear stress, τ, is linearly proportional to the shear strain, γ, and the constant of proportionality is

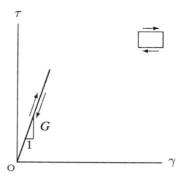

Fig. 13.16 Shear stress versus shear strain diagram for a linearly elastic material

called the *shear modulus* or the *modulus of rigidity*, which is commonly denoted with the symbol G:

$$\tau = G\,\gamma. \tag{13.7}$$

The shear modulus of a given linear material is equal to the slope of the $\tau - \gamma$ curve in the elastic region. The higher the shear modulus, the more rigid the material.

Note that Eqs. (13.5) and (13.6) relate stresses to strains for linearly elastic materials and are called *material functions*. Obviously, for a given material, there may exist different material functions for different modes of deformation. There are also *constitutive equations* that incorporate all material functions.

13.8 Hooke's Law

The load-bearing characteristics of elastic materials are similar to those of springs, which was first noted by Robert Hooke. Like springs, elastic materials have the ability to store potential energy when they are subjected to externally applied loads. During unloading, it is the release of this energy that causes the material to resume its undeformed configuration. A linear spring subjected to a tensile load will elongate, the amount of elongation being linearly proportional to the applied load (Fig. 13.17). The constant of proportionality between the load and the deformation is usually denoted with the symbol k, which is called the *spring constant* or *stiffness* of the spring. For a linear spring with a spring constant k, the relationship between the applied load F and the amount of elongation d is:

$$F = kd. \tag{13.8}$$

By comparing Eqs. (13.5) and (13.8), it can be observed that stress in an elastic material is analogous to the force applied to a spring, strain in an elastic material is analogous to the amount of deformation of a spring, and the elastic modulus of an elastic material is analogous to the spring constant of a spring. This analogy between elastic materials and springs is known as *Hooke's law*.

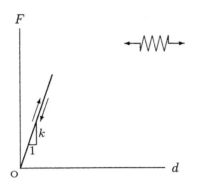

Fig. 13.17 Load-elongation diagram for a linear spring

13.9 Plastic Deformations

We have defined elasticity as the ability of a material to regain completely its original dimensions upon removal of the applied forces. Elastic behavior implies the absence of permanent deformation. On the other hand, *plasticity* implies permanent deformation. In general, materials undergo plastic deformations following elastic deformations when they are loaded beyond their elastic limits or yield points.

Consider the stress–strain diagram of a material shown in Fig. 13.18. Assume that a specimen made of the same material is subjected to a tensile load and the stress, σ, in the specimen is brought to such a level that $\sigma > \sigma_y$. The corresponding strain in the specimen is measured as ε. Upon removal of the applied load, the material will recover the plastic deformation that had taken place by following an unloading path parallel to the initial linearly elastic region (straight line between points O and P). The point where this

path cuts the strain axis is called the plastic strain, ε_p, that signifies the extent of permanent (unrecoverable) shape change that has taken place in the specimen.

The difference in strains between when the specimen is loaded and unloaded ($\varepsilon - \varepsilon_p$) is equal to the amount of elastic strain, ε_e, that had taken place in the specimen and that was recovered upon unloading. Therefore, for a material loaded to a stress level beyond its elastic limit, the total strain is equal to the sum of the elastic and plastic strains:

$$\varepsilon = \varepsilon_e + \varepsilon_p. \tag{13.9}$$

The elastic strain, ε_e, is completely recoverable upon unloading, whereas the plastic strain, ε_p, is a permanent residue of the deformations.

13.10 Necking

As defined in Sect. 13.6, the largest stress a material can endure is called the ultimate strength of that material. Once a material is subjected to a stress level equal to its ultimate strength, an increased rate of deformation can be observed, and in most cases, continued yielding can occur even by reducing the applied load. The material will eventually fail to hold any load and rupture. The stress at failure is called the rupture strength of the material, which may be lower than its ultimate strength. Although this may seem to be unrealistic, the reason is due to a phenomenon called *necking* and because of the manner in which stresses are calculated.

Stresses are usually calculated on the basis of the original cross-sectional area of the material. Such stresses are called *conventional stresses*. Calculating a stress by dividing the applied force with the original cross-sectional area is convenient but not necessarily accurate. The *true* or *actual stress* calculations must be made by taking the cross-sectional area of the deformed material into consideration. As illustrated in Fig. 13.19, under a tensile load, a material may elongate in the direction of the applied load but contract in the transverse directions. At stress levels close to the breaking point, the elongation may occur very rapidly and the material may narrow simultaneously. The cross-sectional area at the narrowed section decreases, and although the force required to further deform the material may decrease, the force per

Fig. 13.18 Plastic deformation

Fig. 13.19 Necking

unit area (stress) may increase. As illustrated with the dotted curve in Fig. 13.20, the actual stress–strain curve may continue having a positive slope, which indicates increasing strain with increasing stress rather than a negative slope, which implies increasing strain with decreasing stress. Also the rupture point and the point corresponding to the ultimate strength of the material may be the same.

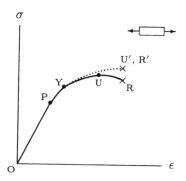

Fig. 13.20 Conventional (*solid curve*) and actual (*dotted curve*) stress–strain diagrams

13.11 Work and Strain Energy

In dynamics, *work done* is defined as the product of force and the distance traveled in the direction of applied force, and *energy* is the capacity of a system to do work. Stress and strain in deformable body mechanics are, respectively, related to force and displacement. Stress multiplied by area is equal to force, and strain multiplied by length is displacement. Therefore, the product of stress and strain is equal to the work done on a body per unit volume of that body, or the *internal work* done on the body by the externally applied forces. For an elastic body, this work is stored as an internal *elastic strain energy*, and it is the release of this energy that brings the body back to its original shape upon unloading. The maximum elastic strain energy (per unit volume) that can be stored in a body is equal to the total area under the $\sigma - \varepsilon$ diagram in the elastic region (Fig. 13.21). There is also a *plastic strain energy* that is dissipated as heat while deforming the body.

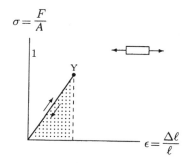

$$\sigma = \frac{F}{A}$$

$$\varepsilon = \frac{\Delta \ell}{\ell}$$

Fig. 13.21 Internal work done and elastic strain energy per unit volume

13.12 Strain Hardening

Figure 13.22 represents the $\sigma - \varepsilon$ diagram of a material. Assume that the material is subjected to a tensile force such that the stress generated is beyond the elastic limit (yield point) of the material. The stress level in the material is indicated with point A on the $\sigma - \varepsilon$ diagram. Upon removal of the applied force, the material will follow the path AB which is almost parallel to the initial, linear section OP of the $\sigma - \varepsilon$ diagram. The strain at B corresponds to the amount of plastic deformation in the material. If the material is reloaded, it will exhibit elastic behavior between B and A, the stress at A being the new yield strength of the material. This technique of changing the yield point of a material is called *strain hardening*. Since the stress at A is greater than the original yield strength of the material, strain hardening increases the yield strength of the material. Upon reloading, if the material is stressed beyond A, then the material will deform according to the original $\sigma - \varepsilon$ curve for the material.

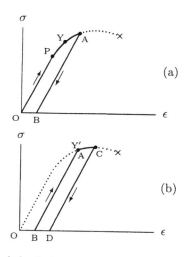

Fig. 13.22 Strain hardening

13.13 Hysteresis Loop

Consider the $\sigma - \varepsilon$ diagram shown in Fig. 13.23. Between points O and A, a tensile force is applied on the material and the material is deformed beyond its elastic limit. At A, the tensile force is removed, and the line AB represents the unloading path. At B, the material is reloaded, this time with a compressive force. At C, the compressive force applied on the material is removed. Between C and O, a second unloading occurs, and finally the material resumes its original shape. The loop OABCO is called the *hysteresis loop*, and the area enclosed by this loop is equal to the total strain energy dissipated as heat to deform the body in tension and compression.

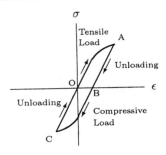

Fig. 13.23 Hysteresis loop

13.14 Properties Based on Stress–Strain Diagrams

As defined earlier, the elastic modulus of a material is equal to the slope of its stress–strain diagram in the elastic region. The elastic modulus is a relative measure of the *stiffness* of one material with respect to another. The higher the elastic modulus, the stiffer the material and the higher the resistance to deformation. For example, material 1 in Fig. 13.24 is stiffer than material 2.

Fig. 13.24 Material 1 is stiffer than material 2

A *ductile* material is one that exhibits a large plastic deformation prior to failure. For example, material 1 in Fig. 13.25 is more ductile than material 2. A *brittle* material, on the other hand, shows a sudden failure (rupture) without undergoing a considerable plastic deformation. Glass is a typical example of a brittle material.

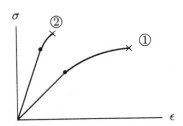

Fig. 13.25 Material 1 is more ductile and less brittle than 2

Toughness is a measure of the capacity of a material to sustain permanent deformation. The toughness of a material is measured by considering the total area under its stress–strain diagram. The larger this area, the tougher the material. For example, material 1 in Fig. 13.26 is tougher than material 2.

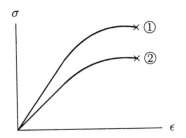

Fig. 13.26 Material 1 is tougher than material 2

The ability of a material to store or absorb energy without permanent deformation is called the *resilience* of the material. The resilience of a material is measured by its *modulus of resilience* which is equal to the area under the stress–strain curve in the elastic region. The modulus of resilience is equal to $\sigma_y \varepsilon_y / 2$ or $\sigma_y^2 / 2E$ for linearly elastic materials.

Although they are not directly related to the stress–strain diagrams, there are other important concepts used to describe material properties. A material is called *homogeneous* if its properties do not vary from location to location within the material. A material is called *isotropic* if its properties are independent of direction or orientation. A material is called *incompressible* if it has a constant density.

13.15 Idealized Models of Material Behavior

Stress–strain diagrams are most useful when they are represented by mathematical functions. The stress–strain diagrams of materials may come in various forms, and it may not be possible to find a single mathematical function to represent them. For the sake of mathematical modeling and the analytical treatment of material behavior, these diagrams can be simplified. Some of these diagrams representing certain idealized material behavior are illustrated in Fig. 13.27.

A *rigid* material is one that cannot be deformed even under very large loads (Fig. 13.27a). A *linearly elastic* material is one for which the stress and strain are linearly proportional, with the modulus of elasticity being the constant of proportionality (Fig. 13.27b). A *rigid-perfectly plastic* material does not exhibit any elastic behavior, and once a critical stress level is reached, it will deform continuously and permanently until failure (Fig. 13.27c). After a linearly elastic response, a *linearly elastic-perfectly plastic* material is one that deforms continuously at a constant stress level (Fig. 13.27d). Figure 13.27e represents the stress–strain diagram for *rigid-linearly plastic* behavior.

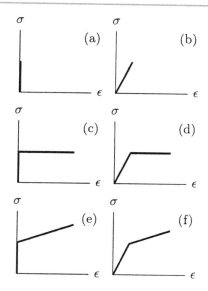

Fig. 13.27 Idealized models of material behavior

The stress–strain diagram of a *linearly elastic-linearly plastic* material has two distinct regions with two different slopes (bilinear), in which stresses and strains are linearly proportional (Fig. 13.27f).

13.16 Mechanical Properties of Materials

Table 13.2 lists properties of selected materials in terms of their tensile yield strengths (σ_y), tensile ultimate strengths (σ_u), elastic moduli (E), shear moduli (G), and Poisson's ratios (v). The significance of the Poisson's ratio will be discussed in the following chapter. Note that mechanical properties of a material can vary depending on many factors including its content (e.g., in the case of steel, its carbon content) and the processes used to manufacture the material (e.g., hard rolled, strain hardened, or heat treated). As will be discussed in later chapters, biological materials exhibit time-dependent properties. That is, their response to external forces depends on the rate at which the forces are applied. Bone, for example, is an anisotropic material. Its response to tensile loading in different directions is different, and

different elastic moduli are established to account for its response in different directions. Figure 13.28 illustrates the tensile stress-strain diagrams of a few materials. Steel and aluminum are relatively stiff and have high ultimate strengths. Glass and dry bone are brittle. Wet bone has the lowest ultimate strength and elastic modulus. Therefore, the values listed in Table 13.2 are some averages and ranges, and are aimed to provide a sense of the orders of magnitude of numbers involved.

Fig. 13.28 Gross comparison of stress–strain diagrams of selected materials

13.17 Example Problems

The following examples will demonstrate some of the uses of the concepts introduced in this chapter.

Example 13.1 A circular cylindrical rod with radius $r = 1.26$ cm is tested in a uniaxial tension test (Fig. 13.29). Before applying a tensile force of $F = 1{,}000$ N, two points A and B that are at a distance $l_0 = 30$ cm (gage length) are marked on the rod. After the force is applied, the distance between A and B is measured as $l_1 = 31.5$ cm.

Determine the tensile strain and average tensile stress generated in the rod.

Table 13.2 Average mechanical properties of selected materials

Material	Yield strength σ_y (MPa)	Ultimate strength σ_u (MPa)	Elastic modulus E (GPa)	Shear modulus G (MPa)	Poisson's ratio v
Muscle	–	0.2	–	–	0.49
Tendon	–	70	0.4	–	0.40
Skin	–	8	0.5	–	0.49
Cortical bone	80	130	17	3.3	0.40
Glass	35–70	–	70–80	–	–
Cast iron	40–260	140–380	100–190	42–90	0.29
Aluminum	60–220	90–390	70	28	0.33
Steel	200–700	400–850	200	80	0.30
Titanium	400–800	500–900	100	45	0.34

Fig. 13.29 Example 13.1

Solution

By definition, the tensile strain is equal to the ratio of the amount of elongation to the original length. The amount of elongation, Δl, is the difference between the gage lengths before and after deformation:

$$\Delta l = l_1 - l_0 = 31.5 - 30.0 = 1.5 \text{ cm}.$$

Therefore, the tensile strain ε generated in the rod is:

$$\varepsilon = \frac{\Delta l}{l_0} = \frac{1.5}{30} = 0.05 \text{ cm/cm}.$$

The bar has a circular cross-section with radius $r = 1.26$ cm or $r = 0.0126$ m. The cross-sectional area A of the bar is:

$$A = \pi r^2 = (3.1416)(0.0126)^2 = 5 \times 10^{-4} \text{m}^2$$

The average tensile stress is equal to the applied force per unit area of the surface that cuts the line of action of the force at right angles. In this case, it is the cross-sectional area of the rod. Therefore:

$$\sigma = \frac{F}{A} = \frac{1,000}{0.0005} = 2,000,000 = 2 \times 10^6 \text{ } Pa = 2 \text{ MPa}.$$

Example 13.2 Two specimens made of two different materials are tested in a uniaxial tension test by applying a force of $F = 20$ kN (20×10^3 N) on each specimen (Fig. 13.30). Specimen 1 is an aluminum bar with elastic modulus E1 = 70 GPa (70×10^9 Pa) and a rectangular cross-section (1 cm × 2 cm). Specimen 2 is a steel rod

SPECIMEN 1: ALUMINUM BAR

SPECIMEN 2: STEEL ROD

Fig. 13.30 Example 13.2

with elastic modulus $E_2 = 200$ GPa (200×10^9 Pa) and a circular cross-section (radius 1 cm).

Calculate tensile stresses developed in each specimen. Assuming that the tensile stress in each specimen is below the proportionality limit of the material, calculate the tensile strain for each specimen. Also, if the original length of each specimen was 30 cm, what are their lengths after deformation?

Solution

The tensile stress is equal to the ratio of the applied force and the cross-sectional area of the specimen. The cross-sectional areas of the specimens are:

$$A_1 = (1 \text{ cm})(2 \text{ cm}) = 2 \text{ cm}^2 = 2 \times 10^{-4} \text{m}^2,$$
$$A_2 = \pi(1 \text{ cm})^2 = 3.14 \text{ cm}^2 = 3.14 \times 10^{-4} \text{m}^2.$$

Therefore, the tensile stresses developed in each specimen are:

$$\sigma_1 = \frac{F}{A_1} = \frac{20 \times 10^3}{2 \times 10^{-4}} = 100 \times 10^6 \text{ } Pa = 100 MPa,$$

$$\sigma_2 = \frac{F}{A_2} = \frac{20 \times 10^3}{3.14 \times 10^{-4}} = 63.7 \times 10^6 \text{ } Pa = 63.7 MPa.$$

That is, the aluminum bar (specimen 1) is stressed more than the steel rod (specimen 2).

To calculate the tensile strains corresponding to tensile stresses σ_1 and σ_2, we can assume that the deformations are elastic and that the stresses σ_1 and σ_2 are below the proportionality limits for aluminum and steel. In other words, the stresses are linearly proportional to strains and the elastic moduli E_1 and E_2 are the constants of proportionality. Hence:

$$\varepsilon_1 = \frac{\sigma_1}{E_1} = \frac{100 \times 10^6}{70 \times 10^9} = 1.43 \times 10^{-3},$$

$$\varepsilon_2 = \frac{\sigma_2}{E_2} = \frac{63.7 \times 10^6}{200 \times 10^9} = 0.32 \times 10^{-3}.$$

These results suggest that the aluminum bar is stretched more than the steel rod.

The original length of each specimen was $l_0 = 30$ cm. After deformation, assume that the aluminum bar is elongated by Δl_1 to length l_1, and the steel rod is elongated by Δl_2 to length l_2. By definition, tensile strain is $\varepsilon = \Delta l / l_0$, or the amount of elongation is $\Delta l = \varepsilon l_0$. On the other hand, the length of the specimen after deformation is equal to the original length plus the amount of elongation. For example, for the aluminum bar, $l_1 = l_0 + \Delta l$, or $l_1 = l_0(1 + \varepsilon_1)$. Therefore:

$$l_1 = l_0(1 + \varepsilon_1) = 30(1 + 0.00143) = 30.0429 \text{ cm},$$
$$l_2 = l_0(1 + \varepsilon_2) = 30(1 + 0.00320) = 30.0960 \text{ cm}.$$
$$0.00032 = 30.01 cm$$

In other words, the increase in length of the aluminum bar and the steel rod is less than 1 mm.

Example 13.3 An experiment was designed to determine the elastic modulus of the human bone (cortical) tissue. Three almost identical bone specimens were prepared. The specimen size and shape used are shown in Fig. 13.31, which has a square (2×2 mm) cross-section. Two sections, A and B, are marked on each specimen at a fixed distance apart. Each specimen was then subjected to tensile loading of varying magnitudes, and the lengths between the marked sections were again measured electronically. The following data was obtained:

Applied force, F (N)	Measured gage length, l (mm)
0	5.000
240	5.017
480	5.033
720	5.050

Determine the tensile stresses and strains developed in each specimen, plot a stress–strain diagram for the bone, and determine the elastic modulus (E) for the bone.

Solution

The cross-sectional area of each specimen is $A = 4$ mm^2 or 4×10^{-6} m^2. When the applied load is zero, the gage length is 5 mm, which is the original (undeformed) gage length, l_0. Therefore, the stress and strain developed in each specimen can be calculated using:

$$\sigma = \frac{F}{A}, \quad \varepsilon = \frac{l - l_0}{l_0}.$$

The following table lists stresses and strains calculated using the above formulas:

F (N)	$\sigma \times 10^6$ (Pa)	l (mm)	ε (mm/mm)
0	0	5.000	0.0
240	60	5.017	0.0034
480	120	5.033	0.0066
720	180	5.050	0.0100

In Fig. 13.32, the stress and strain values computed are plotted to obtain a $\sigma - \varepsilon$ graph for the bone. Notice that the relationship between the stress and strain is almost linear, which is indicated in Fig. 13.32 by a straight line.

Fig. 13.31 The top and side views of the specimen

Fig. 13.32 Stress–strain diagram for the bone

Recall that the elastic modulus of a linearly elastic material is equal to the slope of the straight line representing the $\sigma - \varepsilon$ relationship for that material. Therefore:

$$E = \frac{\sigma}{\varepsilon} = \frac{180 \times 10^6}{0.0100} = 18 \times 10^9 \, Pa = 18 \, GPa.$$

Example 13.4 Figure 13.33 illustrates a fixation device consisting of a plate and two screws, which can be used to stabilize fractured bones. During a single leg stance, a person can apply his/her entire weight to the ground via a single foot. In such situations, the total weight of the person is applied back on the person through the same foot, which has a compressive effect on the leg, its bones, and joints. In the case of a patient with a fractured leg bone (in this case, the femur), this force is transferred from below to above (distal to proximal) the fracture through the screws of the fixation device.

Fig. 13.33 Example 13.4

If the diameter of the screws is $D = 5$ mm and the weight of the patient is $W = 700$ N, determine the shear stress exerted on the screws of a two-screw fixation device during a single leg stance on the leg with a fractured bone.

Solution
Free-body diagrams of the fixation device and the screws are shown in Fig. 13.34. Note that the screw above the fracture is pushing the plate downward, whereas the screw below the fracture is pushing the plate upward. Each screw is applying a force on the plate equal to the weight of the person. The same

Fig. 13.34 Forces applied and the plate and screws

magnitude force is also acting on the screws but in the opposite directions. For example, for the screw above the fracture, the plate is exerting an upward force on the head of the screw and the bone is applying a downward force. The effects of the forces applied on the screws are such that they are trying to shear the screws in a plane perpendicular to the centerline of the screws. With respect to the cross-sectional areas of the screws, these are shearing forces.

The shear stress τ generated in the screws can be calculated by using the following relationship between the shear force F and area A over which the shear stress is to be determined:

$$\tau = \frac{F}{A}.$$

In this case, F is equal to the weight ($W = 700$ N) of the patient. Since the diameters of the screws are given, the cross-sectional area of each screw can be calculated as $A = \pi D^2 / 4 = 19.6 \, mm^2$ or $A = 19.6 \times 10^{-6} \, m^2$ Substituting the numerical value of A and $F = W = 700$ N into the above formula and carrying out the computation will yield $\tau = 35.7 \times 10^6$ Pa.

Note that if we had a four-screw rather than a two-screw fixation device as shown in Fig. 13.35, then each screw would be subjected to a shearing force equal to half of the total weight of the patient.

Fig. 13.35 A four-screw fixation device

Example 13.5 Specimens of human cortical bone tissue were subjected to simple tension test until fracture. The test results revealed a stress–strain diagram shown in Fig. 13.36, which has three distinct regions. These regions are an initial linearly elastic region (between O and A), an intermediate nonlinear elasto-plastic region (between A and B), and a final linearly plastic region (between B and C). The average stresses and corresponding strains at points O, A, B, and C are measured as:

Point	Stress σ (MPa)	Strain ε (mm/mm)
O	0	0.0
A	85	0.005
B	114	0.010
C	128	0.026

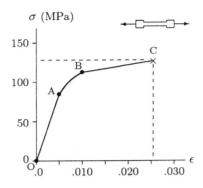

Fig. 13.36 Tensile stress–strain diagram for cortical bone (1 MPa = 10^6 Pa)

Using this information, determine the elastic and strain hardening moduli of the bone tissue in the linear regions of its $\sigma - \varepsilon$ diagram. Note that the *strain hardening modulus* is the slope of the $\sigma - \varepsilon$ curve in the plastic region. Also, find mathematical expressions relating stresses to strains in the linearly elastic and linearly plastic regions.

Solution
The elastic modulus E is the slope of the $\sigma - \varepsilon$ curve in the elastic region. Between O and A, the bone exhibits linearly elastic material behavior, and the $\sigma - \varepsilon$ curve is a straight line. The slope of this line is:

$$E = \frac{\sigma_A - \sigma_0}{\varepsilon_A - \varepsilon_0} = \frac{85 \times 10^6 - 0}{0.005 - 0.0} = 17 \times 10^9 \, \text{Pa} = 17 \, \text{GPa}.$$

The strain hardening modulus E' is equal to the slope of the $\sigma - \varepsilon$ curve in the plastic region. Between B and C, the bone exhibits a linearly plastic material behavior, and its $\sigma - \varepsilon$ curve is a straight line. Therefore:

$$E' = \frac{\sigma_C - \sigma_B}{\varepsilon_C - \varepsilon_B} = \frac{128 \times 10^6 - 114 \times 10^6}{0.026 - 0.010}$$
$$= 0.875 \times 10^9 \, \text{Pa}.$$

The stress–strain relationship between O and A is:

$$\sigma = E\varepsilon \quad or \quad \varepsilon = \frac{\sigma}{E}.$$

The relationship between σ and ε in the linearly plastic region between B and C can be expressed as:

$$\sigma = \sigma_B + E'(\varepsilon - \varepsilon_B) \quad or \quad \varepsilon = \varepsilon_B + \frac{1}{E'}(\sigma - \sigma_B).$$

For example, the strain corresponding to a tensile stress of $\sigma = 120$ MPa can be calculated as:

$$\varepsilon = 0.010 + \frac{120 \times 10^6 - 114 \times 10^6}{0.875 \times 10^9} = 0.017.$$

Example 13.6 Consider the structure shown in Fig. 13.37. The horizontal beam AB has a length $l = 4$ m, weight $W = 500$ N, and is hinged to the wall at A. The beam is supported by two identical steel bars of length $h = 3$ m, cross-sectional area $A = 2 \, \text{cm}^2$, and elastic modulus $E = 200$ GPa. The steel bars are attached to the beam at B and C, where C is equidistant from both ends of the beam.

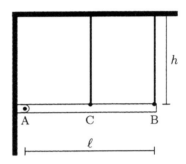

Fig. 13.37 A statically indeterminate system

Determine the forces applied by the beam on the steel bars, the reaction force at A, the amount of elongation in each steel bar, and the stresses generated in each bar. Assume that the beam material is much stiffer than the steel bars.

Solution
This problem will be analyzed in three stages.

Static analysis
The free-body diagram of the beam is shown in Fig. 13.38. The total weight, W, of the beam is assumed to act at its geometric center located at C. R_A is the magnitude of the ground reaction force applied on the beam through the hinge joint at A, and T_1 and T_2 are the forces applied by the steel bars on the beam. Note that only a vertical reaction force is considered at A since there is no horizontal force acting on the beam.

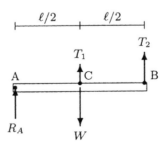

Fig. 13.38 The free-body diagram of the beam

We have a coplanar force system with three unknowns: R_A, T_1, and T_2. There are three equations of equilibrium available from statics. Since there is no horizontal force, the horizontal equilibrium of the beam is automatically satisfied. Therefore, we have two equations but three unknowns. In other words, we have a statically indeterminate system and the equations of equilibrium are not sufficient to fully analyze this problem. We need an additional equation that will be derived by taking into consideration the deformability of the parts constituting the system.

The vertical equilibrium of the beam requires that:

$$\sum F_y = 0: \quad T_1 + T_2 + R_A = W. \tag{i}$$

For the rotational equilibrium of the beam about point A:

$$\sum M_A = 0: \quad \frac{1}{2}T_1 + T_2 = \frac{1}{2}W. \tag{ii}$$

Geometric compatibility
The horizontal beam is hinged to the wall at A, and the weight of the beam tends to rotate the beam about A in the clockwise direction. Because of the weight of the beam, which is applied as tensile forces T_1 and T_2 on the bars, the steel bars will deflect (elongate) and the beam will slightly swing about A (Fig. 13.39). Because of the forces acting on the beam, the beam may bend a little as well. Since it is stated that the beam material is much stiffer than the steel bars, we can ignore the deformability of the beam and assume that it maintains its straight configuration. If δ_1 and δ_2 refer to the amount of deflections in the steel bars, then from Fig. 13.39:

$$\tan \alpha \simeq \frac{\delta_1}{l/2} \simeq \frac{\delta_2}{l}. \tag{iii}$$

Fig. 13.39 Deflection of the beam

Note that this relationship is correct when deflections (δ_1 and δ_2) are small for which angle α is small.

Next we need take into consideration the relationship between applied forces and corresponding deformations.

Stress–strain (force–deformation) analyses
The steel bars with length h elongate by δ_1 and δ_2. Therefore, the tensile strains in the bars are:

$$\varepsilon_1 = \frac{\delta_1}{h}, \quad \varepsilon_2 = \frac{\delta_2}{h}. \tag{iv}$$

The bars are subjected to tensile forces T_1 and T_2. The cross-sectional area of each bar is given as A. Therefore, the tensile stresses exerted by the beam on the steel rods are:

$$\sigma_1 = \frac{T_1}{A}, \quad \sigma_2 = \frac{T_2}{A}. \tag{v}$$

Assuming that the stresses involved are within the proportionality limit for steel, we can apply the Hooke's law to relate stresses to strains:

$$\sigma_1 = E\,\varepsilon_1, \quad \sigma_2 = E\,\varepsilon_2. \tag{vi}$$

Now, we can substitute Eqs. (*iv*) and (*v*) into Eq. (*vi*) so as to eliminate stresses and strains. This will yield:

$$\delta_1 = \frac{T_1\,h}{E\,A}, \quad \delta_2 = \frac{T_2\,h}{E\,A}. \tag{vii}$$

Note that from Eq. (*iii*):

$$\delta_2 = 2\delta_1. \tag{viii}$$

Substituting Eq. (*vii*) into Eq. (*viii*) will yield:

$$T_2 = 2T_1. \tag{ix}$$

Now, we have a total of three equations, Eqs. (*i*), (*ii*), and (*ix*), with three unknowns, R_A, T_1, and R_2. Solving these equations simultaneously will yield:

$$T_1 = \frac{1}{5}W = 100\,N,$$

$$T_2 = \frac{2}{5}W = 200\,N,$$

$$R_A = \frac{2}{5}W = 200\,N.$$

Once tensile forces T_1 and T_2, are determined, Eqs. (*vii*) can be used to calculate the amount of elongations, Eqs. (*v*) can be used to calculate the tensile stresses, and Eqs. (*iv*) can be used to calculate the tensile strains developed in the steel bars:

$$\sigma_1 = \frac{T_1}{A} = 0.5 \times 10^6\,Pa = 0.5\,MPa,$$

$$\sigma_2 = \frac{T_2}{A} = 1.0 \times 10^6\,Pa = 1.0\,MPa,$$

$$\varepsilon_1 = \frac{\sigma_1}{E} = 2.5 \times 10^{-6},$$

$$\varepsilon_2 = \frac{\sigma_2}{E} = 5.0 \times 10^{-6}.$$

Note that the calculated strains are very small. Correspondingly, the deformations are very small as predicted earlier while deriving the relationship in Eqs. (*iii*). Also note that the stresses developed in the steel bars are much lower than the proportionality limit for steel. Therefore, the assumption made to relate stresses and strains in Eqs. (*vi*) was correct as well.

13.18 Exercise Problems

Problem 13.1 Complete the following definitions with appropriate expressions.

(a) Unit deformation of a material as a result of an applied load is called _____.

(b) The internal resistance of a material to deformation due to externally applied forces is called _____.

(c) _____ is a measure of the intensity of internal forces acting parallel or tangent to a plane of cut, while _____ are associated with the intensity of internal forces that are perpendicular to the plane of cut.

(d) On the stress–strain diagram, the stress corresponding to the _____ is the highest stress that can be applied to the material without causing permanent deformation.

(e) On the stress–strain diagram, the highest stress level corresponds to the _____ of the material.

(f) For some materials, it may not be easy to distinguish the yield point. The yield strength of such materials may be determined by the _____.

(g) _____ is defined as the ability of a material to resume its original (stress-free) size and shape upon removal of applied loads.

(h) For linearly elastic materials, stress is linearly proportional to strain and the constant of proportionality is called the_____ of the material.

(i) The distinguishing factor in linearly elastic materials is their _____.

(j) Materials for which the stress–strain curve in the elastic region is not a straight line are known as_____ materials.

(k) _____ is the constant of proportionality between shear stress and shear strain for linearly elastic materials.

(l) A mathematical equation that relates stresses to strains is called a _____.

(m) The analogy between elastic materials and springs is known as _____.

(n) _____ implies permanent (unrecoverable) deformations.

(o) The area under the stress–strain diagram in the elastic region corresponds to the _____ energy stored in the material while deforming the material.

(p) _____ energy is dissipated as heat while deforming the material.

(q) The area enclosed by the _____ signifies the total strain energy dissipated as heat while loading and unloading a material.

(r) The technique of changing the yield point of a material by loading the material beyond its yield point is called _____.

(s) The elastic modulus of a material is a relative measure of the _____ of one material with respect to another.

(t) A _____ material is one that exhibits a large plastic deformation prior to failure.

(u) A _____ material is one that shows a sudden failure (rupture) without undergoing a considerable plastic deformation.

(v) _____ is a measure of the capacity of a material to sustain permanent deformation. The toughness of a material is measured by considering the total area under its stress–strain diagram.

(w) The ability of a material to store or absorb energy without permanent deformation is called the _____ of the material.

(x) If the mechanical properties of a material do not vary from location to location within the material, then the material is called _____.

(y) If a material has constant density, then the material is called _____.

(z) If the mechanical properties of a material are independent of direction or orientation, then the material is called _____.

Answers: Given at the end of the chapter.

Problem 13.2 Curves in Fig. 13.40 represent the relationship between tensile stress and tensile strain for five different materials. The "dot" on each curve indicates the yield point

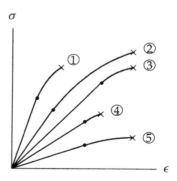

Fig. 13.40 Problem 13.2

and the "cross" represents the rupture point. Fill in the blank spaces below with the correct number referring to a material.

Material _____ has the highest elastic modulus.
Material _____ is the most ductile.
Material _____ is the most brittle.
Material _____ has the lowest yield strength.
Material _____ has the highest strength.
Material _____ is the toughest.
Material _____ is the most resilient.
Material _____ is the most stiff.

Answers: Given at the end of the chapter.

Problem 13.3 Consider two bars, 1 and 2, made of two different materials. Assume that these bars were tested in a uniaxial tension test. Let F_1 and F_2 be the magnitudes of tensile forces applied on bars 1 and 2, respectively. E_1 and E_2 are the elastic moduli and A_1 and A_2 are the cross-sectional areas perpendicular to the applied forces for bar 1 and 2, respectively. For the conditions indicated below, determine the correct symbol relating tensile stresses σ_1 and σ_2 and tensile strains ε_1 and ε_2. Note that ">" indicates greater than, "<" indicates less than, "=" indicates equal to, and "?" indicates that the information provided is not sufficient to make a judgment.

(a) If $A_1 > A_2$ and $F_1 = F_2$ then $\sigma_1 > = ? < \sigma_2$ and $\varepsilon_1 > = ? < \varepsilon_2$.
(b) If $E_1 > E_2$, $A_1 = A_2$, and $F_1 = F_2$, then $\sigma_1 > = ? < \sigma_2$ and $\varepsilon_1 > = ? < \varepsilon_2$.

Answers: Given at the end of the chapter.

Problem 13.4 Figure 13.41 illustrates a bone specimen with a circular cross-section. Two sections, A and B, that are $l_0 = 6$ mm distance apart are marked on the specimen. The radius of the specimen in the region between A and B is $r_0 = 1$ mm.

This specimen was subjected to a series of uniaxial tension tests until fracture by gradually increasing the magnitude of the applied force and measuring corresponding deformations. As a result of these tests, the following data were recorded:

Record no.	Force, F(N)	Deformation, Δl (mm)
1	94	0.009
2	190	0.018
3	284	0.027
4	376	0.050
5	440	0.094

If record 3 corresponds to the end of the linearly elastic region and record 5 corresponds to fracture point, carry out the following:

Fig. 13.41 Problem 13.4

(a) Calculate average tensile stresses and strains for each record.
(b) Draw the tensile stress–strain diagram for the bone specimen.
(c) Calculate the elastic modulus, E, of the bone specimen.
(d) What is the ultimate strength of the bone specimen?
(e) What is the yield strength of the bone specimen? (use the offset method)

Problem 13.5 Human femur bone was subjected to a uniaxial tension test. As a result of a series of experiments, the stress–strain curve shown in Fig. 13.42 was obtained. Based on the graph in Fig. 13.42, determine the following parameters:
(a) Elastic modulus, E.
(b) Apparent yield strength, σ_y (use the offset method).
(c) Ultimate strength, σ_u.
(d) Strain, ε_1, corresponding to yield stress.
(e) Strain, ε_2, corresponding to ultimate stress.
(f) Strain energy when stress is at the proportionality limit.

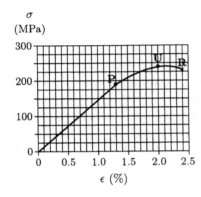

Fig. 13.42 Problem 13.5

Problem 13.6 Consider the uniform, horizontal beam shown in Fig. 13.43a. The beam is hinged to a wall at A and a weight W_2 is attached on the beam at B. Point C represents the center of gravity of the beam, which is equidistant from A and B.

(a)

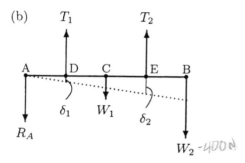

(b)

Fig. 13.43 Problem 13.6

The beam has a weight $W_1 = 100$ N and length $l = 4$ m. The beam is supported by two vertical rods, 1 and 2, attached to the beam at D and E. Rod 1 is made of steel with elastic modulus $E_1 = 200$ GPa and a cross-sectional area $A_1 = 500$ mm^2, and rod 2 is bronze with elastic modulus $E_2 = 80$ GPa and $A_2 = 400$ mm^2. The original (undeformed) lengths of both rods is $h = 2$ m. The distance between A and D is $d_1 = 1$ m and the distance between A and E is $d_2 = 3$ m.

The free-body diagram of the beam and its deflected orientation is shown in Fig. 13.43b, where T_1 and T_2 represent the forces exerted by the rods on the beam. Symbols δ_1 and δ_2 represent the amount of deflection the steel and bronze rods undergo, respectively. Note that the beam material is assumed to be very stiff (almost rigid) as compared to the rods so that it maintains its straight shape.

(a) Calculate tensions T_1 and T_2, and the reactive force R_A on the beam at A.
(b) Calculate the average tensile stresses σ_1 and σ_2 generated in the rods.

Answers to Problem 13.1:

(a) strain	(n) Plasticity
(b) stress	(o) elastic strain
(c) Shear stress, normal stress	(p) Plastic strain
(d) elastic limit	(q) hysteresis loop
(e) ultimate strength	(r) strain hardening
(f) offset method	(s) stiffness
(g) Elasticity	(t) ductile
(h) elastic (Young's) modulus	(u) brittle
(i) elastic (Young's) modulus	(v) Toughness
(j) nonlinear elastic	(w) resilience
(k) Shear modulus	(x) homogeneous
(l) material function	(y) incompressible
(m) Hooke's Law	(z) isotropic

Answers to Problem 13.2: 1, 2, 4, 5, 2, 2, 3, 1

Answers to Problem 13.3:
(a) $\sigma_1 < \sigma_2, \varepsilon_1 ? \varepsilon_2$.
(b) $\sigma_1 = \sigma_2, \varepsilon_1 < \varepsilon_2$.

Answers to Problem 13.4:
Answers (approximately):
(a) $E = 20$ GPa.
(b) $\sigma_u = 140$ MPa.
(c) $\sigma_y = 118$ MPa.

Answers to Problem 13.5:
Answers (approximately):
(a) $E = 14.6$ GPa.
(b) $\sigma_y = 235$ MPa.
(c) $\sigma_u = 240$ MPa.
(d) $\varepsilon_1 = 0.018$.
(e) $\varepsilon_2 = 0.20$. . 02
(f) 1.235 MPa.

Answers to Problem 13.6:
Answers:
(a) $T_1 = 464$ N, $T_2 = 445$ N, $R_A = 409$ N.
(b) $\sigma_1 = 0.928$ GPa, $\sigma_2 = 1.113$ GPa.

14.1 Poisson's Ratio

When a structure is subjected to uniaxial tension, the transverse dimensions decrease (the structure undergoes lateral contractions) while simultaneously elongating in the direction of the applied load. This is illustrated in the previous chapter through the phenomenon called necking. For stresses within the proportionality limit, the results of uniaxial tension and compression experiments suggest that the ratio of deformations occurring in the axial and lateral directions is constant. For a given material, this constant is called the *Poisson's ratio* and is commonly denoted by the symbol v (nu):

$$v = -\frac{\text{lateral strain}}{\text{axial strain}}.$$

Consider the rectangular bar with dimensions a, b, and c shown in Fig. 14.1. To be able to differentiate strains involved in different directions, a rectangular coordinate system is adopted. The bar is subjected to tensile forces of magnitude F_x in the x direction that induces a tensile stress σ_x (Fig. 14.2). Assuming that this stress is uniformly distributed over the cross-sectional area ($A = ab$) of the bar, its magnitude can be determined using:

$$\sigma_x = \frac{F_x}{A}. \tag{14.1}$$

Under the effect of F_x, the bar elongates in the x direction, and contracts in the y and z directions. If the elastic modulus E of the bar material is known and the deformations involved are within the proportionality limit, then the stress and strain in the x direction are related through the Hooke's law:

$$\varepsilon_x = \frac{\sigma_x}{E}. \tag{14.2}$$

Equation (14.2) yields the unit deformation of the bar in the direction of the applied forces. Strains in the lateral directions can now be determined by utilizing the definition of the Poisson's ratio. If ε_y and ε_z are unit contractions in the y and z directions due to the uniaxial loading in the x direction, then:

$$v = -\frac{\varepsilon_y}{\varepsilon_x} = -\frac{\varepsilon_z}{\varepsilon_x}. \tag{14.3}$$

In other words, if the Poisson's ratio of the bar material is known, then the strains in the lateral directions can be determined:

$$\varepsilon_y = \varepsilon_z = -v\,\varepsilon_x = -v\frac{\sigma_x}{E}. \tag{14.4}$$

The minus signs in Eqs. (14.3) and (14.4) indicate a decrease in the lateral dimensions when there is an increase in the axial dimension. Strains ε_y and ε_z are negative when ε_x is positive, which is the case for tensile loading. These equations are also valid for compressive loading in the x direction for which σ_x and ε_x are negative, and ε_y and ε_z are positive.

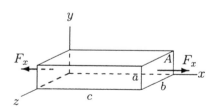

Fig. 14.1 A rectangular bar subjected to uniaxial tension

Fig. 14.2 Stress distribution is uniform over the cross-sectional area $A = ab$ of the bar

N. Özkaya et al., *Fundamentals of Biomechanics: Equilibrium, Motion, and Deformation*,
DOI 10.1007/978-1-4614-1150-5_14, © Springer Science+Business Media, LLC 2012

Once the strains in all three directions are determined, then the deformed dimensions a', b', and c' (Fig. 14.3) of the bar can also be calculated. By definition, strain is equal to the ratio of the change in length and the original length. Therefore,

$$\varepsilon_x = \frac{c' - c}{c}.$$

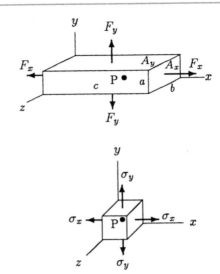

Fig. 14.3 The bar elongates and undergoes lateral contractions simultaneously

Solving this equation for the deformed length c' of the object in the x direction will yield $c' = (1 + \varepsilon_x)c$. Similarly, $a' = (1 + \varepsilon_y)a$ and $b' = (1 + \varepsilon_z)b$.

Note that the stress–strain relationships provided here are valid only for linearly elastic materials, or within the proportionality limits of any elastic–plastic material.

For a given elastic material, the elastic modulus, shear modulus, and Poisson's ratio are related through the expression:

$$G = \frac{E}{2(1 + v)} \quad \text{or} \quad v = \frac{E}{2G} - 1. \quad (14.5)$$

This formula can be used to calculate the Poisson's ratio of a material if the elastic and shear moduli are known.

14.2   Biaxial and Triaxial Stresses

As discussed in the previous section, when an object is subjected to uniaxial loading, strains can occur in all three directions. The strains in the lateral directions can be calculated by utilizing the definition of Poisson's ratio. Poisson's ratio also makes it possible to analyze situations in which there is more than one normal stress acting in more than one direction.

Consider the rectangular bar shown in Fig. 14.4. The bar is subjected to biaxial loading in the xy-plane. Let P be a point in the bar. Stresses induced at point P can be analyzed by constructing a cubical material element around the point. A cubical material element with sides parallel to the sides of the bar itself is shown in Fig. 14.4, along with the stresses acting on it. σ_x and σ_y are the normal stresses due to the tensile forces applied on the bar in the x and y directions, respectively. If $A_x = ab$ and $A_y = bc$ are the areas of the rectangular bar with normals in the x and y directions, respectively, then σ_x and σ_y can be calculated as:

$$\sigma_x = \frac{F_x}{A_x} = \frac{F_x}{ab},$$

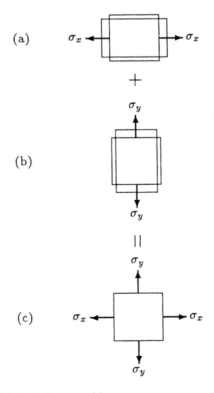

Fig. 14.4 A rectangular bar subjected to tensile forces in the x and y directions and a material element under biaxial stresses

$$\sigma_y = \frac{F_y}{A_y} = \frac{F_y}{bc}.$$

The effects of these biaxial stresses are illustrated graphically in Fig. 14.5. Stress σ_x elongates the material in

(a)

$+$

(b)

$\|$

(c)

Fig. 14.5 Method of superposition

the x direction and causes a contraction in the y (also z) direction. Strains due to σ_x in the x and y directions are:

$$\varepsilon_{x1} = \frac{\sigma_x}{E},$$

$$\varepsilon_{y1} = -v\varepsilon_{x1} = -v\frac{\sigma_x}{E}.$$

Similarly, σ_y elongates the material in the y direction and causes a contraction in the x direction (Fig. 14.5b). Therefore, strains in the x and y directions due to σ_y are:

$$\varepsilon_{y2} = \frac{\sigma_y}{E},$$

$$\varepsilon_{x2} = -v\varepsilon_{y2} = -v\frac{\sigma_x}{E}.$$

The combined effect of σ_x and σ_y on the plane material element is shown in Fig. 14.5c. The same effect can be represented mathematically by adding the individual effects of σ_x and σ_y. The resultant strains in the x and y directions can be determined as:

$$\varepsilon_x = \varepsilon_{x1} + \varepsilon_{x2} = \frac{\sigma_x}{E} - v\frac{\sigma_y}{E}$$
$$\varepsilon_y = \varepsilon_{y1} + \varepsilon_{y2} = \frac{\sigma_y}{E} - v\frac{\sigma_x}{E}. \tag{14.6}$$

If required, these equations can be solved simultaneously to express stresses in terms of strains:

$$\sigma_x = \frac{(\varepsilon_x + v\varepsilon_y)E}{1 - v^2}$$
$$\sigma_y = \frac{(\varepsilon_y + v\varepsilon_x)E}{1 - v^2}. \tag{14.7}$$

This discussion can be extended to derive the following stress–strain relationships for the case of triaxial loading (Fig. 14.6):

$$\varepsilon_x = \frac{1}{E}[\sigma_x - v(\sigma_y + \sigma_z)]$$
$$\varepsilon_y = \frac{1}{E}[\sigma_y - v(\sigma_z + \sigma_x)] \tag{14.8}$$
$$\varepsilon_z = \frac{1}{E}[\sigma_z - v(\sigma_x + \sigma_y)].$$

These formulas are valid for linearly elastic materials, and they can be used when the stresses induced are tensile or compressive, by adapting the convention that tensile stresses are positive and compressive stresses are negative.

Further generalization of stress–strain relationships for linearly elastic materials should take into consideration the relationships between shear stresses and shear strains. As illustrated in Fig. 14.7, the most general case of material loading occurs when an object is subjected to normal and shear stresses in three mutually perpendicular directions. Corresponding to these stresses there are normal and shear strains. In Fig. 14.7, shear stresses are identified by double subscripted symbols. The first subscript indicates the direction normal to the surface over which the shear stress is acting and the second subscript indicates the direction in which the stress is acting. A total of six shear stresses ($\tau_{xy}, \tau_{yx}, \tau_{yz}, \tau_{zy}, \tau_{zx}, \tau_{xz}$) are defined. However, the condition of static equilibrium requires that $\tau_{xy} = \tau_{yx}$, $\tau_{yz} = \tau_{zy}$, and $\tau_{zx} = \tau_{xz}$. Therefore, there are only three independent shear stresses (τ_{xy}, τ_{yz}, and τ_{zx}), and three corresponding shear strains (γ_{xy}, γ_{yz}, and γ_{zx}). For linearly elastic materials, shear stresses are linearly proportional to shear strains and the shear modulus, G, is the constant of proportionality:

$$\tau_{xy} = G\gamma_{xy}$$
$$\tau_{yz} = G\gamma_{yz} \tag{14.9}$$
$$\tau_{zx} = G\gamma_{zx}.$$

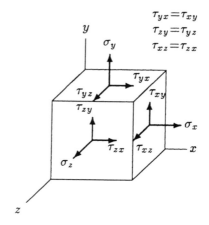

Fig. 14.7 Normal and shear components of the stress tensor

As illustrated in Fig. 14.8, for two-dimensional problems in the xy-plane, only one shear stress (τ_{xy}) and two normal stresses (σ_x and σ_y) need to be considered.

The above discussion indicates that stress and strain have a total of nine components, only six of which are independent. Stress and strain are known as *second order tensors*. Recall that a scalar quantity has a magnitude only. A vector quantity has both a magnitude and a direction, and can be represented in terms of its three components. Scalar quantities are also known as *zero order tensors*, while vectors are *first order*

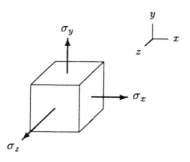

Fig. 14.6 Triaxial normal stresses

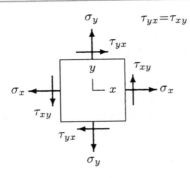

Fig. 14.8 Plane (biaxial) stress components

tensors. Second order tensor components not only have magnitudes and directions associated with them, but they are also dependent upon the plane over which they are determined. For example, in the case of the stress tensor, magnitudes and directions of the stress tensor components at a material point may vary depending upon the orientation of the cubical material element constructed around it. However, if the components of the stress tensor with respect to one material element are known, then the components of the stress tensor with respect to another material element can be determined through appropriate coordinate transformations, which are discussed in the following section. It is also important to note that although the magnitude and direction of the stress tensor components at a material may vary with the orientation of the material element constructed around it, under a given load condition, the overall state of stress at the material point is always the same.

Example 14.1 Consider the cube with sides $a = 10$ cm shown in Fig. 14.9. This block is tested under biaxial forces that are applied in the x and y directions. Assume that the forces applied have equal magnitudes of $F_x = F_y = F = 2 \times 10^6$ N, and that the elastic modulus and Poisson's ratio of the block material are given as $E = 2 \times 10^{11}$ Pa and $v = 0.3$.

Fig. 14.9 Example 14.1

Determine the strains in the x, y, and z directions, and the deformed dimensions of the block if:
(a) Both F_x and F_y are tensile (Fig. 14.10a)
(b) F_x is tensile and F_y is compressive (Fig. 14.10b)
(c) Both F_x and F_y are compressive (Fig. 14.10c)

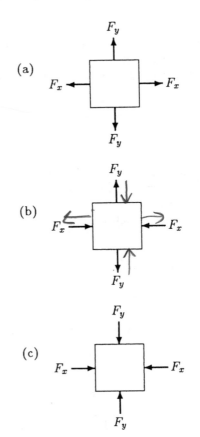

Fig. 14.10 Various modes of biaxial loading of the block

Solution
To be able to calculate the stresses involved, we need to know the areas to which forces F_x and F_y are applied. Let A_x and A_y be the areas of the sides of the cube with normals in the x and y directions, respectively. Since the object is a cube, these areas are equal:

$$A_x = A_y = A = a^2 = 100 \text{ cm}^2 = 1 \times 10^{-2} \text{ m}^2.$$

We can now calculate the normal stresses in the x and y directions. Notice that the magnitudes of the applied forces in the x and y directions and the areas upon which they are applied are equal. Therefore, the magnitudes of the stresses in the x and y directions are equal as well:

$$\sigma_x = \sigma_y = \sigma = \frac{F}{A} = \frac{2 \times 10^6 \, N}{1 \times 10^{-2} \, m^2} = 2 \times 10^8 \, Pa.$$

The normal stress component in the z direction is zero, since there is no force applied on the block in that direction.

Case (a) Both forces are tensile, and therefore, both σ_x and σ_y are tensile and positive (Fig. 14.10a). We can utilize Eqs. (14.8) to calculate the strains involved. Since $\sigma_x = \sigma_y = \sigma$ and $\sigma_z - 0$, these equations can be simplified as:

$$\varepsilon_x = \frac{1}{E}(\sigma_x - v\sigma_y) = \frac{1-v}{E}\sigma,$$

$$\varepsilon_y = \frac{1}{E}(\sigma_y - v\sigma_x) = \frac{1-v}{E}\sigma,$$

$$\varepsilon_z = \frac{-v}{E}(\sigma_x + \sigma_y) = -\frac{2v}{E}\sigma.$$

Substituting the numerical values of E, v, and σ into these equations, and carrying out the computations will yield:

$$\varepsilon_x = 0.7 \times 10^{-3},$$

$$\varepsilon_y = 0.7 \times 10^{-3},$$

$$\varepsilon_z = -0.6 \times 10^{-3}.$$

Note that both ε_x and ε_y are positive, while ε_z is negative. As a result of the applied forces, the dimensions of the block in the x and y directions increase, while its dimension in the z direction decreases. If $a_x, a_y,$ and a_z are the new (deformed) dimensions of the block, then:

$$a_x = (1 + \varepsilon_x)a = (1 + 0.0007)(10\,\text{cm}) = 10.007\,\text{cm}$$
$$a_y = (1 + \varepsilon_y)a = (1 + 0.0007)(10\,\text{cm}) = 10.007\,\text{cm}$$
$$a_z = (1 + \varepsilon_z)a = (1 - 0.0006)(10\,\text{cm}) = 9.994\,\text{cm}.$$

Case (b) In this case, σ_x is tensile and positive while σ_y is compressive and negative (Fig. 14.10b). Therefore, the stress–strain relationships take the following forms:

$$\varepsilon_x = \frac{1}{E}(\sigma_x + v\sigma_y) = \frac{1+v}{E}\sigma,$$

$$\varepsilon_y = \frac{1}{E}(-\sigma_y - v\sigma_x) = -\frac{1+v}{E}\sigma,$$

$$\varepsilon_z = -\frac{v}{E}(\sigma_x + \sigma_y) = 0.$$

Substituting the numerical values of the parameters involved into these equations, and carrying out the computations will yield:

$$\varepsilon_x = 1.3 \times 10^{-3},$$

$$\varepsilon_y = -1.3 \times 10^{-3},$$

$$\varepsilon_z = 0.$$

The tensile force applied in the x direction and the compressive force applied in the y direction elongates the block in the x direction and reduces its dimension in the y direction. In the z direction, the contraction caused by the tensile stress σ_x is counterbalanced by the expansion caused by the compressive stress σ_y.

Case (c) In this case, both forces are compressive. Therefore, both σ_x and σ_y are compressive and negative (Fig. 14.10c). The simplified equations relating normal strains to normal stresses for this case are:

$$\varepsilon_x = \frac{1}{E}(-\sigma_x + v\sigma_y) = \frac{1-v}{E}\sigma,$$

$$\varepsilon_y = \frac{1}{E}(-\sigma_y - v\sigma_x) = -\frac{1-v}{E}\sigma,$$

$$\varepsilon_z = -\frac{v}{E}(-\sigma_x - \sigma_y) = \frac{2v}{E}\sigma.$$

Substituting the numerical values and carrying out the computations will yield:

$$\varepsilon_x = -0.7 \times 10^{-3}$$
$$\varepsilon_y = -0.7 \times 10^{-3}$$
$$\varepsilon_z = 0.6 \times 10^{-3}.$$

These results indicate that the dimensions of the block in the x and y directions decrease, while its dimension in the z direction increases. The deformed dimensions of the block can be determined in a similar manner as employed for Case (a).

14.3 Stress Transformation

Consider the rectangular bar shown in Fig. 14.11. The bar is subjected to externally applied forces that cause various modes of deformation within the bar. Let P be a point within the structure. Assume that a small cubical material element at point P with sides parallel to the sides of the bar is cut out and analyzed. As illustrated in Fig. 14.12, this material element is subjected to a combination of normal (σ_x and σ_y) and shear (τ_{xy}) stresses in the xy-plane. Now, consider a second element at the same material point but with a different orientation than the first element (Fig. 14.13). One can assume that the second material element is obtained simply by rotating the first in the counterclockwise direction

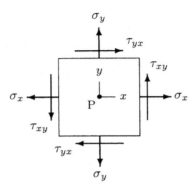

Fig. 14.11 A bar subjected to external forces applied in the *xy*-plane

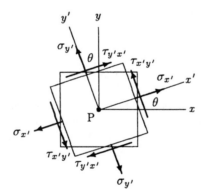

Fig. 14.12 Stress tensor components in the *xy*-plane

Fig. 14.13 Transformation of stress tensor components

$$\sigma_{y'} = \frac{\sigma_y + \sigma_x}{2} - \frac{\sigma_x - \sigma_y}{2} \cos(2\theta) - \tau_{xy} \sin(2\theta), \quad (14.11)$$

$$\tau_{x'y'} = -\frac{\sigma_x - \sigma_y}{2} \sin(2\theta) + \tau_{xy} \cos(2\theta). \quad (14.12)$$

These equations can be used for transforming stresses from one set of coordinates (x, y) to another (x', y').

14.4   Principal Stresses

There are infinitely many possibilities of constructing elements around a given point within a structure. Among these possibilities, there may be one element for which the normal stresses are maximum and minimum. These maximum and minimum normal stresses are called the *principal stresses*, and the planes with normals collinear with the directions of the maximum and minimum stresses are called the *principal planes*. On a principal plane, the shear stress is zero (Fig. 14.14). This condition of zero shear stress and Eqs. (14.10) through (14.12) can be utilized to determine the principal normal stresses and the orientation of the principal planes. This can be achieved by setting Eq. (14.12) equal to zero and solving it for angle θ. If the angle thus determined is denoted as θ_1, then:

$$\theta_1 = \frac{1}{2} \tan^{-1} \left(\frac{2\tau_{xy}}{\sigma_x - \sigma_y} \right). \quad (14.13)$$

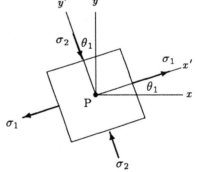

Fig. 14.14 Principal stresses

Here, θ_1 represents the angle of orientation of the principal planes relative to the x and y axes. If θ in Eqs. (14.10) and (14.11) is replaced by θ_1, then the following expressions can be derived for the principal stresses σ_1 and σ_2:

through an angle θ. Let x' and y' be two mutually perpendicular directions representing the normals to the surfaces of the transformed material element. The stress distribution on the transformed material element would be different than that of the first. In general, the second element may be subjected to normal stresses ($\sigma_{x'}$ and $\sigma_{y'}$) and shear stress ($\tau_{x'y'}$) as well. If stresses σ_x, σ_y, and τ_{xy}, and the angle of rotation θ are given, then stresses $\sigma_{x'}$, $\sigma_{y'}$, and $\tau_{x'y'}$ can be calculated using the following formulas:

$$\sigma_{x'} = \frac{\sigma_x + \sigma_y}{2} + \frac{\sigma_x - \sigma_y}{2} \cos(2\theta) + \tau_{xy} \sin(2\theta), \quad (14.10)$$

$$\sigma_1 = \frac{\sigma_x + \sigma_y}{2} + \sqrt{\left(\frac{\sigma_x - \sigma_y}{2} \right)^2 + \tau_{xy}^2}, \quad (14.14)$$

$$\sigma_2 = \frac{\sigma_x + \sigma_y}{2} - \sqrt{\left(\frac{\sigma_x - \sigma_y}{2}\right)^2 + \tau_{xy}^2}. \qquad (14.15)$$

The concept of principal stresses is important in stress analyses. It is known that fracture or material failure occurs along the planes of maximum stresses, and structures must be designed by taking into consideration the maximum stresses involved.

Remember that the response of a material to different modes of loading are different, and different physical properties of a given material must be considered while analyzing its behavior under shear, tension, and compression. Notice that Eqs. (14.14) and (14.15) are useful for calculating the maximum and minimum normal stresses. For a given structure and loading conditions, the maximum normal stress computed using Eq. (14.14) may be well within the limits of operational safety. However, the structure must also be checked for a critical shearing stress, called the *maximum shear stress*. The maximum shear stress, τ_{max}, occurs on a material element for which the normal stresses are equal (Fig. 14.15). Therefore, Eqs. (14.10) and (14.11) can be set equal, and the resulting equation can be solved for the angle of orientation, θ_2, of the element on which the shear stress is maximum. This will yield:

14.5 Mohr's Circle

An effective way of visualizing the state of stress at a material point and calculating the principal stresses can be achieved by means of *Mohr's circle*. A typical Mohr's circle is illustrated in Fig. 14.17 for the plane stress element shown in Fig. 14.16. The procedure for constructing such a diagram and finding the maximum and minimum stresses is outlined below.

- As in Fig. 14.16, make a sketch of the element for which the stresses are known and indicate on this element the proper directions of the stresses involved. The stresses shown in Fig. 14.16 are all positive. The sign convention for positive and negative stresses is such that a tensile stress is positive while a compressive stress is negative. A shear stress on the right-hand surface that tends to rotate the material element in the counterclockwise direction and a shear stress on the upper surface that tends to rotate the element in the clockwise direction are positive.
- Set up a rectangular coordinate system in which the horizontal axis represents the normal stresses and the vertical axis represents the shear stresses. On the τ versus σ diagram, positive normal stresses are plotted to the right

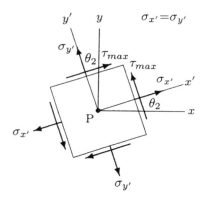

Fig. 14.15 Maximum shear stress

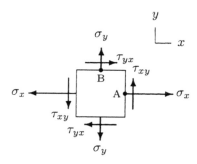

Fig. 14.16 Positive stresses

$$\theta_2 = \frac{1}{2}\tan^{-1}\left(\frac{\sigma_y - \sigma_x}{2\tau_{xy}}\right). \qquad (14.16)$$

An expression for the maximum shear stress, τ_{max}, can then be derived by replacing θ in Eq. (14.12) with θ_2:

$$\tau_{max} = \sqrt{\left(\frac{\sigma_x - \sigma_y}{2}\right)^2 + \tau_{xy}^2}. \qquad (14.17)$$

A graphical method of finding principal stresses is discussed next.

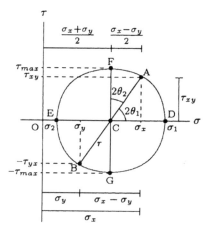

Fig. 14.17 Mohr's circle

of the origin, O, whereas negative normal stresses are plotted to the left of O.

- Let A be a point on the τ–σ diagram with coordinates equal to the normal and shear stresses acting on the right-hand surface of the material element. That is, A has the coordinates σ_x and τ_{xy}. Similarly, B is a point with coordinates equal to the stress components σ_y and $-\tau_{yx}$ acting on the upper surface of the element.

- Connect points A and B with a straight line. Label the point of intersection of line AB and the horizontal axis as C. Point C is the center of the Mohr's circle, and is located at a distance $(\sigma_x+\sigma_y)/2$ from O. Therefore, the stress at C is:

$$\sigma_c = \frac{\sigma_x+\sigma_y}{2}.$$

- The distance between A and C (or B and C) is the radius, r, of the Mohr's circle, which can be calculated as:

$$r = \sqrt{\left(\frac{\sigma_x-\sigma_y}{2}\right)^2 + \tau_{xy}{}^2}.$$

- Draw a circle with radius r and center at C. Intersections of this circle with the horizontal axis (where $\tau = 0$) correspond to the maximum and minimum (principal) normal stresses, which can be calculated as:

$$\sigma_1 = \sigma_c + r = \frac{\sigma_x+\sigma_y}{2} + \sqrt{\left(\frac{\sigma_x-\sigma_y}{2}\right)^2 + \tau_{xy}{}^2},$$

$$\sigma_2 = \sigma_c - r = \frac{\sigma_x+\sigma_y}{2} - \sqrt{\left(\frac{\sigma_x-\sigma_y}{2}\right)^2 + \tau_{xy}{}^2}.$$

If the Mohr's circle is drawn carefully, then σ_1 and σ_2 can also be measured directly from the τ–σ diagram. Notice that the above equations are essentially Eqs. (14.14) and (14.15).

- On the τ–σ diagram, F and G are the points of intersection of the Mohr's circle and a vertical line passing through C. At F and G, normal stresses are both equal to σ_c and the magnitude of the shear stress is maximum. Therefore, points F and G on the Mohr's circle correspond to the state of maximum shear stress. The maximum shear stress is simply equal to the radius of the Mohr's circle:

$$\tau_{max} = \sqrt{\left(\frac{\sigma_x-\sigma_y}{2}\right)^2 + \tau_{xy}{}^2}.$$

Again, this is the same equation provided in the previous section for τ_{max}.

- Mohr's circle has its own way of interpreting angles. On the plane stress element shown in Fig. 14.16, the normals of surfaces A and B are at right angles. On the τ–σ diagram, A and B make an angle of 180°. Therefore, a rotation of $\theta°$ corresponds to an angle of 2θ on the Mohr's circle. Point D on the τ–σ diagram is related to the maximum normal stress, and therefore, to one of the principal directions. On the τ–σ diagram, point D is located at an angle of $2\theta_1$ clockwise from point A. The direction normal to the principal plane (direction of σ_1) can be obtained by rotating the x axis through an angle of θ_1 counterclockwise. A similar procedure is valid for finding the orientation of the element for which the shear stress is maximum.

Example 14.2 Consider the bar shown in Fig. 14.18a, which is subjected to tension in the x direction. Let F be the magnitude of the applied force and A is the cross-sectional area of the bar. The state of uniaxial stress is shown in Fig. 14.18b on a plane material element. The sides of this element have normals in the x and y directions.

Fig. 14.18 Example 14.2

Using Mohr's circle, determine the maximum shear stress induced and the plane of maximum shear stress.

Solution
For the given magnitude of the externally applied force and the cross-sectional area of the bar, the normal stress induced in the bar in the x direction can be determined as $\sigma_x = F/A$. As illustrated in Fig. 14.18b, σ_x is the only component of the stress tensor on a material element with sides parallel to the x and y directions.

Based on the plane stress element of Fig. 14.18b, Mohr's circle is drawn in Fig. 14.19a. Notice that there is only a tensile stress of magnitude σ_x on surface A, and there is no stress on surface B. Therefore, on the τ–σ diagram, point A is located along the σ-axis at a distance σ_x from the origin, and point B is essentially the origin of the τ–σ diagram. The center C of the Mohr's circle lies along the σ-axis between B and A, at a distance $\sigma_x/2$ from both A and B. Therefore, the radius of the Mohr's circle is $\sigma_x/2$.

Fig. 14.19 Analyses of the material element in Fig. 14.18

Point F on the Mohr's circle represents the orientation of the material element for which the shear stress is maximum. The magnitude of the maximum shear stress is equal to the radius of the Mohr's circle:

$$\tau_{\max} = \frac{\sigma_x}{2} = \frac{F}{2A}.$$

On the Mohr's circle, point F is located 90° counterclockwise from A. Therefore, as illustrated in Fig. 14.19b, the material element for which the shear stress is maximum can be obtained by rotating the material element given in Fig. 14.18b in the clockwise direction through an angle $\theta_2 = 45°$.

Note that σ_x is the maximum normal stress on the Mohr's circle. Therefore, point A on the Mohr's circle represents the orientation of the material element for which the normal stresses are maximum and minimum, and the material element in Fig. 14.18b represents the state of principal stresses.

Example 14.3 Consider the material element shown in Fig. 14.20. In the xy-plane, the element is subjected to compressive stress σ_x and shear stress τ_{xy}.

Using Mohr's circle, determine the principal stresses, maximum shear stress, and the principal planes.

Fig. 14.20 Example 14.3

Solution

Mohr's circle in Fig. 14.21a is drawn by using the plane stress element in Fig. 14.20. There is a negative (compressive) normal stress with magnitude σ_x and a negative shear stress with magnitude τ_{xy} on surface A of the material element in Fig. 14.20. Therefore, point A on Mohr's circle has coordinates $-\sigma_x$ and $-\tau_{xy}$. Surface B on the stress element has only a negative shear stress with magnitude τ_{xy}. Therefore, point B on the τ–σ diagram lies along the τ-axis at a τ_{xy} distance above the origin (positive). The center C of Mohr's circle can be determined as the point of intersection of the line connecting A and B with the σ-axis. The radius of the Mohr's circle can also be determined by utilizing the properties of right triangles. In this case,

$$r = \sqrt{\left(\frac{\sigma_x}{2}\right)^2 + \tau_{xy}^2}.$$

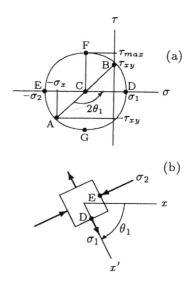

Fig. 14.21 Analyses of the material element in Fig. 14.20

Once the radius of the Mohr's circle is established, it is easy to find the principal stresses σ_1 and σ_2 and the maximum shear stress $\tau_{\max}$:

$$\sigma_1 = r - \frac{\sigma_x}{2} \quad (\text{tensile}),$$

$$\sigma_2 = r + \frac{\sigma_x}{2} \quad (\text{compressive}),$$

$$\tau_{\max} = r.$$

To determine the angle of rotation, θ_1, of the plane for which the stresses are maximum and minimum, we need to read the angle between lines CA and CD in Fig. 14.21a, which is equal to $2\theta_1$. From the geometry of the problem:

$$2\theta_1 = 180° - \tan^{-1}\left(\frac{2\tau_{xy}}{\sigma_x}\right).$$

As illustrated in Fig. 14.21b, to construct the material element for which normal stresses are maximum and minimum, calculate angle θ_1 and rotate the material element in Fig. 14.20 through an angle θ_1 in the clockwise direction.

14.6 Failure Theories

To assure both safety and reliability, a structure must be designed and a proper material must be selected so that the strength of the structure is considerably greater than the stresses to which it will be subjected when it is put in service. For example, if a material is subjected to tensile loads only, then the strength of the material must be judged by its ultimate strength (or yield strength). As discussed in the previous chapter, there are well-established normal stress–strain and shear stress–strain diagrams for most of the common materials. Therefore, it is a relatively straightforward task to predict the response of a material subjected to uniaxial stress or pure shearing stress. However, such a direct approach is not available for a complex state of stress occurring under a combined loading.

Several criteria have been established to predict the conditions under which material failure may occur when it is subjected to combined loading. By material or structural failure, it is meant that the material either ruptures so that it can no longer support any load or undergoes excessive permanent deformation (yielding). Unfortunately, there is no single complete failure criterion that can be used to predict the material response under any type of loading. The purpose of these theories is to relate the stresses to the strength of the material. However, the available data are usually expressed in terms of the yield and ultimate strengths of the material as established in simple tension and pure shear experiments. Therefore, the idea is to utilize this relatively limited information to analyze complex situations in which there may be more than one stress component. A few of these failure criteria are reviewed next.

The *maximum shear stress theory* is used to predict yielding and therefore is applicable to ductile materials. It is also known as *Coulomb theory* or *Tresca theory*. This theory assumes that yielding occurs when the maximum shear stress in a material element reaches the value of the maximum shear stress that would be observed at the instant when yielding occurred if the material were subjected to uniaxial tension.

Assume that a material is subjected to a simple (uniaxial) tension test until yielding. The stress level at yielding is

recorded as σ_{yp}. The maximum shear stress to which this material is subjected can be determined by constructing a Mohr's circle (see Example 14.17) which is illustrated in Fig. 14.22. It is clear that the maximum shear stress in simple tension is equal to half of the normal stress. In this case, the normal stress is also the yield strength of the material in tension, and therefore,

$$\tau_{\max} = \frac{\sigma_x}{2} = \frac{\sigma_{yp}}{2}. \tag{14.18}$$

(a)

(b)

Fig. 14.22 Explaining the maximum shear stress theory

The maximum stress theory states that if the same material is subjected to any combination of normal and shear stresses and the maximum shear stress is calculated, the yielding will start when the maximum shear stress is equal to $\tau_{\max}$. For example, consider that the material is subjected to a combination of normal and shear stresses in the xy-plane as illustrated in Fig. 14.23a. For this state of stress, the maximum shear stress $\tau_{\max}$ can be determined by constructing

(a)

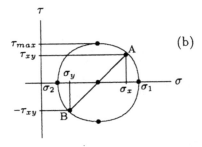

(b)

Fig. 14.23 Combined state of stress and its Mohr's circle a Mohr's circle (Fig. 14.23b) or using Eq. (14.17). Yielding will occur if τ_{max} is equal to or greater than $\sigma_{yp}/2$.

The *maximum distortion energy theory* is a widely accepted criterion that states the conditions for yielding of ductile materials. It is also known as the *von Mises yield theory* or the *Mises–Hencky theory*. This theory assumes that yielding can occur when the root mean square of the differences between the principal stresses is equal to the yield strength of the material established in a simple tension test.

Let σ_1 and σ_2 be the principal stresses in a plane state of stress, and σ_{yp} be the yield strength of the material. According to the distortion energy theory, the failure by yielding is predicted by the condition:

$$\sqrt{\sigma_1{}^2 - \sigma_1\sigma_2 + \sigma_2{}^2} = \sigma_{yp}. \qquad (14.19)$$

The *maximum normal stress theory* is based on the assumption that failure by yielding occurs whenever the largest principal stress is equal to the yield strength σ_{yp}, or that failure by rupture occurs whenever the largest principal stress is equal to the ultimate strength σ_u of the material. This is a relatively simple theory to implement, and it applies both for ductile and brittle materials.

The failure theories reviewed above can be compared by representing them on a common σ_1 versus σ_2 graph, as illustrated in Fig. 14.24. For a given theory, failure occurs if the stress level falls on or outside the closed boundary representing that theory. Among these three failure theories, the maximum distortion energy theory predicts yielding with highest accuracy and shows the best agreement with the experimental results when it is applied to ductile materials. For brittle materials, the maximum normal stress theory is more suitable. If the distortion energy theory is accepted as the basis of comparison, then the maximum shear stress theory is always more conservative and safer, whereas the maximum normal stress theory is conservative whenever the signs of the principal stresses are alike.

The failure theories reviewed here are valid for static loading conditions. These theories must be modified to account for dynamic or repeatedly applied loads that can cause fatigue.

14.7 Allowable Stress and Factor of Safety

Stress and failure analyses constitute the fundamental components in the design of structures. A structure must be designed to withstand the maximum possible stress level, called *working stress*, to which it will be subjected when it is put in service. However, the exact magnitudes of the loads that will be acting upon the structure may not be known. The structure may be subjected to unexpectedly high loads, dynamic loading conditions, or a corrosive environment that can alter the physical properties of the structural material. To account for the effects of uncertainties, a stress level called the *allowable stress* must be set considerably lower than the ultimate strength of the material. The allowable stress must be low enough to provide a margin of safety. It must also allow for an efficient use of the material.

Safety against unpredictable conditions can be achieved by considering a *factor of safety*. The factor of safety n is usually determined as the ratio of the ultimate strength of the material to the allowable stress. The factor of safety is a number greater than one, and may vary depending on whether the material is loaded under tension, compression, or shear. Instead of the ultimate strength, the factor of safety can also be based on the yield strength of the material. This is particularly important for operational conditions in which excessive yielding or plastic deformations cannot be tolerated. In the case of fatigue loading, endurance limit or the fatigue strength of the material must be used. Once the factor of safety is established, the allowable stress σ_{all} can be determined. For example, based on the ultimate strength σ_u criteria:

$$\sigma_{all} = \frac{\sigma_u}{n}. \qquad (14.20)$$

Fig. 14.24 Comparison of the failure theories

14.8 Factors Affecting the Strength of Materials

As discussed in the previous chapter, elastic and shear moduli, yield strength, ultimate strength, stress at failure, and the area under the stress–strain diagram for a material are indications of the strength of the material. There are many physical and environmental factors that may influence the properties of a material. For example, temperature can alter the strength of a material by altering its physical properties.

Common materials expand when heated and contract when cooled. An increase in temperature will lower the ultimate strength of a material. The stresses in structures caused by temperature changes can be quite considerable.

Friction occurs when two surfaces roll or slide over one another. Friction dissipates energy primarily as heat. Another consequence of the sliding action of two surfaces is the removal of material from the surfaces, which is called *wear*. Wear can alter the surface quality of structures, expose them to corrosive environments, and consequently reduce their mechanical strength. *Corrosion* is one of the primary causes of mechanical failure. Failure by corrosion can be accelerated by the presence of stresses. Corrosion can cause the development of minute cracks in the material, which can propagate in a stressed environment. In general, friction, wear, corrosion, and the presence of discontinuities in a material can lower its strength.

14.9 Fatigue and Endurance

The failure theories reviewed in Sect. 14.6 are based on the material properties established under static loading configurations. They attempt to predict the response of a material to a loading configuration that is applied once in a specific direction. Many structures, including machine parts and muscles and bones in the human body, are subjected to repeated loading and unloading. Loads that may not cause the failure of a structure in a single application may cause fracture when applied repeatedly. Failure may occur after a few cycles of loading and unloading, or after millions of cycles, depending on the amplitude of the applied load, the physical properties of the material, the size of the structure, the surface quality of the structure, and the operational conditions. Fracture due to repeated loading is called *fatigue*, and in mechanics, fatigue implies a condition of complete structural failure.

Fatigue analysis of structures is quite complicated. There are several experimental techniques developed to understand the fatigue behavior of materials. The fatigue behavior of a material can be determined in a fatigue test using tensile, compressive, bending, or torsional forces. Here, fatigue due to a combination of tension and compression is explained.

Consider the bar shown in Fig. 14.25. Assume that the bar is made of a material whose ultimate strength is σ_u. This bar is first stressed to a level σ_m (a mean stress) that is considerably lower than σ_u. The bar is then subjected to a stress fluctuating over time, sometimes tensile and other times compressive. The amplitude σ_a of the stress is such that the bar is subjected to a maximum tensile stress of $\sigma_{max} = \sigma_m + \sigma_a$, which is less than the ultimate strength σ_u of the material. This reversible and periodic stress is applied until the bar fractures and the number of cycles (N) to fracture is recorded. This experiment is repeated on specimens having the same geometric and

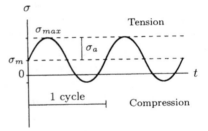

Fig. 14.25 Uniaxial fatigue test

material properties by applying sinusoidal stresses of varying amplitude. The results show that the number of cycles to failure depends on the stress amplitude σ_a. The higher the σ_a, the lower the N.

A typical result of a fatigue test is plotted in Fig. 14.26 on a diagram showing stress amplitude versus number of cycles to failure (σ–N). For a given N, the corresponding stress value is called the *fatigue strength* of the material at that number of cycles. For a given stress level, N represents the fatigue life of the material, which increases rapidly with decreasing stress amplitude. In Fig. 14.26, the experimental data are represented by a single curve. For some materials, the σ–N curve levels off and becomes essentially a horizontal line. The stress at which the fatigue curve levels off is called the *endurance limit* of the material, which is denoted by σ_e in Fig. 14.26. Below the endurance limit, the material has a high probability of not failing in fatigue no matter how many cycles of stress are imposed upon the material.

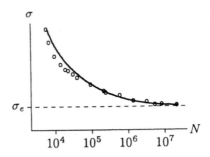

Fig. 14.26 Stress amplitude versus number of cycles to failure

A brittle material such as glass or ceramic will undergo elastic deformation when it is subjected to a gradually increasing load. *Brittle fracture* occurs suddenly without exhibiting considerable plastic deformation (Fig. 14.27). *Ductile fracture*, on the other hand, is characterized by failure accompanied by considerable elastic and plastic deformations. When a ductile material is subjected to fatigue loading, the failure occurs suddenly without showing

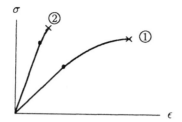

Fig. 14.27 Comparing ductile (*1*) and brittle (*2*) fractures

considerable plastic deformation (yielding). The fatigue failure of a ductile material occurs in a manner similar to the static failure of a brittle material.

The fatigue behavior of a material depends upon several factors. The higher the temperature, the lower the fatigue strength. The fatigue behavior is very sensitive to surface imperfections and presence of discontinuities within the material that cause stress concentrations. The fatigue failure starts with the creation of a small crack on the surface of the material, which propagates under the effect of repeated loads, resulting in the rupture of the material. A good surface finish can improve the fatigue life of a material.

Orthopedic devices undergo repeated loading and unloading due to the activities of the patients and the actions of their muscles. Over a period of years, a weight-bearing prosthetic device or a fixation device can be subjected to a considerable number of cycles of stress reversals due to normal daily activity. This cyclic loading and unloading can cause fatigue failure of the components of a prosthetic device.

14.10 Stress Concentration

Consider the rectangular bar shown in Fig. 14.28. The bar has a cross-sectional area A and is subjected to a tensile force F. The internal reaction force per unit cross-sectional area is defined as the stress, and in this case,

$$\sigma = \frac{F}{A}. \qquad (14.21)$$

Fig. 14.28 Uniform stress distribution

The classic definition of stress is based on the assumption that the external force F is applied over a relatively large area rather than at a single point, and that the cross-sectional area of the bar is constant throughout the length of the bar. Consequently, as illustrated in Fig. 14.28, the stress distribution is uniform over the cross-sectional area of the bar throughout the length of the bar. If the uniformity of the cross-sectional area of the bar is disturbed by the presence of holes, cracks, fillets, scratches, or notches, or if the force is applied over a very small area, then the stress distribution will no longer be uniform at the section where the discontinuity is present, or around the region where the force is applied.

Consider the plate with a circular hole of diameter d shown in Fig. 14.29a. The plate is subjected to a tensile load of F. The equilibrium considerations of the plate require that the resultant internal reaction force is equal to F at any section of the plate. As shown in Fig. 14.29b, at a section away from the hole (e.g., at section aa), the stress distribution is assumed to be uniform. If the cross-sectional area of the plate is A, the magnitude of this uniform stress can be calculated as $\sigma = F/a$. Now, consider the section cc passing through the center of the hole. The magnitude of the average stress at section cc is $\bar{\sigma} = F/(A - A_h)$ where A_h is the hollow area of the cross-section of the plate at section cc. Since $A - A_h$ is always less than A, the magnitude of the average stress at section cc is greater than the magnitude of the uniform stress at section aa. Furthermore, the distribution of stress at section cc is not uniform, but the stress is maximum along the edges of the hole (Fig. 14.29c). That is, the stress is concentrated around the hole. This phenomenon is known as the *stress concentration*.

Fig. 14.29 Effects of stress concentration

Based on experimental observations, there are empirical formulas established to calculate the maximum stresses developed due to the presence of stress concentrators. The general relationship between the maximum stress σ_{max} (or τ_{max}) and the average stress $\bar{\sigma}$ (or $\bar{\tau}$) is such that:

$$\sigma_{max} = k\bar{\sigma}. \qquad (14.22)$$

In Eq. (14.22), k is known as the *stress concentration factor*. The value of the stress concentration factor is greater than one and varies depending on many factors such as the size of the stress concentrator relative to the size of the structure (e.g., the ratio of the diameter of the hole and the width of the structure), the type of applied load (tension, compression, shear, bending, torsion, or combined), and the physical properties of the material (ductility, brittleness, hardness).

Although the stress levels measured by considering the uniform cross-sectional area of the structure may be below the fracture strength of the material, the structure may fail unexpectedly due to stress concentration effects. The fracture or ultimate strength of a material may be exceeded locally due to the presence of a stress concentrator. Note here that the fatigue failure of structures is explained by the localized stress theory due to the stress concentration effects. There may be very small imperfections or discontinuities inside or on the surface of a structure. These small holes or notches may not cause any serious problem when the structure is subjected to static loading configurations. However, repeatedly applied loads can start minute cracks in the material at the locations of discontinuities. With each application of the load these cracks may propagate, and eventually cause the material to rupture.

The effect of stress concentrations on the lives of bones and orthopedic devices is very important. It is noted that after the removal of orthopedic screws from a bone, the screw holes remain in the bone for many months. During the first few months after the removal of screws, the bones may fracture through the sections of one of the screw holes. A screw hole in the bone causes stress concentration effects and makes the bone weaker, particularly in bending and torsion.

The effects of stress concentrations can be reduced by good surface finish and by avoiding unnecessary holes or any other sudden shape changes in the structure.

14.11 Torsion

Torsion is one of the fundamental modes of loading resulting from the twisting action of applied forces. Here, torsional analyses will be limited to circular shafts. The analyses of structures with noncircular cross-sections subjected to torsional loading are complex and beyond the scope of this text.

Consider the solid circular shaft shown in Fig. 14.30. The shaft has a length l and a radius r_o. AB represents a straight line on the outer surface of the shaft that is parallel to its centerline. Note that a plane passing through the centerline and cutting the shaft into two semicylinders is called a *longitudinal plane*, and a plane perpendicular to the longitudinal planes is called a *transverse plane* (plane *abcd* in Fig. 14.31). In this case, line AB lies along one of the longitudinal planes. The shaft is mounted to the wall at one end, and a twisting torque with magnitude M is applied at the other end (shown in Fig. 14.30 with a double-headed arrow). Due to the externally applied torque, the shaft deforms in such a way that the straight line AB is twisted into a helix AB'. The deformation at A is zero because the shaft is fixed at that end. The extent of deformation increases in the direction from the fixed end toward the free end. Angle γ in Fig. 14.30 is a measure of the deformation of the shaft, and it represents the shear strain due to the shear stresses induced in the transverse planes. The tangent of angle γ is approximately equal to the ratio of the arc length BB' and the length l of the shaft. For small deformations, the tangent of this angle is approximately equal to the angle itself measured in radians. Therefore,

$$\gamma = \frac{\text{arc length BB}'}{l}. \qquad (14.23)$$

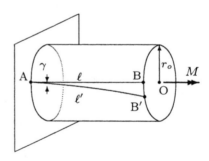

Fig. 14.30 A circular shaft subjected to torsion

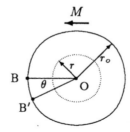

Fig. 14.31 A plane perpendicular to the centerline cuts the shaft into two

As illustrated in Fig. 14.32, the amount of deformation within the shaft also varies with respect to the radial distance r measured from the centerline of the shaft. This variation is such that the deformation is zero at the center, increases toward the rim, and reaches a maximum on the outer surface. Angle θ in Fig. 14.32 is called the *angle of twist* and it is a measure of the extent of the twisting action that the shaft suffers. From the geometry of the problem,

$$\theta = \frac{\text{arc length BB}'}{r_o}. \quad (14.24)$$

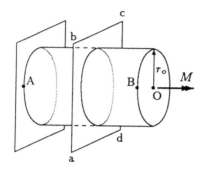

Fig. 14.32 θ is the angle of twist

Equations (14.23) and (14.24) can be combined together by eliminating the arc length BB'. Solving the resulting equation for the angle of twist will yield:

$$\theta = \frac{l}{r_o}\gamma. \quad (14.25)$$

Now consider a plane perpendicular to the centerline of the shaft (plane *abcd* in Fig. 14.31) that cuts the shaft into two segments. Since the shaft as a whole is in static equilibrium, its individual parts have to be in static equilibrium as well. This condition requires the presence of internal shearing forces distributed over the cross-sectional area of the shaft (Fig. 14.33). The intensity of these internal forces (force per unit area) is the shear stress τ. The magnitude of the shear stress is related to the magnitude M of the applied torque, the cross-sectional area of the shaft, and the radial distance r between the centerline and the point at which the shear stress is to be determined. This relationship can be determined by satisfying the rotational equilibrium of either the right-hand or the left-hand segment of the shaft, which will yield:

$$\tau = \frac{Mr}{J}. \quad (14.26)$$

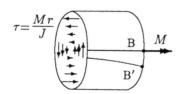

Fig. 14.33 Shear stress, τ, distribution over the cross-sectional area of the shaft

This is known as the *torsion formula*. In Eq. (14.26), J is the *polar moment of inertia* of the cross-sectional area about the centerline of the shaft. The polar moment of inertia for a solid circular shaft with radius r_o (Fig. 14.34a) about its centerline is:

$$J = \frac{\pi r_o^4}{2}.$$

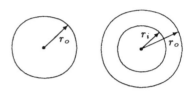

(a) Solid shaft   (b) Hollow shaft

$$J = \frac{\pi r_o^4}{2} \qquad J = \frac{\pi(r_o^4 - r_i^4)}{2}$$

Fig. 14.34 Polar moments of inertia for circular cross-sections

The polar moment of inertia for a hollow circular shaft (Fig. 14.34b) with inner radius r_i and outer radius r_o is:

$$J = \frac{\pi}{2}(r_o^4 - r_i^4).$$

The polar moment of inertia has the dimension of length to the power four, and therefore, has a unit m^4 in SI.

If the shaft material is linearly elastic or the deformations are within the proportionality limit, then the stress and strain must be linearly proportional. In the case of shear loading, the constant of proportionality is the shear modulus, G, of the material:

$$\tau = G\gamma. \quad (14.27)$$

Solving Eq. (14.27) for γ and substituting (14.26) into the resulting equation will yield:

$$\gamma = \frac{\tau}{G} = \frac{Mr}{GJ}. \quad (14.28)$$

On the circumference of the shaft, r in Eq. (14.28) is equal to r_o. Substituting Eq. (14.28) into Eq. (14.25), a more useful expression for the angle of twist can be obtained:

$$\theta = \frac{Ml}{GJ}. \quad (14.29)$$

The following are some important remarks about torsion, torsion formula, and torsional stresses.

- To derive the torsion formula, several assumptions and idealizations were made. For example, it is assumed that the material is isotropic, homogeneous, and linearly elastic.

- For a given shaft and applied torque, the torsional shear stress τ is a linear function of the radial distance r measured from the center of the shaft. The shear stress is distributed nonuniformly over the cross-sectional area of the shaft. At the center of the shaft, $r = 0$ and $\tau = 0$. The stress-free centerline of the solid circular shaft is called the *neutral axis*. The magnitude of the torsional shear stress increases in the direction from the center toward the rim, and reaches a maximum on the circumference of the shaft where $r = r_o$ and $\tau = Mr_o/J$ (Fig. 14.35).

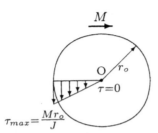

Fig. 14.35 Shear stress varies linearly with radial distance

- Torsion formula takes a special form, $\tau = 2M/\pi r_o{}^3$, at the rim of a solid circular shaft for which $J = \pi r_o{}^4/2$. This equation indicates that the larger the radius of the shaft, the harder it is to deform it in torsion.

- Since $\gamma = \tau/G = Mr/GJ$, the greater the magnitude of the applied torque, the larger the shear stress and shear deformation. The greater the shear modulus of the shaft material, the stiffer the material and the more difficult to deform it in torsion.

- The shear stress discussed herein is that induced in the transverse planes. For a shaft subjected to torsional loading, shear stresses are also developed along the longitudinal planes. This is illustrated in Fig. 14.36 on a material element that is obtained by cutting the shaft with two transverse and two longitudinal planes. The transverse and longitudinal stresses are denoted with τ_t and τ_l, respectively, and the equilibrium of the material element requires that τ_t and τ_l are numerically equal.

Fig. 14.36 Transverse (τ_t) and longitudinal (τ) stresses

- A shaft subjected to torsion not only deforms in shear but is also subjected to normal stresses. This can be explained by the fact that the straight line AB deforms into a helix AB′, as illustrated in Fig. 14.30. The length l before deformation is increased to length l' after the deformation, and an increase in length indicates the presence of tensile stresses along the direction of elongation.

- Consider the material element in Fig. 14.37. The normals of the sides of this material element make an angle 45° with the centerline of the shaft. It can be illustrated by proper coordinate transformations that the only stresses induced on the sides of such an element are normal stresses (tensile stress σ_1 and compressive stress σ_2). The absence of shear stresses on a material element indicates that the normal stresses present are the principal (maximum and minimum) stresses, and that the planes on which these stresses act are the principal planes. (These concepts are discussed in the following chapter.) For structures subjected to pure torsion, material failure occurs along one of the principal planes. This can be demonstrated by twisting a piece of chalk until it breaks into two pieces. A careful examination of the chalk will reveal the occurrence of the fracture along a spiral line normal to the direction of maximum tension. For circular shafts, the spiral lines make an angle of 45° with the neutral axis (centerline). The same fracture pattern has been observed for bones subjected to pure torsion (Fig. 14.38).

Fig. 14.37 Under pure torsion, principal normal stresses (τ_1, τ_2) occur on planes whose normals are at 45° with the centerline

Fig. 14.38 Spiral fracture pattern for a bone subjected to pure torsion

There are various experimental methods to analyze the behavior of structures under torsion. Figure 14.39 illustrates a simplified, schematic drawing of a standard torsion testing machine. The important components of this machine are an adjustable pendulum (A), an angular displacement transducer (B), a torque transducer (C), a rotating grip (D), and a stationary grip (E). This machine can be used to determine

Fig. 14.39 Standard torsion testing machine

the torsional characteristics of specimens, in this case, of bone (F).

The pendulum generates a twisting torque about its shaft, which is also connected to the rotating grip of the machine, and forces an angular deformation in the specimen. The magnitude of the torque applied on the specimen can be controlled by adjusting the position of the mass of the pendulum relative to its center of rotation. The closer the mass to the center of rotation of the pendulum, the smaller the length of the moment arm as measured from the center and, therefore, the smaller the torque generated. Conversely, the more distal the mass of the pendulum from the center, the larger the length of the moment arm and the larger the twisting action (torque) of the weight of the pendulum. The torque generated by the weight of the pendulum is transmitted through a shaft that is connected to the rotating grip, thereby applying the same torque to the specimen firmly placed between the two grips. The torque and angular displacement transducers measure the amount of torque applied on the specimen and the corresponding angular deformation of the specimen. The fracture occurs when the torque applied is sufficiently high so that the stresses generated in the specimen are beyond the ultimate strength of the material.

The data collected by the transducers of the torsion test machine can be plotted on a torque (M) versus angular deformation (angle of twist, θ) graph. A typical M–θ graph is shown in Fig. 14.40. This graph can be analyzed to gather information about the material properties of the specimen.

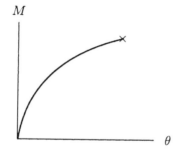

Fig. 14.40 Applied torque versus angular displacement

Example 14.4 A human femur is mounted in the grips of the torsion testing machine (Fig. 14.41). The length of the bone at sections between the rotating (D) and stationary (E) grips is measured as $L = 37$ cm. The femur is subjected to a torsional loading until fracture, and the applied torque versus angular displacement (deflection) graph shown in Fig. 14.42 is obtained. The femur is fractured at a section (section *aa* in Fig. 14.41) that is $l = 25$ cm distance away from the stationary grip. The geometry of the bony tissue at the fractured section is observed to be a circular ring with inner radius $r_i = 7$ mm and outer radius $r_o = 13$ mm (Fig. 14.41).

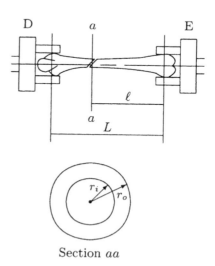

Section *aa*

Fig. 14.41 Fractured bone and its cross-sectional geometry

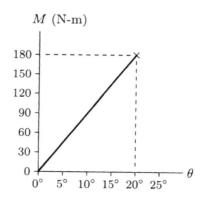

Fig. 14.42 Troque versus angle of twist diagram

Calculate the maximum shear strain and shear stress at the fractured section of the femur, and determine the shear modulus of elasticity of the femur.

Solution

In Fig. 14.41, $\theta = 20°$ is the maximum angle of deformation (angle of twist) measured at the rotating grip at the instant when fracture occurred. The total length of the bone between the rotating and stationary grips is measured as $L = 37$ cm. Therefore, the angular deformation is $(20°)/(37 \text{ cm}) = 0.54°/\text{cm}$ of bone length as measured from the stationary grip. The fracture occurred at section aa which is $l = 25$ cm away from the stationary grip. Therefore, the angular deflection θ_{aa} at section aa just before fracture is $(0.54°/\text{cm})(25 \text{ cm}) = 13.5°$ or $(13.5°)(\pi/180°) = 0.236$ rad.

The shear strain γ on the surface of the bone at section aa can be determined from Eq. (14.25). In this case, $l = 0.25$ m is the distance between section aa and the section of the femur where the stationary grip holds the bone, $r_0 = 0.013$ m is the outer radius of the cross-section of the bone at the fracture, and $\theta = \theta_{aa} = 0.236$ is the angle of twist measured in radians. Therefore,

$$\gamma = \frac{r_0}{l}\theta_{aa} = \left(\frac{0.013}{0.25}\right)(0.236) = 0.0123 \text{ rad}.$$

The torsion formula Eq. (14.26) relates the applied torque M, radial distance r, polar moment of inertia J of the cross-section, and the shear stress τ. The cross-sectional geometry of the bony structure of the femur at section aa is a ring with inner radius $r_i = 0.007$ m and outer radius $r_0 = 0.013$ m. Therefore, the polar moment of inertia of the cross-section of the femur at section aa is:

$$J = \frac{\pi}{2}(r_0^4 - r_i^4) = \frac{3.14}{2}[(0.013)^4 - (0.007)^4]$$
$$= 41.1 \times 10^{-9} \text{ m}^4.$$

In Fig. 14.42, the magnitude of the applied torque at the instant when the fracture occurred (maximum torque) is $M = 180$ N–m. Hence, the maximum shear stress on the surface of the bone at section aa can be determined using the torsion formula:

$$\tau = \frac{Mr_0}{J} = \frac{(180)(0.013)}{41.1 \times 10^{-9}} = 56.9 \times 10^6 \text{ Pa} = 56.9 \text{ MPa}.$$

Assuming that the deformations are elastic and the relationship between the shear stress and shear strain is linear, the shear modulus G of the bone can be determined from (14.27):

$$G = \frac{\tau}{\gamma} = \frac{56.9 \times 10^6}{0.0123} = 4.6 \times 10^9 \text{ Pa} = 4.6 \text{ GPa}.$$

Remarks
- For a specimen subjected to torsion, the maximum shear stress before fracture is the *torsional strength* of that specimen. In this case, the torsional strength of the femur is 56.9 MPa.

- The *torsional stiffness* is the ratio of the applied torque and the resultant angular deformation. In this case, the torsional stiffness of the femur is $(180 \text{ N–m})/(20° \times \pi/180°) = 515.7$ N–m/rad.
- The *torsional rigidity* is the torsional stiffness multiplied by the length of the specimen. In this case, the torsional rigidity of the femur is $(509.9 \text{ N-m/rad})(0.37 \text{ m}) = 190.8$ N–m²/rad.
- The maximum amount of torque applied to a specimen before fracture is defined as the *torsional load capacity* of the specimen. In this case, the torsional load capacity of the femur is 180 N–m.
- The total area under the torque versus angular displacement diagram represents the *torsional energy storage capacity* of the specimen or the *torsional energy absorbed* by the specimen. In this case, the torsional energy storage capacity of the femur is $1/2(180 \text{ N–m})(20° \times (\pi/180°)) = 31.4$ N–m–rad.
- Torsional fractures are usually initiated at regions of the bones where the cross-sections are the smallest. Some particularly weak sections of human bones are the upper and lower thirds of the humerus, femur, and fibula; the upper third of the radius; and the lower fourth of the ulna and tibia.

Example 14.5 Consider the solid circular cylinder shown in Fig. 14.43a. The cylinder is subjected to pure torsion by an externally applied torque, M. As illustrated in Fig. 14.43b, the state of stress on a material element with sides parallel to the longitudinal and transverse planes of the cylinder is pure shear. For given M and the parameters defining the geometry of the cylinder, the magnitude τ_{xy} of the torsional shear stress can be calculated using the torsion formula provided in Eq. (14.26).

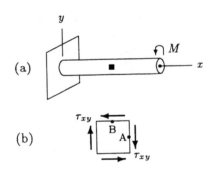

Fig. 14.43 Example 14.5

Using Mohr's circle, investigate the state of stress in the cylinder.

Solution

Mohr's circle in Fig. 14.44a is drawn by using the stress element of Fig. 14.43b. On surfaces A and B of the stress element shown in Fig. 14.43b, there is only a negative shear stress with magnitude τ_{xy}. Therefore, both A and B on the τ–σ diagram lie along the τ axis where $\sigma = 0$. Furthermore, the origin of the τ–σ diagram constitutes the midpoint between A and B, and hence, the center, C, of the Mohr's circle. The distance between C and A is equal to τ_{xy}, which is the radius of the Mohr's circle. Mohr's circle cuts the horizontal axis at two locations, both at a distance τ_{xy} from the origin. Therefore, the principal stresses are such that $\sigma_1 = \tau_{xy}$ (tensile) and $\sigma_2 = \tau_{xy}$ (compressive). Furthermore, τ_{xy} is also the maximum shear stress.

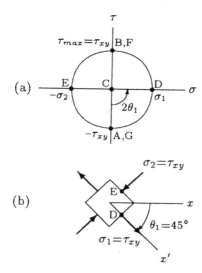

Fig. 14.44 Analyses of the material element in Fig. 14.43

The point where $\sigma = \sigma_1$ on the τ–σ diagram in Fig. 14.44a is labeled as D. The angle between lines CA and CD is 90°, and it is equal to half of the angle of orientation of the plane with normal in one of the principal directions. Therefore, the planes of maximum and minimum normal stresses can be obtained by rotating the element in Fig. 14.43b 45° (clockwise). This is illustrated in Fig. 14.44b.

The lines that follow the directions of principal stresses are called the *stress trajectories*. As illustrated in Fig. 14.45 for a circular cylinder subjected to pure torsion, the stress trajectories are in the form of helices making an angle 45° (clockwise and counterclockwise) with the longitudinal axis of the cylinder. As discussed previously, the significance of these stress trajectories is such that if the material is weakest in tension, the failure occurs along a helix such as *bb* in Fig. 14.45, where the tensile stresses are at a maximum.

Fig. 14.45 Stress trajectories

14.12 Bending

Consider the simply supported, originally straight beam shown in Fig. 14.46. A force with magnitude F applied downward bends the beam, subjecting parts of the beam to shear, tension, and compression. There are several ways to subject structures to bending. The free-body diagram of the beam in Fig. 14.46 is shown in Fig. 14.47a. There are three parallel forces acting on the beam. F is the applied load, and R_1 and R_2 are the reaction forces. The type of bending to which this beam is subjected is called a *three-point bending*. The beam in Fig. 14.47b, on the other hand, is subjected to a *four-point bending*.

Fig. 14.46 Bending

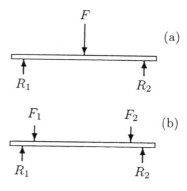

Fig. 14.47 Three- and four-point bending

The stress analyses of structures subjected to bending start with static analyses that can be employed to determine both external reactions and internal resistances. For two-dimensional problems in the xy-plane, the number of equations available from statics is three. Two of these equations are translational equilibrium conditions ($\sum F_x = 0$ and $\sum F_y = 0$) and the third equation guarantees rotational equilibrium ($\sum M_z = 0$). The internal resistance of structures to externally applied loads can be determined by applying the *method of sections*, which is based on the fact that the individual parts of a structure that is itself in static equilibrium must also be in equilibrium. This concept makes it possible to utilize the equations of equilibrium for calculating internal forces and moments, which can then be used to determine stresses. Stresses in a structure subjected to bending may vary from one section to another and from one point to another over a given section. For design and failure analyses, maximum normal and shear stresses must be considered. These critical stresses can be determined by the repeated application of the method of sections throughout the structure.

Consider the simply supported beam shown in Fig. 14.48a. Assume that the weight of the beam is negligible. The length of the beam is l, and the two ends of the beam are labeled as A and B. The beam is subjected to a concentrated load with magnitude F, which is applied vertically downward at point C. The distance between A and C is l_1. The free-body diagram of the beam is shown in Fig. 14.48b. By applying the equations of equilibrium, the reaction forces at A and B can be determined as $R_A = (1 - (l_1/l))F$ and $R_B = (l_1/l)F$.

To determine the internal reactions, the method of sections can be applied. As shown in Fig. 14.48c, this method is applied at sections aa and bb because the nature of the internal reactions on the left-hand and right-hand sides

of the concentrated load are different. Once a hypothetical cut is made, either the left-hand or the right-hand segment of the beam can be analyzed for the internal reactions. In Fig. 14.48c, the free-body diagrams of the left-hand segments of the beam are shown. For the vertical equilibrium of each of these segments, there must be an internal shear force at the cut. The magnitude of this force at section aa can be determined as $V = R_A$ by applying the equilibrium condition $\sum F_y = 0$. This force acts vertically downward and is constant between A and C. The magnitude of the shear force at section bb is $V = F - R_A = R_B$, and since F is greater than R_A, it acts vertically upward. The sign convention adopted in this text for the shear force is illustrated in Figs. 14.49 and 14.50. An upward internal force on the left-hand segment (or the downward internal force on the right-hand segment) of a cut is positive. Otherwise the shear force is negative. Therefore, as illustrated in Fig. 14.48d, the shear force is negative between A and C, and positive between C and B.

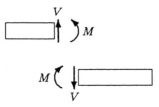

Fig. 14.49 Positive shear forces and bending moments

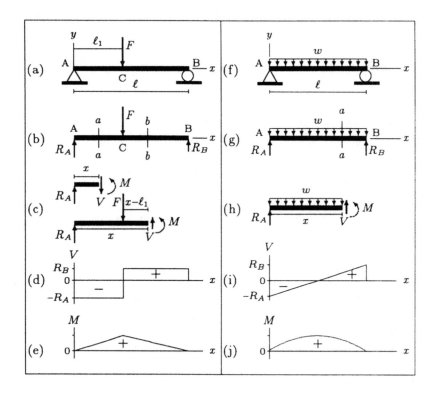

Fig. 14.48 Free-body, shear, and moment diagrams for simply supported beams subjected to concentrated and distributed loads

Fig. 14.50 Negative shear forces and bending moments

In addition to the vertical equilibrium, the segments must also be checked for rotational equilibrium. As illustrated in Fig. 14.48c, this condition is satisfied if there are internal resisting moments at sections *aa* and *bb*. By utilizing the equilibrium condition $\sum M = 0$, the magnitudes of these counterclockwise moments can be determined for sections *aa* and *bb* as $M = xR_A$ and $M = xR_A - (x - l_1)F$, respectively. Note that M is a function of the axial distance x measured from A. The function relating M and x between A and C is different than that between C and B. These functions are plotted on an M versus x graph in Fig. 14.48e. The moment is maximum at C where the load is applied. As illustrated in Figs. 14.49 and 14.50, the sign convention adopted for the moment is such that a counterclockwise moment on the left-hand segment (or the clockwise moment on the right-hand segment) of a cut is positive.

It can also be demonstrated that shear force and bending moment are related through the following equation:

$$V = \frac{dM}{dx}. \tag{14.30}$$

If the variation of M along the length of the structure is known, then the shear force at a given section of the structure can also be determined using Eq. (14.30).

The procedure outlined above is also applied to analyze a simply supported beam subjected to a distributed load of w (force per unit length of the beam), which is illustrated in Fig. 14.48f–j. The procedure for determining the internal shear forces and resisting moments in cantilever beams subjected to concentrated and distributed loads is outlined graphically in Fig. 14.51.

When a structure is subjected to bending, both normal and shear stresses are generated in the structure. For example, consider the beam shown in Fig. 14.52. The beam is bent by a downward force. If the beam is assumed to consist of layers of material, then the upper layers of the beam are compressed while the layers on the lower portion of the beam are subjected to tension. The extent of compression or the amount of contraction is maximum at the uppermost layer, and the amount of elongation is maximum on the bottom layer. There is a layer somewhere in the middle of the beam, where a transition from tension to compression occurs. For such a layer, there is no tension or compression, and therefore, no deformation in the longitudinal direction. This stress-free layer that separates the zones of tension and compression is called the *neutral plane* of the beam. The line of intersection of the neutral plane with a plane (transverse) cutting the longitudinal axis of the beam at right angles is called the *neutral axis*. The neutral axis passes through the

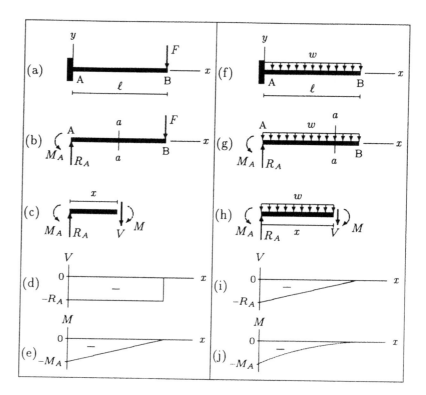

Fig. 14.51 Free-body, shear, and moment diagrams for cantilever beams subjected to concentrated and distributed loads

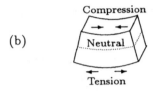

(a)

(b)

Fig. 14.52 Tension and compression in bending

centroid. Centroids of symmetrical cross-sections are located at their geometric centers (Fig. 14.53).

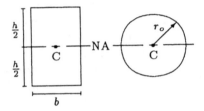

Fig. 14.53 Centroids (C) and neutral axes (NA) for a rectangle and a circle

The above discussion indicates that when a beam is bent, it is subjected to stresses occurring in the longitudinal direction or in a direction normal to the cross-section of the beam. Furthermore, for the loading configuration shown in Fig. 14.52, the distribution of these normal stresses over the cross-section of the beam is such that it is zero on the neutral axis, negative (compressive) above the neutral axis, and positive (tensile) below the neutral axis. For a beam subjected to pure bending, the following equation can be derived from the equilibrium considerations of a beam segment:

$$\sigma_x = \frac{My}{I}. \tag{14.31}$$

This equation is known as the *flexure formula*, and the stress σ_x is called the *flexural stress*. In Eq. (14.31), M is the bending moment, y is the vertical distance between the neutral axis and the point at which the stress is sought, and I is the *area moment of inertia* of the cross-section of the beam about the neutral axis. The area moments of inertia for a number of simple geometries are listed in Table 14.1.

Table 14.1 Area (A), area moment of inertia (I) about the neutral axis (NA), first moment of the area (Q) about the neutral axis, and maximum normal (σ_{max}) and shear (τ_{max}) stresses for beams subjected to bending and with cross-section as shown

h — NA, b	$A = bh$ $I = \frac{bh^3}{12}$	$Q = \frac{bh^2}{8}$ $\sigma_{max} = \frac{Mh}{2I}$ $\tau_{max} = \frac{3V}{2A}$
r_o — NA	$A = \pi r_o^2$ $I = \frac{\pi r_o^4}{4}$	$Q = \frac{2r_o^3}{3}$ $\sigma_{max} = \frac{Mr_o}{I}$ $\tau_{max} = \frac{4V}{3A}$
r_o, r_i — NA	$A = \pi(r_o^2 - r_i^2)$ $I = \frac{\pi(r_o^4 - r_i^4)}{4}$	$Q = r_o(r_o^2 + r_i^2)$ $\sigma_{max} = \frac{Mr_o}{I}$ $\tau_{max} = \frac{2V}{A}$

The stress distribution at a section of a beam subjected to pure bending is shown in Fig. 14.54. At a given section of the beam, both the bending moment and the area moment of inertia of the cross-section are constant. According to the flexure formula, the flexural stress σ_x is a linear function of the vertical distance y measured from the neutral axis, which can take both positive and negative values. At the neutral axis, $y = 0$ and σ_x is zero. For points above the neutral axis, y is positive and σ_x is negative, indicating compression. For points below the neutral axis, y is negative and σ_x is positive and tensile. At a given section, the stress reaches its absolute maximum value either at the top or the bottom of the beam where y is maximum. It is a common practice to indicate the maximum value of y with c, eliminate the negative sign indicating direction (which can be found by inspection), and write the flexure formula as:

$$\sigma_{max} = \frac{Mc}{I}. \tag{14.32}$$

Fig. 14.54 Normal (flexural) stress distribution over the cross-section of the beam

For example, for a beam with a rectangular cross-section, width b, height h, and bending moment M, $c = h/2$ and the magnitude of the maximum flexural stress is:

$$\sigma_{max} = \frac{M}{I}\frac{h}{2} \quad with \quad I = \frac{bh^3}{12}.$$

Note that while deriving the flexure formula, a number of assumptions and idealizations are made. For example, it is assumed that the beam is subjected to pure bending. That is, it is assumed that shear, torsional, or axial forces are not present. The beam is initially straight with a uniform, symmetric cross-section. The beam material is isotropic and homogeneous, and linearly elastic. Therefore, Hooke's law ($\sigma_x = E\varepsilon_x$) can be used to determine the strains due to flexural stresses. Furthermore, Poisson's ratio of the beam material can be used to calculate lateral contractions and/or elongations.

When the internal shear force in a beam subjected to bending is not zero, a shear stress is also developed in the beam. The distribution of this shear stress on the cross-section of the beam is such that it is greatest on the neutral axis and zero on the top and bottom surfaces of the beam. The following formula is established to calculate shear stresses in bending:

$$\tau_{xy} = \frac{VQ}{Ib}. \tag{14.33}$$

In Eq. (14.33), V is the shear force at a section where the shear stress τ_{xy} is sought, I is the moment of inertia of the cross-sectional area about the neutral axis, and b is the width of the cross-section. As shown in Fig. 14.55, let y_1 be the vertical distance between the neutral axis and the point at which τ_{xy} is to be determined. Then Q is the *first moment* of the area $abcd$ about the neutral axis. The first moment of area $abcd$ can be calculated as $Q = A\bar{y}$ where A is the area enclosed by $abcd$ and $\bar{y}$ is the distance between the neutral axis and the centroid of area $abcd$. Note that both A and $\bar{y}$ are maximum at the neutral axis, and A is zero at the top and bottom surfaces. Therefore, Q is greatest at the neutral axis,

and zero at the top and bottom surfaces. Maximum Q's for different cross-sections are listed in Table 14.1.

The shear stress distribution over the cross-section of a beam is shown in Fig. 14.56. The shear stress is constant along lines parallel to the neutral axis. The shear stress is maximum at the neutral axis where $y_1 = 0$ and Q is maximum. Maximum shear stresses for different cross-sections can be obtained by substituting the values of I and Q into Eq. (14.19), which are listed in Table 14.1. For example, for a rectangular cross-section:

$$\tau_{max} = \frac{3V}{2bh}.$$

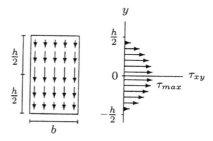

Fig. 14.56 Shear stress distribution over a section of the beam

Consider a cubical material element in a beam subjected to shear force and bending moment as illustrated in Fig. 14.57. On this material element, the effect of bending moment M is represented by a normal (flexural) stress σ_x, and the effect of shear force V is represented by the shear stress τ_{xy} acting on the surfaces with normals in the positive

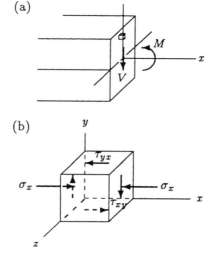

Fig. 14.57 Both normal and shear stresses occur on a material element subjected to bending

Fig. 14.55 Definitions of the parameters involved

and negative x (longitudinal) directions. As shown in Fig. 14.57b, for the rotational equilibrium of this material element, there have to be additional shear stresses (τ_{xy}) on the upper and lower surfaces of the cube (with normals in the positive and negative y directions) such that numerically $\tau_{yx} = \tau_{xy}$. The occurrence of τ_{yx} can be understood by assuming that the beam is made of layers of material, and that these layers tend to slide over one another when the beam is subjected to bending (Fig. 14.58).

Fig. 14.58 Shear stresses in the longitudinal direction

There are various experiments that may be conducted to analyze the behavior of specimens subjected to bending forces. Some of these experiments are introduced within the context of the following examples.

Example 14.6 Figure 14.59 illustrates an apparatus that may be used to conduct three-point bending experiments. This apparatus consists of a stationary head (A) to which the specimen (B) is attached, two rings (C and D), and a mass (E) with weight W applied to the specimen through the rings.

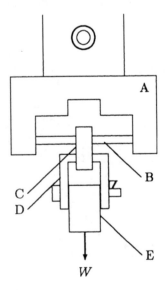

Fig. 14.59 Three-point bending apparatus

For a weight $W = 1,000$ N applied to the middle of the specimen and for a support length of $l = 16$ cm (the distance between the left and the right supports), determine the maximum flexural and shear stresses generated at section bb of a specimen. The specimen has a square ($a = 1$ cm) cross-section, and the distance between the left support and section bb is $d = 4$ cm (Fig. 14.60a).

Fig. 14.60 The free-body diagrams

Solution
The free-body diagram of the specimen is shown in Fig. 14.60a. The force (W) is applied to the middle of the specimen. The rotational and translational equilibrium of the specimen require that the magnitude R of the reaction forces generated at the supports must be equal to half of W. That is, $R = 500$ N.

The specimen has a square cross-section, and its neutral axis is located at a vertical distance $a/2$ measured from both the top and bottom surfaces of the specimen. The normal (flexural) stresses generated at section bb of the specimen depend on the magnitude of the bending moment M at section bb and the area moment of inertia I of the specimen at section bb about the neutral axis. At section bb, the magnitude of the flexural stress is maximum ($\sigma_{\max}$) at the top (compressive) and the bottom (tensile) surfaces of the specimen:

$$\sigma_{\max} = \frac{M}{I}\frac{a}{2}.$$

The internal resistances at section bb of the specimen are shown in Fig. 14.60b. For the rotational equilibrium of the specimen,

$$M = dR = (0.04)(500) = 20\,\text{N–m}.$$

The area moment of inertia of a square with sides a is:

$$I = \frac{a^4}{12} = \frac{(0.01)^4}{12} = 8.33 \times 10^{-10}\,\text{m}^4.$$

Substituting M and I into the flexure formula will yield:

$$\sigma_{\text{max}} = \left(\frac{20}{8.33 \times 10^{-10}}\right) \frac{0.01}{2} = 120 \times 10^6\,\text{Pa} = 120\,\text{MPa}.$$

The shear stress generated at section bb of the specimen is a function of the shear force V at section bb, and the first moment Q and the area moment of inertia I of the cross-section of the specimen at section bb. The shear stress is maximum (τ_{max}) along the neutral axis, such that:

$$\tau_{\text{max}} = \frac{VQ}{Ia}.$$

For the vertical equilibrium of the specimen (Fig. 14.60b), the magnitude V of the internal shear force at section bb must be equal to the magnitude R of the reaction force; $V = R = 500$ N. The first moment of the cross-sectional area of the specimen about the neutral axis is:

$$Q = \frac{a^3}{8} = \frac{(0.01)^3}{8} = 0.125 \times 10^{-6}\,\text{m}^3.$$

Therefore, the maximum shear stress occurring at section bb along the neutral axis is:

$$\tau_{\text{max}} = \frac{(500)(0.125 \times 10^{-6})}{(8.33 \times 10^{-10})(0.01)} = 7.5 \times 10^6\,\text{Pa} = 7.5\,\text{MPa}.$$

Remarks Note that magnitudes and directions of internal shear force and bending moment are different at different sections of the specimen. At a given section, the internal shear force and bending moment are functions of the horizontal distance between the left support (or the right support) and the section. This is illustrated in Fig. 14.61a for a section that lies somewhere between the left support and where the load W is applied on the specimen. If the distance between the left support and the section is x, then the magnitude of the internal bending moment at the section is:

$$M = xR.$$

Fig. 14.61 Stress distribution at a section of the specimen in Fig. 14.60a

In other words, M is a function of x. The stress distribution at the same section is shown in Fig. 14.61b. The upper layers of the specimen are under compression (negative σ_x), while the lower layers are in tension (positive σ_x). Also, a downward shear stress (τ_{xy}) acts throughout the section. Also indicated in Fig. 14.61b are two material elements, A and B. The nonzero stress tensor components acting on the surfaces of these material elements are shown in Fig. 14.62. It is clear that the state of stress at the upper layers of the specimen is different than the state of stress at the lower layers.

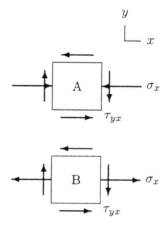

Fig. 14.62 Stress components on material elements. *A* and *B* in Fig. 14.61b

Example 14.7 Consider the beam shown in Fig. 14.63a. The externally applied force and the reactions at the supports bend the beam (three-force bending), subjecting it to shear stresses. In addition to shear, the upper layers of the beam

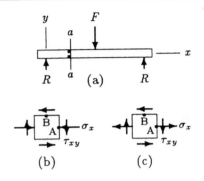

Fig. 14.63 Example 14.7

are subjected to compression and the lower layers to tension. Figure 14.63b shows the state of stress occurring at a material point along a section (section aa) on the left-hand side of the applied force F and above the neutral plane of the beam. Figure 14.63c illustrates the state of stress at the same section below the neutral plane.

Using Mohr's circle, determine the principal stresses and maximum shear stresses occurring in the beam for the states of stress shown in Fig. 14.63b, c.

Solution
At a given section, magnitude F of the externally applied force and parameters defining the geometry of the beam, the normal (flexural) stress σ_x and the shear stress τ_{xy} distributions can be determined in a manner similar to that described in Example 14.6. Once the state of stress at a material point is known, Mohr's circle can be used for further analyses of the stresses involved.

Mohr's circle in Fig. 14.64a is drawn by using the plane stress element of Fig. 14.63b. There is a negative normal stress

with magnitude σ_x and a negative shear stress with magnitude τ_{xy} on surface A of the stress element in Fig. 14.63b. Therefore, point A on Mohr's circle has coordinates $-\sigma_x$ and $-\tau_{xy}$. Surface B on the stress element has only a negative shear stress with magnitude τ_{xy}. Therefore, point B on the τ–σ diagram lies along the σ-axis, τ_{xy} distance above the origin. The center C of Mohr's circle can be determined as the point of intersection of the line connecting A and B with the σ-axis. The radius of the Mohr's circle can also be determined by utilizing the properties of right triangles. In this case,

$$r = \sqrt{(\sigma_x/2)^2 + \tau_{xy}^2}.$$

Once the radius of the Mohr's circle is established, principal stresses and maximum shear stress can be determined as $\sigma_1 = r - \sigma_x/2$ (tensile), $\sigma_2 = r + \sigma_x/2$ (compressive), and $\tau_{max} = r$. To determine the angle of rotation, θ_1, of the plane for which the stresses are maximum and minimum, we need to read the angle between lines CA and CD (in this case, it is $2\theta_1$, counterclockwise), divide it by two, and rotate the stress element in Fig. 14.63b in the clockwise direction through an angle θ_1. This is illustrated in Fig. 14.64b.

Analyses of the material element in Fig. 14.63c are illustrated graphically in Fig. 14.65.

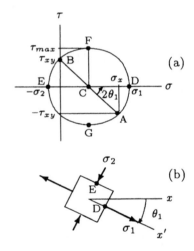

Fig. 14.65 Analyses of the element in Fig. 14.63c

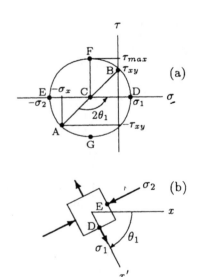

Fig. 14.64 Analyses of the element in Fig. 14.63b

Example 14.8 Figure 14.66 illustrates a bench test that may be used to subject bones to bending. In the case shown, the distal end of a human femur is firmly clamped to the bench and a horizontal force with magnitude $F = 500$ N is applied to the head of the femur at point P.

Determine the maximum normal and shear stresses generated at section aa of the femur that is located at a vertical distance $h = 16$ cm measured from point P. Assume that the geometry of the femur at section aa is circular with inner radius $r_i = 6$ mm and outer radius $r_o = 13$ mm.

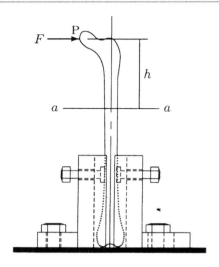

Fig. 14.66 A bench test

Solution

The femur is hypothetically cut into two parts by a plane passing through section aa, and the free-body diagram of the proximal part of the femur is shown in Fig. 14.67. The internal resistances at section aa are the shear force V and bending moment M. The translational and rotational equilibrium of the proximal part of the femur require that:

$$V = F = 500\,N,$$

$$M = hF = (0.16)(500) = 80\,N\text{--}m.$$

Fig. 14.67 The free-body diagram

Formulas to determine the cross-sectional area, A, the area moment of inertia, I, and the first moment of cross-sectional area about the neutral axis of a structure with a hollow circular cross-section, Q, are provided in Table 14.1. Accordingly,

$$A = \pi(r_o^2 - r_i^2) = 4.18 \times 10^{-4}\,\text{m}^2,$$

$$I = \frac{\pi}{4}(r_o^4 - r_i^4) = 2.14 \times 10^{-8}\,\text{m}^4,$$

$$Q = r_o(r_o^2 + r_i^2) = 2.67 \times 10^{-6}\,\text{m}^3.$$

Formulas to determine the maximum normal stress σ_{max} and maximum shear stress τ_{max} for a structure with a hollow circular cross-section and those subjected to bending forces are also provided in Table 14.1:

$$\sigma_{max} = \frac{Mr_o}{I} = \frac{(80)(0.013)}{2.14 \times 10^{-8}} = 48.6 \times 10^6\,\text{Pa} = 48.6\,\text{MPa},$$

$$\tau_{max} = \frac{2V}{A} = \frac{2(500)}{4.18 \times 10^{-4}} = 2.4 \times 10^6\,\text{Pa} = 2.4\,\text{MPa}.$$

Due to the direction of the applied force, the flexural stress is greatest on the medial and lateral sides of the femur. The flexural stress is tensile on the medial side and compressive on the lateral side. The shear stress is greatest along the inner surface of the bony structure of the femur on the ventral and dorsal sides. The loading configuration of the bone is such that it behaves like a cantilever beam.

14.13 Combined Loading

The stress analyses discussed so far were concerned with axial (tension or compression), pure shear, torsional, and flexural (bending) loading of structures based on the assumption that these loads were applied on a structure one at a time. The stresses due to these basic types of loading configurations can be calculated using the following formulas:

Axial Loading: $\qquad \sigma_a = \dfrac{F_a}{A_a}$

Pure shear loading: $\qquad \tau_s = \dfrac{F_s}{A_s}$

Torsional loading: $\qquad \tau_t = \dfrac{M_t r}{J}$

Flexural loading: $\qquad \sigma_b = \dfrac{M_b y}{I}$

Here, σ_a is the normal stress due to axial load F_a applied on an area A_a, τ_s is the shear stress due to shear load F_s applied on an area A_s, τ_t is the shear stress due to applied twisting torque M_t at a point r from the centerline of a cylindrical shaft and at a section with polar moment of inertia J, and σ_b is the normal stress due to bending moment M_b at a distance y from the neutral axis of the structure at a section with area moment of inertia of I.

A structure may be subjected to two or more of these loads simultaneously. To analyze the overall effects of such combined loading configurations, first the stresses generated at a given section of the structure due to each load are determined individually. Next, the normal stresses are combined (added or subtracted) together, and the shear stresses are combined together. The following example is aimed to illustrate how combined stresses can be handled.

Example 14.9 Figure 14.68 illustrates a bench test performed on an intertrochanteric nail. The nail is firmly clamped to the bench and a downward force with magnitude $F = 1,000$ N is applied.

Fig. 14.68 A bench test

Determine the stresses generated at section bb of the nail which is located at a horizontal distance $d = 6$ cm measured from the point of application of the force on the nail. The geometry of the nail at section bb is a square with sides $a = 15$ mm.

Solution
The nail is hypothetically cut into two parts by a plane passing through section bb, and the free-body diagram of the proximal part of the nail is shown in Fig. 14.69.

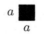

Fig. 14.69 The free-body diagram

The translational equilibrium of the nail in the vertical direction requires the presence of a compressive force at section bb with a magnitude equal to the magnitude $F = 1,000$ N of the external force applied on the nail. The rotational equilibrium condition requires that there is a clockwise internal bending moment at section bb with magnitude:

$$M = dF = (0.06)(1,000) = 60\,\text{N–m}.$$

The compressive force at section bb gives rise to an axial stress σ_a. The nail has a square geometry at section bb, and its cross-sectional area is:

$$A = a^2 = (0.015)^2 = 2.25 \times 10^{-4}\text{m}^2.$$

Therefore, the magnitude of the axial stress at section bb due to the compressive force is:

$$\sigma_\text{a} = \frac{F}{A} = \frac{1,000}{2.25 \times 10^{-4}} = 4.4 \times 10^6\,\text{Pa} = 4.4\,\text{MPa}.$$

The area moment of inertia I of the nail at section bb is:

$$I = \frac{a^4}{12} = \frac{(0.015)^4}{12} = 4.2 \times 10^{-9}\,\text{m}^4.$$

The bending moment M at section bb gives rise to a flexural stress σ_b. The flexural stress is maximum on the medial and lateral sides of the nail, which are indicated as M and L in Fig. 14.69. The maximum flexural stress is:

$$\sigma_{b_\text{max}} = \frac{Ma}{2I} = \frac{(60)(0.015)}{2(4.2 \times 10^{-9})} = 107.1 \times 10^6\,\text{Pa}$$
$$= 107.1\,\text{MPa}.$$

The flexural stress σ_b varies linearly over section bb. It is compressive on the medial half of the nail, zero in the middle, and tensile on the lateral half of the nail. The distribution of normal stresses σ_a and σ_b due to the compressive force F and bending moment M at section bb are shown in Fig. 14.70a, b, respectively. The combined effect of these stresses is shown in Fig. 14.70c. Note from Fig. 14.70c that the resultant normal stress generated at section bb is maximum at the medial (M) side of the nail. This maximum stress is compressive and has a magnitude:

$$\sigma_\text{max} = \sigma_\text{a} + \sigma_{b_\text{max}} = 111.5\,\text{MPa}.$$

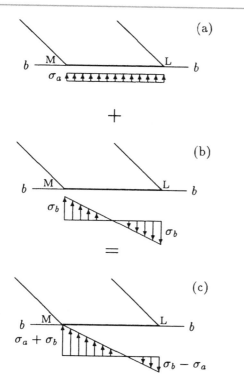

Fig. 14.70 Combined stresses

Since $\sigma_{b_{max}}$ (tensile) is greater than σ_a (compressive), the resultant stress σ_L on the lateral end of section bb is tensile, and its magnitude is equal to:

$$\sigma_L = \sigma_{b_{max}} - \sigma_a = 102.7 \text{ MPa}.$$

14.14 Exercise Problems

Problem 14.1 Complete the following definitions with appropriate expressions.

(a) For a given material and for stresses within the proportionality limit, the ratio of deformations occurring in the axial and lateral directions is constant, and this constant is called the _____.

(b) Stress and strain are known as _____, while force is a vector quantity.

(c) Maximum and minimum normal stresses at a material point are called the _____.

(d) Planes with normals collinear with the directions of the maximum and minimum normal stresses are called the _____.

(e) The _____ occurs on a material element for which normal stresses are equal.

(f) _____ is an effective graphical method of investigating the state of stress at a material point.

(g) By material _____, it is meant that the material either ruptures or undergoes excessive permanent deformation.

(h) Fracture due to repeated loading and unloading is called _____.

(i) The stress at which the fatigue curve levels off is called the _____ of the material.

(j) _____ fracture occurs suddenly without exhibiting considerable plastic deformation.

(k) _____ fracture is characterized by failure accompanied by considerable elastic and plastic deformations.

(l) Presence of holes, cracks, fillets, scratches, and notches in a structure can cause _____ that may lead to unexpected structural failure.

(m) For a circular cylindrical shaft, the _____ formula relates shear stresses to applied torque, polar moment of inertia, and the radial distance from the centerline of the shaft.

(n) The shear stress is zero at the center of a circular shaft subject to pure torsion. The stress-free centerline of the shaft is called the _____.

(o) For structures subjected to pure torsion, material failure occurs along one of the principal planes that make an angle _____ degrees with the centerline.

(p) The stress analyses of structures subjected to bending start with _____ analyses that can be employed to determine both external reactions and internal resistances.

(q) The internal resistance of structures to externally applied forces can be determined by applying the _____.

(r) For a structure subjected to bending, _____ formula relates normal stresses to bending moment and the geometric parameters of the structure.

Answers: Given at the end of the chapter.

Problem 14.2 Consider the rectangular bar shown in Fig. 14.71, with original (undeformed) dimensions $a = b = 2$ cm and $c = 20$ cm. The elastic modulus of the bar material is $E = 100$ GPa and its Poisson's ratio is $v = 0.30$. The bar is subjected to biaxial forces in the x and y directions such that $F_x = 4 \times 10^6$ N and that $\underline{F}_x$ is tensile while $\underline{F}_y$ is compressive.

Assuming that the bar material is linearly elastic, determine:

(a) Average normal stresses σ_x, σ_y, and σ_z developed in the bar

(b) Average normal strains $\varepsilon_x, \varepsilon_y$, and ε_z

(c) Dimension, c', of the bar in the x direction after deformation

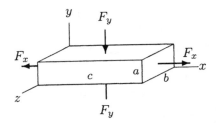

Fig. 14.71 Problem 14.2 (dimensions of the bar are not drawn to scale)

Answers
(a) $\sigma_x = 10\,\text{GPa}$, $\sigma_y = 1\,\text{GPa}$, $\sigma_z = 0$
(b) $\varepsilon_x = 0.1030$, $\varepsilon_y = -0.0400$, $\varepsilon_z = -0.0027$
(c) $c' = 22.06$ cm 0.027

Problem 14.3 Consider the solid circular cylinder shown in
Fig. 14.72. The cylinder has a length $l = 10$ cm and radius
$r_0 = 2$ cm. The cylinder is made of a linearly elastic material
with shear modulus $G = 10\,\text{GPa}$. If the cylinder is subjected to
a twisting torque with magnitude $M = 3{,}000$ N–m, calculate:
(a) The polar moment of inertia, J, of the cross-section of
    the cylinder
(b) The maximum angle of twist, θ, in degrees
(c) The maximum shear stress, τ, in the transverse plane
(d) The maximum shear strain, γ, in the transverse plane

Answers
(a) $J = 25.13 \times 10^{-8}\,\text{m}^4$
(b) $\theta = 6.84°$
(c) $\tau = 0.23876\,\text{GPa}$
(d) $\gamma = 0.02388$

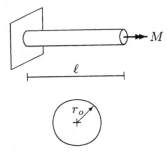

Fig. 14.72 Problem 14.3

Problem 14.4 Consider the uniform horizontal beam shown
in Fig. 14.73. Also shown is the cross-section of the beam.
The beam has a length $l = 4$ m, width $b = 10$ cm, and height
$h = 20$ cm. The beam is hinged to the ground at A,
supported by a roller at B, and a downward force with
magnitude $F = 400$ N is applied on the beam at C which is
located at a distance $d = 1$ m from A.

Fig. 14.73 Problem 14.4

Assuming that the weight of the beam is negligible,
calculate:
(a) The reactive forces on the beam at A and B
(b) The internal shearing force, V_{aa}, and bending moment,
    M_{aa}, at section aa of the beam which is 2 m from A
(c) The internal shearing force, V_{bb}, and bending moment,
    M_{bb}, at section bb of the beam which is 3 m from A
(d) The first moment, Q, of the cross-sectional area of the
    beam
(e) The maximum shear stress, τ_{aa}, generated in the beam at
    section aa

Answers
(a) $R_A = 300\,\text{N}$ $(\uparrow)$, $R_B = 100\,\text{N}$ $(\uparrow)$
(b) $V_{aa} = 100\,\text{N}$, $M_{aa} = 200\,\text{N–m}$
(c) $V_{bb} = 100\,\text{N}$, $M_{bb} = 100\,\text{N–m}$
(d) $Q = 0.0005\,\text{m}^3$
(e) $\tau_{aa} = 7{,}500\,\text{Pa}$

Problem 14.5 Consider the plane stress element shown in
Fig. 14.74 representing the state of stress at a material point.
If the magnitudes of the stresses are such that $\sigma_x = 200\,\text{Pa}$,
$\sigma_y = 100\,\text{Pa}$, and $\tau_{xy} = 50\,\text{Pa}$, calculate the magnitudes of

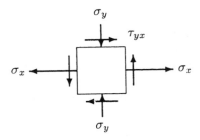

Fig. 14.74 Problem 14.5

principal stresses σ_1 and σ_2, and the maximum shear stress, τ_{max}, generated at this material point.

Answers

$\sigma_1 = 208\,\text{Pa}$ (tensile), $\sigma_2 = 108\,\text{Pa}$ (compressive),

$\tau_{max} = 158\,\text{Pa}$

Answers to Problem 14.1:

(a) Poisson's ratio	(j) Brittle
(b) second-order tensor	(k) Ductile
(c) principal stresses	(l) stress concentration effects
(d) principal planes	(m) torsion
(e) maximum shear stress	(n) neutral axis
(f) Mohr's circle	(o) 45
(g) failure	(p) static
(h) fatigue	(q) method of sections
(i) endurance limit	(r) flexure

15.1 Viscoelasticity

The material response discussed in the previous chapters was limited to the response of elastic materials, in particular to linearly elastic materials. Most metals, for example, exhibit linearly elastic behavior when they are subjected to relatively low stresses at room temperature. They undergo plastic deformations at high stress levels. For an elastic material, the relationship between stress and strain can be expressed in the following general form:

$$\sigma = \sigma(\varepsilon). \tag{15.1}$$

Equation (15.1) states that the normal stress σ is a function of normal strain ε only. The relationship between the shear stress τ and shear strain γ can be expressed in a similar manner. For a linearly elastic material, stress is linearly proportional to strain, and in the case of normal stress and strain, the constant of proportionality is the elastic modulus E of the material (Fig. 15.1):

$$\sigma = E\varepsilon. \tag{15.2}$$

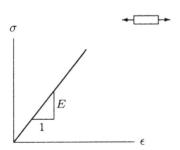

Fig. 15.1 Linearly elastic material behavior

While investigating the response of an elastic material, the concept of time does not enter into the discussions. Elastic materials show time-independent material behavior. Elastic materials deform instantaneously when they are subjected to externally applied loads. They resume their original (unstressed) shapes almost instantly when the applied loads are removed.

There is a different group of materials—such as polymer plastics, almost all biological materials, and metals at high temperatures—that exhibits gradual deformation and recovery when they are subjected to loading and unloading. The response of such materials is dependent upon how quickly the load is applied or removed, the extent of deformation being dependent upon the rate at which the deformation-causing loads are applied. This time-dependent material behavior is called *viscoelasticity*. Viscoelasticity is made up of two words: viscosity and elasticity. *Viscosity* is a fluid property and is a measure of resistance to flow. *Elasticity*, on the other hand, is a solid material property. Therefore, a viscoelastic material is one that possesses both fluid and solid properties.

For viscoelastic materials, the relationship between stress and strain can be expressed as:

$$\sigma = \sigma(\varepsilon, \dot{\varepsilon}). \tag{15.3}$$

Equation (15.3) states that stress, σ, is not only a function of strain, ε, but is also a function of the *strain rate*, $\dot{\varepsilon} = d\varepsilon/dt$, where t is time. A more general form of Eq. (15.3) can be obtained by including higher order time derivatives of strain. Equation (15.3) indicates that the stress–strain diagram of a viscoelastic material is not unique but is dependent upon the rate at which the strain is developed in the material (Fig. 15.2).

N. Özkaya et al., *Fundamentals of Biomechanics: Equilibrium, Motion, and Deformation,*
DOI 10.1007/978-1-4614-1150-5_15, © Springer Science+Business Media, LLC 2012

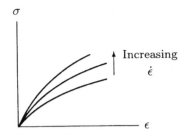

Fig. 15.2 Strain rate ($\dot{\epsilon}$) dependent viscoelastic behavior

In Eq. (15.4), η (eta) is the constant of proportionality between the stress σ and the strain rate $\dot{\epsilon}$, and is called the *coefficient of viscosity* of the fluid. As illustrated in Fig. 15.4, the coefficient of viscosity is the slope of the $\sigma - \dot{\epsilon}$ graph of a Newtonian fluid. The physical significance of this coefficient is similar to that of the coefficient of friction between the contact surfaces of solid bodies. The higher the coefficient of viscosity, the "thicker" the fluid and the more difficult it is to deform. The coefficient of viscosity for water is about 1 centipoise at room temperature, while it is about 1.2 centipoise for blood plasma.

15.2   Analogies Based on Springs and Dashpots

In Sect. 13.8, while covering Hooke's Law, an analogy was made between linearly elastic materials and linear springs. An elastic material deforms, stores potential energy, and recovers deformations in a manner similar to that of a spring. The elastic modulus E for a linearly elastic material relates stresses and strains, whereas the constant k for a linear spring relates applied forces and corresponding deformations (Fig. 15.3). Both E and k are measures of stiffness. The similarities between elastic materials and springs suggest that springs can be used to represent elastic material behavior. Since these similarities were first noted by Robert Hooke, elastic materials are also known as *Hookean solids*.

When subjected to external loads, fluids deform as well. Fluids deform continuously, or *flow*. For fluids, stresses are not dependent upon the strains but on the strain rates. If the stresses and strain rates in a fluid are linearly proportional, then the fluid is called a *linearly viscous fluid* or a *Newtonian fluid*. Examples of linearly viscous fluids include water and blood plasma. For a linearly viscous fluid,

$$\sigma = \eta(\dot{\epsilon}). \tag{15.4}$$

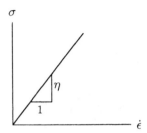

Fig. 15.4 Stress–strain rate diagram for a linearly viscous fluid

The spring is one of the two basic mechanical elements used to simulate the mechanical behavior of materials. The second basic mechanical element is called the *dashpot*, which is used to simulate fluid behavior. As illustrated in Fig. 15.5, a dashpot is a simple piston–cylinder or a syringe type of arrangement. A force applied on the piston will advance the piston in the direction of the applied force. The speed of the piston is dependent upon the magnitude of the applied force and the friction occurring between the contact surfaces of the piston and cylinder. For a linear dashpot, the applied force and speed (rate of displacement) are linearly proportional, the *coefficient of friction* μ (mu) being the constant of proportionality. If the applied force and the displacement are both in the x direction, then,

$$F = \mu\dot{x}. \tag{15.5}$$

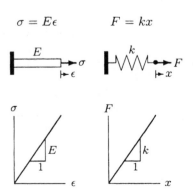

Fig. 15.3 Analogy between a linear spring and an elastic solid

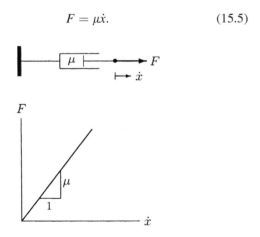

Fig. 15.5 A linear dashpot and its force–displacement rate diagram

In Eq. (15.5), $\dot{x} = dx/dt$ is the time rate of change of displacement or the speed.

By comparing Eqs. (15.4) and (15.5), an analogy can be made between linearly viscous fluids and linear dashpots. The stress and the strain rate for a linearly viscous fluid are, respectively, analogous to the force and the displacement rate for a dashpot; and the coefficient of viscosity is analogous to the coefficient of viscous friction for a dashpot. These analogies suggest that dashpots can be used to represent fluid behavior.

15.3   Empirical Models of Viscoelasticity

Springs and dashpots constitute the building blocks of model analyses in viscoelasticity. Springs and dashpots connected to one another in various forms are used to construct empirical viscoelastic models. Springs are used to account for the elastic solid behavior and dashpots are used to describe the viscous fluid behavior (Fig. 15.6). It is assumed that a constantly applied force (stress) produces a constant deformation (strain) in a spring and a constant rate of deformation (strain rate) in a dashpot. The deformation in a spring is completely recoverable upon release of applied forces, whereas the deformation that the dashpot undergoes is permanent.

SPRING: ELASTIC SOLID

$$\sigma = E\,\epsilon$$

DASHPOT: VISCOUS FLUID

$$\sigma = \eta\,\dot{\epsilon}$$

Fig. 15.6 Spring represents elastic and dashpot represents viscous material behaviors

15.3.1 Kelvin–Voight Model

The simplest forms of empirical models are obtained by connecting a spring and a dashpot together in parallel and in series configurations. As illustrated in Fig. 15.7, the *Kelvin–Voight model* is a system consisting of a spring and a dashpot connected in a parallel arrangement. If subscripts "s" and "d" denote the spring and dashpot, respectively, then a stress σ applied to the entire system will produce stresses σ_s and σ_d in the spring and the dashpot. The total stress

Fig. 15.7 Kelvin–Voight model

applied to the system will be shared by the spring and the dashpot such that:

$$\sigma = \sigma_s + \sigma_d. \qquad (15.6)$$

As the stress σ is applied, the spring and dashpot will deform by an equal amount because of their parallel arrangement. Therefore, the strain ε of the system will be equal to the strains ε_s and ε_d occurring in the spring and the dashpot:

$$\varepsilon = \varepsilon_s = \varepsilon_d. \qquad (15.7)$$

The stress–strain relationship for the spring and the stress–strain rate relationship for the dashpot are:

$$\sigma_s = E\varepsilon_s, \qquad (15.8)$$

$$\sigma_d = \eta\dot{\varepsilon}_d. \qquad (15.9)$$

Substituting Eqs. (15.8) and (15.9) into Eq. (15.6) will yield:

$$\sigma = E\varepsilon_s + \eta\dot{\varepsilon}_d. \qquad (15.10)$$

From (15.7), $\varepsilon_s = \varepsilon_d = \varepsilon$. Therefore,

$$\sigma = E\varepsilon + \eta\dot{\varepsilon}. \qquad (15.11)$$

Note that the strain rate $\dot{\varepsilon}$ can alternatively be written as $d\varepsilon/dt$. Consequently,

$$\sigma = E\varepsilon + \eta\frac{d\varepsilon}{dt}. \qquad (15.12)$$

Equation (15.12) relates stress to strain and the strain rate for the Kelvin–Voight model, which is a two-parameter (E and η) viscoelastic model. Equation (15.12) is an *ordinary differential equation*. More specifically, it is a first order, linear ordinary differential equation. For a given stress σ, Eq. (15.12) can be solved for the corresponding strain ε. For prescribed strain ε, it can be solved for stress σ.

Note that the review of how to handle ordinary differential equations is beyond the scope of this text. The interested reader is encouraged to review textbooks in "differential equations."

15.3.2 Maxwell Model

As shown in Fig. 15.8, the *Maxwell model* is constructed by connecting a spring and a dashpot in a series. In this case, a stress σ applied to the entire system is applied equally on the spring and the dashpot ($\sigma = \sigma_s = \sigma_d$), and the resulting strain ε is the sum of the strains in the spring and the dashpot ($\varepsilon = \varepsilon_s + \varepsilon_d$). Through stress–strain analyses similar to those carried out for the Kelvin–Voight model, a differential equation relating stresses and strains for the Maxwell model can be derived in the following form:

$$\eta\dot{\sigma} + E\sigma = E\eta\dot{\varepsilon}. \tag{15.13}$$

Fig. 15.8 Maxwell model

This is also a first order, linear ordinary differential equation representing a two-parameter (E and η) viscoelastic behavior. For a given stress (or strain), Eq. (15.13) can be solved for the corresponding strain (or stress).

Notice that springs are used to represent the elastic solid behavior, and there is a limit to how much a spring can deform. On the other hand, dashpots are used to represent fluid behavior and are assumed to deform continuously (flow) as long as there is a force to deform them. For example, in the case of a Maxwell model, a force applied will cause both the spring and the dashpot to deform. The deformation of the spring will be finite. The dashpot will keep deforming as long as the force is maintained. Therefore, the overall behavior of the Maxwell model is more like a fluid than a solid, and is known to be a *viscoelastic fluid* model. The deformation of a dashpot connected in parallel to a spring, as in the Kelvin–Voight model, is restricted by the response of the spring to the applied loads. The dashpot in the Kelvin–Voight model cannot undergo continuous deformations. Therefore, the Kelvin–Voight model represents a *viscoelastic solid* behavior.

15.3.3 Standard Solid Model

The Kelvin–Voight solid and Maxwell fluid are the basic viscoelastic models constructed by connecting a spring and a dashpot together. They do not represent any known real material. However, in addition to springs and dashpots,

they can be used to construct more complex viscoelastic models, such as the standard solid model. As illustrated in Fig. 15.9, the *standard solid model* is composed of a spring and a Kelvin–Voight solid connected in a series. The standard solid model is a three-parameter (E_1, E_2, and η) model and is used to describe the viscoelastic behavior of a number of biological materials such as the cartilage and the white blood cell membrane. The material function relating the stress, strain, and their rates for this model is:

$$(E_1 + E_2)\sigma + \eta\dot{\sigma} = (E_1 E_2 \varepsilon + E_1 \eta\dot{\varepsilon}). \tag{15.14}$$

Fig. 15.9 Standard solid model

In Eq. (15.14), $\dot{\sigma} = d\sigma/dt$ is the stress rate and $\dot{\varepsilon} = d\varepsilon/dt$ is the strain rate. This equation can be derived as follows. As illustrated in Fig. 15.10, the model can be represented by two units, A and B, connected in a series such that unit A is an elastic solid and unit B is a Kelvin–Voight solid. If σ_A and ε_A represent stress and strain in unit A, and σ_B and ε_B are stress and strain in unit B, then,

$$\sigma_A = E_1 \varepsilon_A, \tag{i}$$

$$\sigma_B = E_2 \varepsilon_B + \eta\frac{d\varepsilon_B}{dt} = \left(E_2 + \eta\frac{d}{dt}\right)\varepsilon_B. \tag{ii}$$

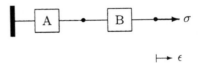

Fig. 15.10 Standard solid model is represented by units A and B

Since units A and B are connected in a series:

$$\varepsilon_A + \varepsilon_B = \varepsilon, \tag{iii}$$

$$\sigma_A = \sigma_B = \sigma. \tag{iv}$$

Substitute Eq. (*iv*) into Eqs. (*i*) and (*ii*) and express them in terms of strains ε_A and ε_B:

$$\varepsilon_A = \frac{\sigma}{E_1}, \tag{v}$$

$$\varepsilon_B = \frac{\sigma}{E_2 + \eta(d/dt)}. \tag{vi}$$

Substitute Eqs. (v) and (vi) into Eq. (iii):

$$\frac{\sigma}{E_1} + \frac{\sigma}{E_2 + \eta(d/dT)} = \varepsilon.$$

Employ cross multiplication and rearrange the order of terms to obtain

$$(E_1 + E_2)\sigma + \eta\frac{d\sigma}{dt} = E_1 E_2 \varepsilon + E_1 \eta \frac{d\varepsilon}{dt}$$

15.4 Time-Dependent Material Response

An empirical model for a given viscoelastic material can be established through a series of experiments. There are several experimental techniques designed to analyze the time-dependent aspects of material behavior. As illustrated in Fig. 15.11a, a *creep and recovery (recoil)* test is conducted by applying a load (stress σ_o) on the material at time t_0, maintaining the load at a constant level until time t_1, suddenly removing the load at t_1, and observing the material response. As illustrated in Fig. 15.11b, the *stress relaxation* experiment is done by straining the material to a level ε_o and maintaining the constant strain while observing the stress response of the material. In an *oscillatory response* test, a harmonic stress is applied and the strain response of the material is measured (Fig. 15.11c).

Fig. 15.11 (a) Creep and recovery, (b) stress relaxation, and (c) oscillatory response tests

Consider a viscoelastic material. Assume that the material is subjected to a creep test. The results of the creep test can be represented by plotting the measured strain as a function of time. An empirical viscoelastic model for the material behavior can be established through a series of trials. For this purpose, an empirical model is constructed by connecting a number of springs and dashpots together. A differential equation relating stress, strain, and their rates is derived through the procedure outlined in Sect. 15.3 for the Kelvin–Voight model. The imposed condition in a creep test is $\sigma = \sigma_0$. This condition of constant stress is substituted into the differential equation, which is then solved (integrated) for strain ε. The result obtained is another equation relating strain to stress constant σ_0, the elastic moduli and coefficients of viscosity of the empirical model, and time. For a given σ_0 and assigned elastic and viscous moduli, this equation is reduced to a function relating strain to time. This function is then used to plot a strain versus time graph and is compared to the experimentally obtained graph. If the general characteristics of the two (experimental and analytical) curves match, the analyses are furthered to establish the elastic and viscous moduli (material constants) of the material. This is achieved by varying the values of the elastic and viscous moduli in the empirical model until the analytical curve matches the experimental curve as closely as possible. In general, this procedure is called *curve fitting*. If there is no general match between the two curves, the model is abandoned and a new model is constructed and checked.

The result of these mathematical model analyses is an empirical model and a differential equation relating stresses and strains. The stress–strain relationship for the material can be used in conjunction with the fundamental laws of mechanics to analyze the response of the material to different loading conditions.

Note that the deformation processes occurring in viscoelastic materials are quite complex, and it is sometimes necessary to use an array of empirical models to describe the response of a viscoelastic material to different loading conditions. For example, the shear response of a viscoelastic material may be explained with one model and a different model may be needed to explain its response to normal loading. Different models may also be needed to describe the response of a viscoelastic material at low and high strain rates.

15.5 Comparison of Elasticity and Viscoelasticity

There are various criteria with which the elastic and viscoelastic behavior of materials can be compared. Some of these criteria are discussed in this section.

An elastic material has a unique stress–strain relationship that is independent of the time or strain rate. For elastic materials, normal and shear stresses can be expressed as functions of normal and shear strains:

$$\sigma = \sigma(\varepsilon) \quad and \quad \tau = \tau(\gamma).$$

For example, the stress–strain relationships for a linearly elastic solid are $\sigma = E\varepsilon$ and $\tau = G\gamma$, where E and G are constant elastic moduli of the material. As illustrated in Fig. 15.12, a linearly elastic material has a unique normal stress–strain diagram and a unique shear stress–strain diagram.

Fig. 15.12 An elastic material has unique normal and shear stress–strain diagrams

Viscoelastic materials exhibit time-dependent material behavior. The response of a viscoelastic material to an applied stress not only depends upon the magnitude of the stress but also on how fast the stress is applied to or removed from the material. Therefore, the stress–strain relationship for a viscoelastic material is not unique but is a function of the time or the rate at which the stresses and strains are developed in the material:

$$\sigma = \sigma(\varepsilon, \dot{\varepsilon}, \ldots, t) \quad and \quad \tau = \tau(\gamma, \dot{\gamma}, \ldots, t).$$

Consequently, as illustrated in Fig. 15.13, a viscoelastic material does not have a unique stress–strain diagram.

Fig. 15.13 Stress–strain diagram for a viscoelastic material may not be unique

For an elastic body, the energy supplied to deform the body (strain energy) is stored in the body as potential energy. This energy is available to return the body to its original (unstressed) size and shape once the applied stress is removed. As illustrated in Fig. 15.14, the loading and unloading paths for an elastic material coincide. This indicates that there is no loss of energy during loading and unloading.

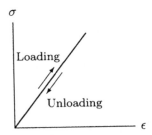

Fig. 15.14 For an elastic material, loading and unloading paths coincide

For a viscoelastic body, some of the strain energy is stored in the body as potential energy and some of it is dissipated as heat. For example, consider the Maxwell model. The energy provided to stretch the spring is stored in the spring while the energy supplied to deform the dashpot is dissipated as heat due to the friction between the moving parts of the dashpot. Once the applied load is removed, the potential energy stored in the spring is available to recover the deformation of the spring, but there is no energy available in the dashpot to regain its original configuration.

Consider the three-parameter standard solid model shown in Fig. 15.9. A typical loading and unloading diagram for this model is shown in Fig. 15.15. The area enclosed by the loading and unloading paths is called the *hysteresis loop*, which represents the energy dissipated as heat during the deformation and recovery phases. This area, and consequently the amount of energy dissipated as heat, is dependent upon the rate of strain employed to deform the body. The presence of the hysteresis loop in the stress–strain diagram for a viscoelastic material indicates that continuous loading and unloading would result in an increase in the temperature of the material.

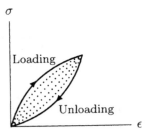

Fig. 15.15 Hysteresis loop

Note here that most of the elastic materials exhibit plastic behavior at stress levels beyond the yield point. For elastic–plastic materials, some of the strain energy is dissipated as heat during plastic deformations. This is indicated with the presence of a hysteresis loop in their loading and unloading diagrams (Fig. 15.16). For such

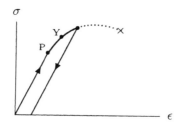

Fig. 15.16 Hysteresis loop for an elastic–plastic material

materials, energy is dissipated as heat only if the plastic region is entered. Viscoelastic materials dissipate energy regardless of whether the strains or stresses are small or large.

Since viscoelastic materials exhibit time-dependent material behavior, the differences between elastic and viscoelastic material responses are most evident under time-dependent loading conditions, such as during the creep and stress relaxation experiments.

As discussed earlier, a creep and recovery test is conducted by observing the response of a material to a constant stress σ_o applied at time t_0 and removed at a later time t_1 (Fig. 15.17a). As illustrated in Fig. 15.17b, such a load will cause a strain $\varepsilon_o = \dfrac{\sigma_o}{E}$ in a linearly elastic material instantly at time t_0. This constant strain will remain in the material until time t_1. At time t_1, the material will instantly and completely recover the deformation. To the same constant loading condition, a viscoelastic material will respond with a strain gradually increasing between times

t_0 and t_1. At time t_1, gradual recovery will start. For a viscoelastic solid material, the recovery will eventually be complete (Fig. 15.17c). For a viscoelastic fluid, complete recovery will never be achieved and there will be a residue of deformation left in the material (Fig. 15.17d).

As illustrated in Fig. 15.18a, the stress relaxation test is performed by straining a material instantaneously, maintaining the constant strain level ε_o in the material, and observing the response of the material. A linearly elastic material response is illustrated in Fig. 15.18b. The constant stress $\sigma_o = E\varepsilon_o$ developed in the material will remain as long as the strain ε_o is maintained. In other words, an elastic material will not exhibit a stress relaxation behavior. A viscoelastic material, on the other hand, will respond with an initial high stress that will decrease over time. If the material is a viscoelastic solid, the stress level will never reduce to zero (Fig. 15.18c). As illustrated in Fig. 15.18d, the stress will eventually reduce to zero for a viscoelastic fluid.

Fig. 15.18 Stress relaxation

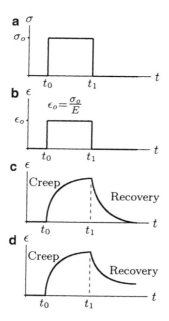

Fig. 15.17 Creep and recovery

Because of their time-dependent material behavior, viscoelastic materials are said to have a "memory." In other words, viscoelastic materials remember the history of deformations they undergo and react accordingly.

Almost all biological materials exhibit viscoelastic properties, and the remainder of this chapter is devoted to the discussion and review of the mechanical properties of biological tissues including bone, tendons, ligaments, muscles, and articular cartilage.

15.6 Common Characteristics of Biological Tissues

One of the objectives of studies in the field of biomechanics is to establish the mechanical properties of biological tissues so as to develop mathematical models that help us describe and further investigate their behavior under various loading conditions. While conducting studies in biomechanics, it has been a common practice to utilize engineering methods and principles, and at the same time to treat biological tissues like engineering materials. However, living tissues have characteristics that are very different than engineering materials. For example, living tissues can be self-adapting and self-repairing. That is, they can adapt to changing mechanical demand by altering their mechanical properties, and they can repair themselves. The mechanical properties of living tissues tend to change with age. Most biological tissues are composite materials (consisting of materials with different properties) with nonhomogeneous and anisotropic properties. In other words, the mechanical properties of living tissues may vary from point to point within the tissue, and their response to forces applied in different directions may be different. For example, values for strength and stiffness of bone may vary between different bones and at different points within the same bone. Furthermore, almost all biological tissues are viscoelastic in nature. Therefore, the strain or loading rate at which a specific test is conducted must also be provided while reporting the results of the strength measurements. These considerations require that most of the mechanical properties reported for living tissues are only approximations and a mathematical model aimed to describe the behavior of a living tissue is usually limited to describing its response under a specific loading configuration.

From a mechanical point of view, all tissues are composite materials. Among the common components of biological tissues, collagen and elastin fibers have the most important mechanical properties affecting the overall mechanical behavior of the tissues in which they appear. *Collagen* is a protein made of crimped fibrils that aggregate into fibers. The mechanical properties of collagen fibrils are such that each fibril can be considered a mechanical spring and each fiber as an assemblage of springs. The primary mechanical function of collagen fibers is to withstand axial tension. Because of their high length-to-diameter ratios (aspect ratio), collagen fibers are not effective under compressive loads. Whenever a fiber is pulled, its crimp straightens, and its length increases. Like a mechanical spring, the energy supplied to stretch the fiber is stored and it is the release of this energy that returns the fiber to its unstretched configuration when the applied load is removed. The individual fibrils of the collagen

fibers are surrounded by a gel-like *ground substance* that consists largely of water. Collagen fibers possess a two-phase, solid–fluid, or viscoelastic material behavior with a relatively high tensile strength and poor resistance to compression.

The geometric configuration of collagen fibers and their interaction with the noncollagenous tissue components form the basis of the mechanical properties of biological tissues. Among the noncollagenous tissue components, *elastin* is another fibrous protein with material properties that resemble the properties of rubber. Elastin and microfibrils form elastic fibers that are highly extensible, and their extension is reversible even at high strains. Elastin fibers behave elastically with low stiffness up to about 200% elongation followed by a short region where the stiffness increases sharply until failure (Fig. 15.19). The elastin fibers do not exhibit considerable plastic deformation before failure, and their loading and unloading paths do not show significant hysteresis. In summary, elastin fibers possess a low-modulus elastic material property, while collagen fibers show a higher modulus viscoelastic material behavior.

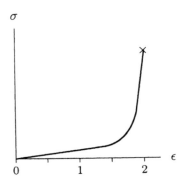

Fig. 15.19 Stress–strain diagram for elastin

15.7 Biomechanics of Bone

Bone is the primary structural element of the human body. Bones form the building blocks of the skeletal system that protects the internal organs, provides kinematic links, provides muscle attachment sites, and facilitates muscle actions and body movements. Bone has unique structural and mechanical properties that allow it to carry out these functions. As compared to other structural materials, bone is also unique in that it is self-repairing. Bone can also alter its shape, mechanical behavior, and mechanical properties to adapt to the changes in mechanical demand. The major factors that influence the mechanical behavior of bone are the composition of bone, the mechanical properties of the tissues comprising the bone, the size and geometry of the bone, and the direction, magnitude, and rate of applied loads.

15.7.1 Composition of Bone

In biological terms, bone is a *connective tissue* that binds together various structural elements of the body. In mechanical terms, bone is a *composite material* with various solid and fluid phases. Bone consists of cells and an organic mineral matrix of fibers and a ground substance surrounding collagen fibers. Bone also contains inorganic substances in the form of mineral salts. The inorganic component of bone makes it hard and relatively rigid, and its organic component provides flexibility and resilience. The composition of bone varies with species, age, sex, type of bone, type of bone tissue, and the presence of bone disease.

At the macroscopic level, all bones consist of two types of tissues (Fig. 15.20). The *cortical* or *compact* bone tissue is a dense material forming the outer shell (cortex) of bones and the diaphysial region of long bones. The *cancellous*, *trabecular*, or *spongy* bone tissue consists of thin plates (trabeculae) in a loose mesh structure that is enclosed by the cortical bone. Bones are surrounded by a dense fibrous membrane called the *periosteum*. The periosteum covers the entire bone except for the joint surfaces that are covered with articular cartilage.

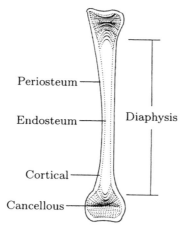

Fig. 15.20 Sectional view of a whole bone showing cortical and cancellous tissues

15.7.2 Mechanical Properties of Bone

Bone is a nonhomogeneous material because it consists of various cells, organic and inorganic substances with different material properties. Bone is an anisotropic material because

its mechanical properties are different in different directions. That is, the mechanical response of bone is dependent upon the direction as well as the magnitude of the applied load. For example, the compressive strength of bone is greater than its tensile strength. Bone possesses viscoelastic (time-dependent) material properties. The mechanical response of bone is dependent on the rate at which the loads are applied. Bone can resist rapidly applied loads much better than slowly applied loads. In other words, bone is stiffer and stronger at higher strain rates.

Bone is a complex structural material. The mechanical response of bone can be observed by subjecting it to tension, compression, bending, and torsion. Various tests to implement these conditions were discussed in the previous chapters. These tests can be performed using uniform bone specimens or whole bones. If the purpose is to investigate the mechanical response of a specific bone tissue (cortical or cancellous), then the tests are performed using bone specimens. Testing a whole bone, on the other hand, attempts to determine the "bulk" properties of that bone.

The tensile stress–strain diagram for the cortical bone is shown in Fig. 15.21. This $\sigma-\varepsilon$ curve is drawn using the averages of the elastic modulus, strain hardening modulus, ultimate stress, and ultimate strain values determined for the human femoral cortical bone tested under tensile and compressive loads applied in the longitudinal direction at a moderate strain rate ($\dot{\varepsilon} = 0.05\,s^{-1}$). The $\sigma-\varepsilon$ curve in Fig. 15.21 has three distinct regions. In the initial linearly elastic region, the $\sigma-\varepsilon$ curve is nearly a straight line and the slope of this line is equal to the elastic modulus (E) of the bone, which is about 17 GPa. In the intermediate region, the bone exhibits nonlinear elasto-plastic material behavior. Material yielding also occurs in this region. By the offset method discussed in Chap. 13, the yield strength of the cortical bone for the $\sigma-\varepsilon$ diagram shown in Fig. 15.21 can be determined to be about 110 MPa. In the final region, the bone exhibits a linearly plastic material behavior and the $\sigma-\varepsilon$ diagram is another straight line. The slope of this line is the

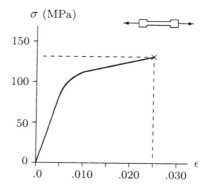

Fig. 15.21 Tensile stress–strain diagram for human cortical bone loaded in the longitudinal direction (strain rate $\dot{\varepsilon} = 0.05\,s^{-1}$)

strain hardening modulus (E') of bone tissue, which is about 0.9 GPa. The bone fractures when the tensile stress is about 128 MPa, for which the tensile strain is about 0.026.

The elastic moduli and strength values for bone are dependent upon many factors including the test conditions such as the rate at which the loads are applied. This viscoelastic nature of bone tissue is demonstrated in Fig. 15.22. The stress–strain diagrams in Fig. 15.22 for different strain rates indicate that a specimen of bone tissue that is subjected to rapid loading (high $\dot{\epsilon}$) has a greater elastic modulus and ultimate strength than a specimen that is loaded more slowly (low $\dot{\epsilon}$). Figure 15.22 also demonstrates that the energy absorbed (which is proportional to the area under the σ–ϵ curve) by the bone tissue increases with an increasing strain rate. Note that during normal daily activities, bone tissues are subjected to a strain rate of about 0.01 s^{-1}.

Fig. 15.22 The strain rate-dependent stress–strain curves for cortical bone tissue

The stress–strain behavior of bone is also dependent upon the orientation of bone with respect to the direction of loading. This anisotropic material behavior of bone is demonstrated in Fig. 15.23. Notice that the cortical bone has a larger ultimate strength (stronger) and a larger elastic modulus (stiffer) in the longitudinal direction than the transverse direction. Furthermore, bone specimens loaded in the

transverse direction fail in a more brittle manner (without showing considerable yielding) as compared to bone specimens loaded in the longitudinal direction. The ultimate strength values for adult femoral cortical bone under various modes of loading, and its elastic and shear moduli are listed in Table 15.1. The ultimate strength values in Table 15.1 demonstrate that the bone strength is highest under compressive loading in the longitudinal direction (the direction of osteon orientation) and lowest under tensile loading in the transverse direction (the direction perpendicular to the longitudinal direction). The elastic modulus of cortical bone in the longitudinal direction is higher than its elastic modulus in the transverse direction. Therefore, cortical bone is stiffer in the longitudinal direction than in the transverse direction.

Table 15.1 Ultimate strength, and elastic and shear moduli for human femoral cortical bone.
(1 GPa = 10^9 Pa, 1 MPa = 10^6 Pa)

LOADING MODE	ULTIMATE STRENGTH
LONGITUDINAL	
Tension	133 MPa
Compression	193 MPa
Shear	68 MPa
TRANSVERSE	
Tension	
Compression	51 MPa
	133 MPa
ELASTIC MODULI, E	
Longitudinal	17.0 GPa
Transverse	11.5 GPa
SHEAR MODULUS, G	3.3 GPa

It should be noted that there is a wide range of variation in values reported for the mechanical properties of bone. It may be useful to remember that the tensile strength of bone is less than 10% of that of stainless steel. Also, the stiffness of bone is about 5% of the stiffness of steel. In other words, for specimens of the same dimension and under the same tensile load, a bone specimen will deform 20 times as much as the steel specimen.

The chemical compositions of cortical and cancellous bone tissues are similar. The distinguishing characteristic of the cancellous bone is its porosity. This physical difference between the two bone tissues is quantified in terms of the *apparent density* of bone, which is defined as the mass of bone tissue present in a unit volume of bone. To a certain degree, both cortical and cancellous bone tissues can be regarded as a single material of variable density. The material properties such as strength and stiffness, and the stress–strain characteristics of cancellous bone depend not only on the apparent density that may be different for

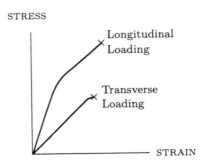

Fig. 15.23 The direction-dependent stress–strain curves for bone tissue

different bone types or at different parts of a single bone, but also on the mode of loading. The compressive stress–strain curves (Fig. 15.24) of cancellous bone contain an initial linearly elastic region up to a strain of about 0.05. The material yielding occurs as the trabeculae begin to fracture. This initial elastic region is followed by a plateau region of almost constant stress until fracture, exhibiting a ductile material behavior. By contrast to compact bone, cancellous bone fractures abruptly under tensile forces, showing a brittle material behavior. Cancellous bone is about 25–30% as dense, 5–10% as stiff, and five times as ductile as cortical bone. The energy absorption capacity of cancellous bone is considerably higher under compressive loads than under tensile loads.

Fig. 15.24 Apparent density-dependent stress–strain curves for cancellous bone tissue

15.7.3 Structural Integrity of Bone

There are several factors that may affect the structural integrity of bones. For example, the size and geometry of a bone determine the distribution of the internal forces throughout the bone, thereby influencing its response to externally applied loads. The larger the bone, the larger the area upon which the internal forces are distributed and the smaller the intensity (stress) of these forces. Consequently, the larger the bone, the more resistant it is to applied loads.

A common characteristic of long bones is their tubular structure in the diaphysial region, which has considerable mechanical advantage over solid circular structures of the same mass. Recall from the previous chapter that the shear stresses in a structure subjected to torsion are inversely proportional with the polar moment of inertia (J) of the cross-sectional area of the structure, and the normal stresses in a structure subjected to bending are inversely proportional to the area moment of inertia (I) of the cross-section of the structure. The larger the polar and area moments of inertia of a structure, the lower the maximum normal stresses due to torsion and bending. Since tubular structures have larger polar and area moments of inertia as compared to solid cylindrical structures of the same volume, tubular structures

are more resistant to torsional and bending loads as compared to solid cylindrical structures. Furthermore, a tubular structure can distribute the internal forces more evenly over its cross-section as compared to a solid cylindrical structure of the same cross-sectional area.

Certain skeletal conditions such as osteoporosis can reduce the structural integrity of bone by reducing its apparent density. Small decreases in bone density can generate large reductions in bone strength and stiffness. As compared to a normal bone with the same geometry, an osteoporotic bone will deform easier and fracture at lower loads. The density of bone can also change with aging, after periods of disuse, or after chronic exercise, thereby changing its overall strength. Certain surgical procedures that alter the normal bone geometry may also reduce the strength of bone. Bone defects such as screw holes reduce the load-bearing ability of bone by causing stress concentrations around the defects.

Bone becomes stiffer and less ductile with age. Also with age, the ability of bone to absorb energy and the maximum strain at failure are reduced, and the bone behaves more like dry bone. Although the properties of dry bone may not have any value in orthopedics, it may be important to note that there are differences between bone in its wet and dry states. Dry bone is stiffer, has a higher ultimate strength, and is more brittle than wet bone (Fig. 15.25).

Fig. 15.25 Stress–strain curves for dry and wet bones

15.7.4 Bone Fractures

When bones are subjected to moderate loading conditions, they respond by small deformations that are only present while the loads are applied. When the loads are removed, bones exhibit elastic material behavior by resuming their original (unstressed) shapes and positions. Large deformations occur when the applied loads are high. Bone fractures when the stresses generated in any region of bone are larger than the ultimate strength of bone.

Fractures caused by pure tensile forces are observed in bones with a large proportion of cancellous bone tissue.

Fractures due to compressive loads are commonly encountered in the vertebrae of the elderly, whose bones are weakened as a result of aging. Bone fractures caused by compression occur in the diaphysial regions of long bones. Compressive fractures are identified by their oblique fracture pattern. Long bone fractures are usually caused by torsion and bending. Torsional fractures are identified by their spiral oblique pattern, whereas bending fractures are usually identified by the formation of "butterfly" fragments. Fatigue fracture of bone occurs when the damage caused by repeated mechanical stress outpaces the bone's ability to repair to prevent failure. Bone fractures caused by fatigue are common among professional athletes and dedicated joggers. Clinically, most bone fractures occur as a result of complex, combined loading situations rather than simple loading mechanisms.

15.8   Tendons and Ligaments

Tendons and ligaments are fibrous connective tissues. Tendons help execute joint motion by transmitting mechanical forces (tensions) from muscles to bones. Ligaments join bones and provide stability to the joints. Unlike muscles, which are active tissues and can produce mechanical forces, tendons and ligaments are passive tissues and cannot actively contract to generate forces.

Around many joints of the human body, there is insufficient space to attach more than one or a few muscles. This requires that to accomplish a certain task, one or a few muscles must share the burden of generating and withstanding large loads with intensities (stress) even larger at regions closer to the bone attachments where the cross-sectional areas of the muscles are small. As compared to muscles, tendons are stiffer, have higher tensile strengths, and can endure larger stresses. Therefore, around the joints where the space is limited, muscle attachments to bones are made by tendons. Tendons are capable of supporting very large loads with very small deformations. This property of tendons enables the muscles to transmit forces to bones without wasting energy to stretch tendons.

The mechanical properties of tendons and ligaments depend upon their composition which can vary considerably. The most common means of evaluating the mechanical response of tendons and ligaments is the uniaxial tension test. Figure 15.26 shows a typical tensile stress–strain diagram for tendons. The shape of this curve is the result of the interaction between elastic elastin fibers and the viscoelastic collagen fibers. At low strains (up to about 0.05), less stiff elastic fibers dominate and the crimp of the collagen fibers straightens, requiring very little force to stretch the tendon. The tendon becomes stiffer when the crimp is straightened. At the same time, the fluid-like ground

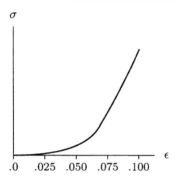

Fig. 15.26   Tensile stress–strain diagram for tendon

substance in the collagen fibers tends to flow. At higher strains, therefore, the stiff and viscoelastic nature of the collagen fibers begins to take an increasing portion of the applied load. Tendons are believed to function in the body at strains of up to about 0.04, which is believed to be their yield strain (ε_y). Tendons rupture at strains of about 0.1 (ultimate strain, ε_u), or stresses of about 60 MPa (ultimate stress, σ_u).

Note that the shape of the stress–strain curve in Fig. 15.26 is such that the area under the curve is considerably small. In other words, the energy stored in the tendon to stretch the tendon to a stress level is much smaller than the energy stored to stretch a linearly elastic material (with a stress–strain diagram that is a straight line) to the same stress level. Therefore, the tendon has higher resilience than linearly elastic materials.

The time-dependent, viscoelastic nature of the tendon is illustrated in Figs. 15.27 and 15.28. When the tendon is stretched rapidly, there is less chance for the ground substance to flow, and consequently, the tendon becomes stiffer. The hysteresis loop shown in Fig. 15.28 demonstrates the time-dependent loading and unloading behavior of the tendon. Note that more work is done in stretching the tendon than is recovered when the tendon is allowed to relax, and therefore, some of the energy is dissipated in the process.

The mechanical role of ligaments is to transmit forces from one bone to another. Ligaments also have a stabilizing

Fig. 15.27   The strain rate-dependent stress–strain curves for tendon

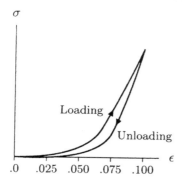

Fig. 15.28 The hysteresis loop of stretching and relaxing modes of the tendon

role for the skeletal joints. The composition and structure of ligaments depend upon their function and position within the body. Like tendons they are composite materials containing crimped collagen fibers surrounded by ground substance. As compared to tendons, they often contain a greater proportion of elastic fibers that accounts for their higher extensibility but lower strength and stiffness. The mechanical properties of ligaments are qualitatively similar to those of tendons. Like tendons, they are viscoelastic and exhibit hysteresis, but deform elastically up to strains of about $\varepsilon_y = 0.25$ (about five times as much as the yield strain of tendons) and stresses of about $\sigma_y = 5$ MPa. They rupture at a stress of about 20 MPa.

Since tendons and ligaments are viscoelastic, some of the energy supplied to stretch them is dissipated by causing the flow of the fluid within the ground substance, and the rest of the energy is stored in the stretched tissue. Tendons and ligaments are tough materials and do not rupture easily. Most common damages to tendons and ligaments occur at their junctions with bones.

15.9 Skeletal Muscles

There are three types of muscles: skeletal, smooth, and cardiac. Smooth muscles line the internal organs, and cardiac muscles form the heart. Here, we are concerned with the characteristics of the skeletal muscles, each of which is attached, via aponeuroses and/or tendons, to at least two bones causing and/or controlling the relative movement of one bone with respect to the other. When its fibers contract under the stimulation of a nerve, the muscle exerts a pulling effect on the bones to which it is attached. *Contraction* is a unique ability of the muscle tissue, which is defined as the development of tension in the muscle. Muscle contraction can occur as a result of muscle shortening (concentric contraction) or muscle lengthening (eccentric contraction), or it can occur without any change in the muscle length (static or isometric contraction).

The skeletal muscle is composed of muscle fibers and myofibrils. Myofibrils in turn are made of contractile elements: *actin* and *myosin* proteins. Actin and myosin appear in bands or filaments. Several relatively thick myosin filaments interact across cross-bridges with relatively thin actin filaments to form the basic structure of the contractile element of the muscle, called the *sarcomere* (Fig. 15.29). Many sarcomere elements connected in a series arrangement form the contractile element (motor unit) of the muscle. It is within the sarcomere that the muscle force (tension) is generated, and where muscle shortening and lengthening takes place. The active contractile elements of the muscle are contained within a fibrous passive connective tissue, called *fascia*. Fascia encloses the muscles, separates them into layers, and connects them to tendons.

Fig. 15.29 Basic structure of the contractile element of muscle (*thick lines* represent myosin filaments, *thin horizontal lines* are actin filaments, and *cross-hatched lines* are cross-bridges)

The force and torque developed by a muscle is dependent on many factors, including the number of motor units within the muscle, the number of motor units recruited, the manner in which the muscle changes its length, the velocity of muscle contraction, and the length of the lever arm of the muscle force. For muscles, two different forces can be distinguished. *Active tension* is the force produced by the contractile elements of the muscle and is a result of voluntary muscle contraction. *Passive tension*, on the other hand, is the force developed within the connective muscle tissue when the muscle length surpasses its resting length. The net tensile force in a muscle is dependent on the force–length characteristics of both the active and passive components of the muscle. A typical tension versus muscle length diagram is shown in Fig. 15.30. The number of cross-bridges between the filaments is maximum, and therefore, the active tension (T_a) is maximum at the resting length (l_o) of the muscle. As the muscle lengthens, the filaments are pulled apart, the number of cross-bridges is reduced and the active tension is decreased. At full length, there are no cross-bridges and the active tension reduces to zero. As the muscle shortens, the cross-bridges overlap and the active tension is again reduced. When the muscle is at its resting length or

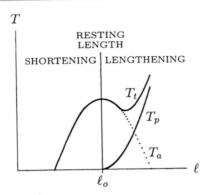

Fig. 15.30 Muscle force (T) versus muscle length (ℓ)

less, the passive (connective) component of the muscle is in a loose state with no tension. As the muscle lengthens, a passive tensile force (T_p) builds up in the connective tissues. The force–length characteristic of this passive component resembles that of a nonlinear spring. Passive tensile force increases at an increasing rate as the length of the muscle increases. The overall, total, or net muscle force (T_t) that is transmitted via tendons is the sum of the forces in the active and passive elements of the muscle. Note here that for a given muscle, the tension–length diagram is not unique but dependent on the number of motor units recruited. The magnitude of the active component of the muscle force can vary depending on how the muscle is excited, and usually expressed as the percentage of the maximum voluntary contraction.

The force generated by a contracting muscle is usually transmitted to a bone through a tendon. There is a functional reason for tendons to make the transfer of forces from muscles to bones. As compared to tendons, muscles have lower tensile strengths. The relatively low ultimate strength requires muscles to have relatively large cross-sectional areas in order to transmit sufficiently high forces without tearing. Tendons are better designed to perform this function.

15.10 Articular Cartilage

Cartilage covers the articulating surfaces of bones at the diarthrodial (synovial) joints. The primary function of cartilage is to facilitate the relative movement of articulating bones. Cartilage reduces stresses applied to bones by increasing the area of contact between the articulating surfaces and reduces bone wear by reducing the effects of friction.

Cartilage is a two-phase material consisting of about 75% water and 25% organic solid. A large portion of the solid phase of the cartilage material is made up of collagen fibers. The remaining ground substance is mainly proteoglycan (hydrophilic molecules). Collagen fibers are relatively strong and stiff in tension, while proteoglycans are strong

in compression. The solid–fluid composition of cartilage makes it a viscoelastic material.

The mechanical properties of cartilage under various loading conditions have been investigated using a number of different techniques. For example, the response of the human patella to compressive loads has been investigated by using an *indentation test* in which a small cylindrical or hemispherical indenter is pressed into the articulating surface, and the resulting deformation is recorded (Fig. 15.31a). A typical result of an indentation test is shown in Fig. 15.31b. When a constant magnitude load is applied, the material initially responds with a relatively large elastic deformation. The applied load causes pressure gradients to occur in the interstitial fluid, and the variations in pressure cause the fluid to flow through and out of the cartilage matrix. As the load is maintained, the amount of deformation increases at a decreasing rate. The deformation tends toward an equilibrium state as the pressure variations within the fluid are dissipated. When the applied load is removed (unloading phase), there is an instantaneous elastic recovery (recoil) that is followed by a more gradual recovery leading to complete recovery. This creep–recovery response of cartilage may be qualitatively represented by the three-parameter viscoelastic solid model (Fig. 15.32), which consists of a linear spring and a Kelvin–Voight unit connected in series.

Fig. 15.31 Indentation test

Fig. 15.32 The standard solid model has been used to represent the creep–recovery behavior of cartilage

Another experiment designed to investigate the response of cartilage to compressive loading conditions is the *confined compression test* illustrated in Fig. 15.33. In this test, the specimen is confined in a rigid cylindrical die and loaded with a rigid permeable block. The compressive load causes pressure variations in the interstitial fluid and consequent fluid exudation. Eventually, the pressure variations dissipate and equilibrium is reached. The state at which the equilibrium is reached is indicative of the compressive stiffness of the cartilage. The compressive stiffness and resistance of cartilage depend upon the water and proteoglycan content of the tissue. The higher the proteoglycan content, the higher the compressive resistance of the tissue.

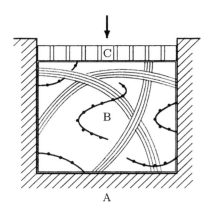

Fig. 15.33 Confined compression test. A is the rigid die, B is the specimen, and C is the permeable block

During daily activities, the articular cartilage is subjected to tensile and shear stresses as well as compressive stresses. Under tension, cartilage responds by realigning the collagen fibers that carry the tensile loads applied to the tissue. The tensile stiffness and strength of cartilage depend on the collagen content of the tissue. The higher the collagen content, the higher the tensile strength of cartilage. Shear stresses on the articular cartilage are due to the frictional forces between the relative movement of articulating surfaces. However, the coefficient of friction for synovial joints is so low (of the order 0.001–0.06) that friction has an insignificant effect on the stress resultants acting on the cartilage.

Both structural (such as intraarticular fracture) and anatomical abnormalities (such as rheumatoid arthritis and acetabular dysplasia) can cause cartilage damage, degeneration, wear, and failure. These abnormalities can change the load-bearing ability of the joint by altering its mechanical properties. The importance of the load-bearing ability of the cartilage and maintaining its mechanical integrity may become clear if we consider that the magnitude of the forces involved at the human hip joint is about five times body weight during ordinary walking (much higher during running or jumping). The hip contact area over which these

forces are applied is about 15 cm^2 (0.0015 m^2). Therefore, the compressive stresses (pressures) involved are of the order 3 MPa for an 85 kg person.

15.11 Discussion

Here we have covered, very briefly, the mechanical properties of selected biological tissues. We believe that the knowledge of the mechanical properties and structural behavior of biological tissues is an essential prerequisite for any experimental or theoretical analysis of their physiological function in the body. We are aware of the fact that the proper coverage of each of these topics deserves at least a full chapter. Our purpose here was to provide a summary, to illustrate how biological phenomena can be described in terms of the mechanical concepts introduced earlier, and hope that the interested reader would refer to more complete sources of information to improve his or her knowledge of the subject matter.

15.12 Exercise Problem

Problem 15.1 Complete the following definitions with appropriate expressions.
(a) Elastic materials show time-independent material behavior. Elastic materials deform _____ when they are subjected to externally applied loads.
(b) Time-dependent material behavior is known as _____.
(c) Elasticity is a solid material behavior, whereas _____ is a fluid property and is a measure of resistance to flow.
(d) For a viscoelatic material, stress is not only a function of strain, but also a function of _____.
(e) _____ and _____ are basic mechanical elements that are used to simulate elastic solid and viscous fluid behaviors, respectively.
(f) The _____ is a viscoelastic model consisting of a spring and a dashpot connected in a parallel arrangement.
(g) The _____ is a viscoelastic model consisting of a spring and a dashpot connected in a series arrangement.
(h) The _____ is a viscoelastic model consisting of a spring and a Kelvin–Voight solid connected in a series.
(i) A _____ test is conducted by applying a load on the material, maintaining the load at a constant level for some time, suddenly removing the load, and observing the material response.

(j) A _____ test is conducted by straining the material at a level and maintaining the strain at a constant level while observing the stress response of the material.

(k) In a _____ test, a harmonic stress is applied on the material and the strain response of the material is observed.

(l) The area enclosed by the loading and unloading paths is called the _____, which represents the energy dissipated as heat.

(m) Because of their time-dependent behavior, viscoelastic materials are said to have a _____.

(n) Living tissues have characteristics that are very different than engineering materials. For example, they are _____ and _____.

(o) Among the common components of biological tissues, _____ and _____ fibers have the most important mechanical properties affecting the overall mechanical behavior of the tissues in which they appear.

(p) _____ is a protein made of crimped fibrils that aggregate into fibers.

(q) _____ is a fibrous protein with material properties that resemble the properties of rubber.

(r) In biological terms, bone is a _____ tissue that binds together various structural elements of the body. In mechanical terms, bone is a _____ material with various solid and fluid phases.

(s) The _____ bone tissue is a dense material forming the outer shell (cortex) of bones and the diaphysial region of long bones.

(t) The _____ bone tissue consists of thin plates (trabeculae) in a loose mesh structure that is enclosed by the cortical bone.

(u) Bone is stiffer and stronger at _____ strain rates.

(v) Cortical bone strength is highest under compressive loading in the _____ direction (direction of osteon orientation) and lowest under tensile loading in the _____ direction (direction perpendicular to the longitudinal direction).

(w) The tensile strength of bone is less than _____ percent of stainless steel, and the stiffness of bone is about _____ percent of the stiffness of steel.

(x) The chemical compositions of cortical and cancellous bone tissues are similar. The distinguishing characteristic of the cancellous bone is its _____.

(y) _____ is a unique ability of the muscle tissue, which is defined as the development of tension in the muscle.

(z) _____ tension is the force produced by the contractile elements of the muscle and is a result of voluntary muscle contraction, and _____ tension is the force developed within the connective muscle tissue when the muscle length surpasses its resting length.

Answers to Problem 15.1:

(a) instantaneously	(n) self-adapting, self-repairing
(b) viscoelasticity	(o) collagen, elastin
(c) viscosity	(p) Collagen
(d) strain rate or time	(q) Elastin
(e) Spring, dashpot	(r) connective, composite
(f) Kelvin–Voight	(s) cortical or compact
(g) Maxwell	(t) cancellous or trabecular
(h) standard solid	(u) higher
(i) creep and recovery	(v) longitudinal, transverse
(j) stress relaxation	(w) 10, 5
(k) oscillatory response	(x) porosity
(l) hysteresis loop	(y) Contraction
(m) memory	(z) Active, passive

Appendix A: Plane Geometry

A.1 Angles

In geometry, the basic building element is *line*, and when we say line, we assume that it is straight and infinite in length. Two points on a line at some distance from one another constitute a line segment, which is finite in length. When lines intersect, they form *angles*.

The angles identified as θ in Fig. A.1 that are formed by two intersecting straight lines *aa* and *bb* are equal and called *opposite angles*. If the two lines are perpendicular to one another, then the angles formed are called *right angles*. A right angle is equal to 90°. Graphically, right angles are usually denoted by small square boxes. An angle is called *acute* if it is smaller than 90° and is called *obtuse* if it is greater than 90°.

Fig. A.1 Opposite angles

The angles identified as θ in Fig. A.2 are equal and are called *alternate angles*. These angles are formed by a straight line *cc* intersecting two parallel straight lines *aa* and *bb*.

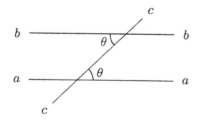

Fig. A.2 Alternate angles

The angles identified as θ in Fig. A.3 are equal. In this case, straight line *cc* is perpendicular to *bb* and *dd* is perpendicular to *aa*. The geometry illustrated in Fig. A.3 is utilized extensively in physics and mechanics, for example, while analyzing motions on inclined surfaces. For such analyses, *aa* represents the horizontal, *bb* represents the inclined surface that makes an angle θ with the horizontal, *cc* is perpendicular to the inclined surface, and *dd* represents the vertical.

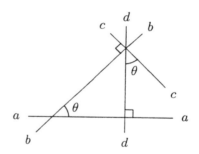

Fig. A.3 Lines *bb* and *cc*, and *aa* and *dd* are perpendicular

A.2 Triangles

A *triangle* is a geometric shape with three sides and three interior angles. There are different kinds of triangles, including equilateral, isosceles, irregular, and right-triangles. An *equilateral* triangle is a triangle with all sides of equal length (Fig. A.4). The three interior angles of the equilateral triangle are also equal to each other. A triangle with at least two sides of equal length is called an *isosceles* triangle (Fig. A.5). In the isosceles triangle, the angles that are opposite to the two equal sides are equal to each other. A triangle with three sides of various lengths is called an *irregular* triangle (Fig. A.6). For any triangle, the sum of

N. Özkaya et al., *Fundamentals of Biomechanics: Equilibrium, Motion, and Deformation,*
DOI 10.1007/978-1-4614-1150-5, © Springer Science+Business Media, LLC 2012

Fig. A.4 Equilateral triangle

Fig. A.5 Isosceles triangle

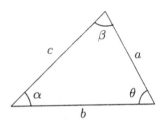

Fig. A.6 An irregular triangle

the three interior angles is equal to 180°. In the case of the irregular triangle shown:

$$\alpha + \beta + \theta = 180°.$$

A.3 Law of Sines

For any triangle, such as the one in Fig. A.4, the angles and sides of the triangle are related through the *law of sines* which states that:

$$\frac{\sin \alpha}{a} = \frac{\sin \beta}{b} = \frac{\sin \theta}{c}. \tag{A.1}$$

The definition of sine (abbreviated as sin) is given in Sect. A.7.

A.4 Law of Cosine

For any triangle, such as the one in Fig. A.6, if two sides of the triangle (e.g., a and b) and an angle between them (θ) are known, the unknown third side (c) can be determined by applying the law of cosine, which states that:

$$c^2 = a^2 + b^2 - 2ab \, \cos \theta. \tag{A.2}$$

The definition of cosine (abbreviated as cos) is given in Sect. A.7.

A.5 The Right-Triangle

A *right-triangle* is formed if one of the angles of a triangle is equal to 90°. For the right-triangle shown in Fig. A.7, angle θ is equal to 90°, and the sum of the remaining angles is also equal to 90°:

$$\theta = \alpha + \beta = 90°.$$

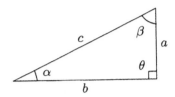

Fig. A.7 The right-triangle

In Fig. A.6, side c of the triangle opposite to the right angle (angle θ) is called the *hypotenuse*, and it is the longest side of the right-triangle. With respect to angle a, b is the length of the *adjacent* side and a is the length of the *opposite* side. The two sides forming the right angle are frequently called the legs of the right-triangle.

A.6 Pythagorean Theorem

The *Pythagorean theorem* states that the square of the length of the hypotenuse of a right-triangle is equal to the sum of the squares of the lengths of the other sides of the right-triangle:

$$c^2 = a^2 + b^2. \tag{A.3}$$

Considering the square root of both sides:

$$c = \sqrt{a^2 + b^2}. \tag{A.4}$$

A.7 Sine, Cosine, and Tangent

The *sine* of an acute angle (angles other than the right angle) in a right-triangle is equal to the ratio of the lengths of the opposite side and the hypotenuse. For the right-triangle in Fig. A.7:

$$\sin \alpha = \frac{a}{c}, \quad \sin \beta = \frac{b}{c}. \tag{A.5}$$

The *cosine* of an acute angle in a right-triangle is the ratio of the lengths of the adjacent side and the hypotenuse:

$$\cos \alpha = \frac{b}{c}, \quad \cos \beta = \frac{a}{c}. \tag{A.6}$$

Note that the sine of angle α in Fig. A.7 is equal to the cosine of angle β, and the cosine of angle α is equal to the sine of angle β. Also note that Eqs. (A.5) and (A.6) for angle α can alternatively be written as:

$$a = c \sin \alpha,$$
$$b = c \cos \alpha. \tag{A.7}$$

Furthermore, we can take the squares of a and b in Eq. (A.7) and substitute them in Eq. (A.2) to obtain:

$$c^2 = a^2 + b^2,$$
$$c^2 = (c \cos \alpha)^2 + (c \sin \alpha)^2,$$
$$c^2 = c^2(\cos^2\alpha + \sin^2\alpha).$$

Dividing both sides of the above equation by c^2 will yield an important trigonometric identity:

$$\cos^2\alpha + \sin^2\alpha = 1. \tag{A.8}$$

The *tangent* of an acute angle in a right-triangle is the ratio of the lengths of the opposite side and the adjacent side:

$$\tan \alpha = \frac{a}{b}, \quad \tan \beta = \frac{b}{a}. \tag{A.9}$$

The sine, cosine, and tangent of a few selected angles are listed in Table A.1. Note that the tangent of an angle is also equal to the ratio of the sine and the cosine of that angle:

$$\tan \alpha = \frac{\sin \alpha}{\cos \alpha}, \quad \tan \beta = \frac{\sin \beta}{\cos \beta}. \tag{A.10}$$

Table A.1 Sine, cosine, and tangent of selected angles

Angle	sin	cos	tan
0°	0.000	1.000	0.000
30°	0.500	0.866	0.577
45°	0.707	0.707	1.000
60°	0.866	0.500	1.732
90°	1.000	0.000	

It should also be noted that a right-triangle can uniquely be defined if the lengths of two sides of the triangle are known, or if one of the acute angles along with one of the sides of the triangle are known. The remaining sides and angles can be determined using trigonometric identities.

A.8 Inverse Sine, Cosine, and Tangent

Sometimes the sine, cosine, or the tangent of an angle is known and the task is to determine the angle itself. For this purpose, *inverse* sine, cosine, and tangent are defined such that:

If $\sin \alpha = A$ then $\alpha = \sin^{-1}(A)$
If $\cos \alpha = B$ then $\alpha = \cos^{-1}(B)$
If $\tan \alpha = C$ then $\alpha = \tan^{-1}(C)$

Inverse sine, cosine, and tangent are alternatively referred as arcsine, arccosine, and arctangent that are abbreviated as arcsin, arccos, and arctan, respectively.

Example A.1
For the right-triangle shown in Fig. A.8, if $a = 4$ and $b = 3$, determine angles α and β, and the length c of the hypotenuse.

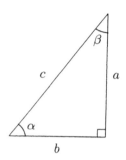

Fig. A.8 Example A.1

Solution
From Eq. (A.4):

$$c = \sqrt{a^2 + b^2}.$$

Substitute the numerical values of a and b, and carry out the calculations:

$$c = \sqrt{4^2 + 3^2} = \sqrt{16 + 9} = \sqrt{25} = 5.$$

The cosine of angle α is:

$$\cos \alpha = \frac{b}{c} = \frac{3}{5} = 0.6.$$

Take the inverse cosine of both sides:

$$\alpha = \cos^{-1}(0.6) = 53.13°.$$

The cosine of angle β is:

$$\cos \beta = \frac{a}{c} = \frac{4}{5} = 0.8.$$

Take the inverse cosine of both sides:

$$\beta = \cos^{-1}(0.8) = 36.87°.$$

Check whether the results are correct:

$$\alpha + \beta \stackrel{?}{=} 53.13° + 36.87° \stackrel{\checkmark}{=} 90°.$$

Example A.2
For the right-triangle shown in Fig. A.9, if $b = 2$ and $\alpha = 30°$, determine angle β, and sides a and c.

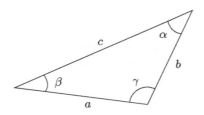

Fig. A.9 Problems A.1, A.2 and A.3

Solution
The sum of the acute angles of a right-triangle must be equal to 90°. Therefore:

$$\beta = 90° - \alpha = 90° - 30° = 60°.$$

Consider the cosine of angle α:

$$\cos \alpha = \frac{b}{c}.$$

Multiply both sides by c and divide by $\cos \alpha$:

$$c = \frac{b}{\cos \alpha}.$$

Substitute the numerical values of b and α:

$$c = \frac{2}{\cos 30°} = \frac{2}{0.866} = 2.31.$$

Consider the tangent of angle α:

$$\tan \alpha = \frac{a}{b}.$$

Multiply both sides by b:

$$a = b \tan \alpha.$$

Substitute the numerical values of b and α:

$$a = (2)(\tan 30°) = (2)(0.577) = 1.15.$$

Check if the results are correct:

$$a^2 + b^2 = c^2,$$
$$(1.15)^2 + (2)^2 \stackrel{?}{=} (2.31)^2,$$
$$5.3 \stackrel{\checkmark}{=} 5.3.$$

A.9 Exercise Problems

Problem A.1
For the irregular triangle shown in Fig. A.9, if $a = 3.5$, $b = 2.0$, and $\gamma = 130°$, determine the side c and angles α and β.
Answers: $c = 5.02$, $\alpha = 32.3°$, $\beta = 17.7°$.

Problem A.2
For the irregular triangle shown in Fig. A.9, if sides $a = 9$, $b = 6$, and $c = 13.4$, determine the interior angles of α, β, and an angle γ of the triangle.
Answers: $\alpha = 32.2°$, $\beta = 21.4°$, $\gamma = 125.4°$.

Problem A.3
For the irregular triangle shown in Fig. A.9, if $a = 15$ and the angles $\beta = 27°$ and $\alpha = 36°$, determine the sides b and c and an angle γ of the triangle.
Answers: $b = 11.6$, $c = 27.7$, $\gamma = 117°$.

Problem A.4
For the right-triangle shown in Fig. A.10, if the hypotenuse $c = 16.3$ and an angle $\alpha = 38°$, determine the two unknown sides of the right-triangle.
Answers: $a = 10.4$, $b = 12.8$.

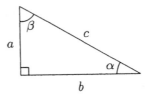

Fig. A.10 Problems A.4, A.5 and A.6

Problem A.5
For the right-triangle shown in Fig. A.10, if sides $a = 3.2$ and $b = 5.6$, determine the hypotenuse c and angles α and β.
Answers: $c = 6.5$, $\alpha = 29.7°$, $\beta = 60.3°$.

Problem A.6
For the right-triangle in Fig. A.10, if the hypotenuse $c = 12$ and angles α and β are equal to each other, determine the sides a and b, and the angles α and β.
Answers: $a = b = 8.5$, $\alpha = \beta = 45°$.

Appendix B: Vector Algebra

B.1 Definitions

Most of the concepts in mechanics are either scalar or vector. A *scalar* quantity has a magnitude only. Concepts such as mass, energy, power, mechanical work, and temperature are scalar quantities. For example, it is sufficient to say that an object has 80 kilograms (kg) of mass. A *vector* quantity, on the other hand, has both a magnitude and a direction associated with it. Force, moment, velocity, and acceleration are examples of vector quantities. To describe a force fully, one must state how much force is applied and in which direction it is applied. The magnitude of a vector is also a scalar quantity which is always a positive number.

It should be noted that both scalars and vectors are special forms of a more general category of all quantities in mechanics called *tensors*. Scalars are also known as "zero-order tensors," whereas vectors are "first-order tensors." Concepts such as stress and strain, on the other hand, are "second-order tensors." In this definition, the order corresponds to the power n in 3^n. For scalars, n is zero, and therefore, $3^0 = 1$, or only one quantity (magnitude) is necessary to define a scalar quantity. For vectors, n is one, and therefore, $3^1 = 3$. That is, three quantities (components in three directions) are necessary to define a vector quantity in three-dimensional space. On the other hand, n is two for second-order tensors, and since $3^2 = 9$, nine quantities (three components in three planes) are needed to define concepts such as stress and strain.

B.2 Notation

There are various notations used to refer to vector quantities. In this text, we shall use letters with underbars. For example, $\underline{A}$ will refer to a vector quantity, whereas A without the underbar will be used to refer to a scalar quantity. In graphical solutions, vectors are commonly represented by arrows, as illustrated in Fig. B.1. The orientation of the arrow indicates the *line of action* of the vector. The arrowhead denotes the *direction* of the vector by defining its *sense* along its line of action. If the vector represents, for example, an applied force, then the base (tail) of the arrow corresponds to the *point of application* of the force vector. If there is a need to show more than one vector in a single drawing, the length of each arrow must be proportional to the magnitude of the vector it is representing. The *magnitude* of a vector quantity is always a positive number corresponding to the numerical measure of that quantity. There are two ways of referring to the magnitude of a vector quantity: either by dropping the underbar (A) or by enclosing the vector quantity with a set of vertical lines known as the absolute sign ($|\underline{A}|$).

B.3 Multiplication of a Vector by a Scalar

Let $\underline{A}$ be a vector quantity with magnitude A and m be a scalar quantity. The product $m\underline{A}$ is equal to a new vector, $\underline{B} = m\underline{A}$, such that it has the same direction as vector $\underline{A}$ but a magnitude equal to m times A. For example, if $m = 2$, then the magnitude of the product vector $\underline{B}$ is twice as large as the magnitude of vector $\underline{A}$.

B.4 Negative Vector

Let $\underline{A}$ be a vector quantity with magnitude A. $-\underline{A}$ is called a *negative vector* and it differs from vector $\underline{A}$ in that $-\underline{A}$ has a direction opposite to that of vector $\underline{A}$ (Fig. B.2). Since magnitudes of vector quantities are positive scalar quantities,

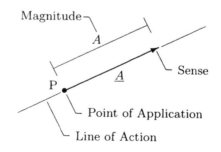

Fig. B.1 Graphical representation of vector $\underline{A}$

N. Özkaya et al., *Fundamentals of Biomechanics: Equilibrium, Motion, and Deformation*,
DOI 10.1007/978-1-4614-1150-5, © Springer Science+Business Media, LLC 2012

Fig. B.2 Negative vector

both $-\underline{A}$ and $\underline{A}$ have the same magnitude equal to A. Therefore, for vector quantities, a negative sign implies change in direction and has nothing to do with the magnitude of the vector.

B.5 Addition of Vectors: Graphical Methods

There are two ways of adding two or more vectors graphically: parallelogram and triangle or tail-to-tip methods. Consider the two vectors $\underline{A}$ and $\underline{B}$ shown in Fig. B.3. Vector $\underline{A}$ is pointing toward point Q and vector $\underline{B}$ is pointing toward point R. The *parallelogram* method of adding two vectors involves the construction of a parallelogram by drawing a line at the tip of one of the vectors parallel to the other vector, and repeating the same thing for the second vector. If S corresponds to the point of intersection of these parallel lines, then an arrow drawn from point P toward S represents a vector that is equal to the vector sum of $\underline{A}$ and $\underline{B}$ (Fig. B.4). The third vector thus obtained is called the *resultant* or the *net* vector. In Fig. B.4, the resultant vector is identified as $\underline{C}$ which can be mathematically expressed as:

$$\underline{A} + \underline{B} = \underline{C}. \qquad (B.1)$$

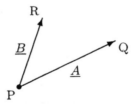

Fig. B.3 Vectors $\underline{A}$ and $\underline{B}$

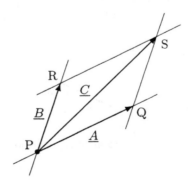

Fig. B.4 Parallelogram

The *triangle* or *tail-to-tip* method of adding two vectors graphically is illustrated in Fig. B.5. In this case, without changing its orientation, one of the vectors to be added is translated to the tip of the other vector in such a way that the tip of one of the vectors coincides with the tail of the other. An arrow drawn from the tail of the first vector toward the tip of the second vector represents the resultant vector.

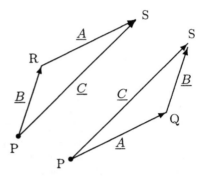

Fig. B.5 Tail-to-tip method

Notice that while performing vector addition, the order of appearance of vectors is arbitrary. That is, the addition of vectors is a *commutative* operation. The sum of $\underline{A}$ and $\underline{B}$ and the sum of $\underline{B}$ and $\underline{A}$ result in the same vector $\underline{C}$:

$$\underline{A} + \underline{B} = \underline{B} + \underline{A} = \underline{C}. \qquad (B.2)$$

B.6 Subtraction of Vectors

The subtraction of one vector from another can easily be done by noting that the negative of any vector is a vector of the same magnitude pointing in the opposite direction. For example, as illustrated in Fig. B.6, to subtract vector $\underline{B}$ from vector $\underline{A}$, vector $-\underline{B}$ can be added to $\underline{A}$ to determine the resultant vector.

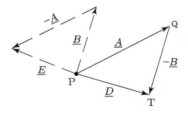

Fig. B.6 $\underline{D} = \underline{A} - \underline{B}, \underline{E} = \underline{B} - \underline{A}$

Note that subtraction of vector $\underline{A}$ from $\underline{B}$ follows a similar approach. If $\underline{D}$ and $\underline{E}$ are two vectors such that

$$\underline{A} - \underline{B} = \underline{D},$$

$$\underline{B} - \underline{A} = \underline{E},$$

then since $\underline{A} - \underline{B} = -(\underline{B} - \underline{A})$, vectors $\underline{D}$ and $\underline{E}$ have an equal magnitude ($D = E$) but opposite directions ($\underline{D} = -\underline{E}$).

B.7 Addition of More Than Two Vectors

As illustrated in Fig. B.7, let $\underline{A}$, $\underline{B}$, and $\underline{C}$ be three vectors such that all three vectors lie on a two-dimensional surface (a *coplanar* system of vectors). There are a number of different ways to add three vectors together. Since the addition of vectors is a commutative operation, the order of appearance of vectors to be added does not influence the resultant vector.

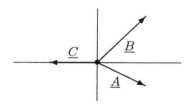

Fig. B.7 Three vectors

The three vectors shown in Fig. B.7 are graphically added together in Fig. B.8 by using the tail-to-tip method. In the case illustrated, vector $\underline{C}$ is added to vector $\underline{B}$ which is added to vector $\underline{A}$ to obtain the resultant vector $\underline{D}$. That is:

$$\underline{A} + \underline{B} + \underline{C} = \underline{D}.$$

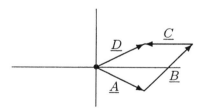

Fig. B.8 Vector sum of $\underline{A}$, $\underline{B}$, and $\underline{C}$

There are other ways to graphically add vectors $\underline{A}$, $\underline{B}$, and $\underline{C}$ using the tail-to-tip method. Some of these are shown in Figs. B.9–B.11.

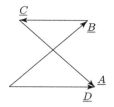

Fig. B.9 Vector sum of $\underline{B}$, $\underline{C}$, and $\underline{A}$

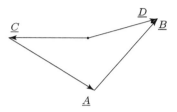

Fig. B.10 Vector sum of $\underline{C}$, $\underline{A}$, and $\underline{B}$

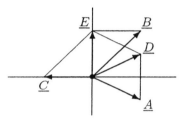

Fig. B.11 Parallelogram method of adding three vectors

In the case illustrated in Fig. B.9, vector $\underline{A}$ is added to vector $\underline{C}$ which is added to vector $\underline{B}$ to obtain the resultant vector $\underline{D}$. That is:

$$\underline{B} + \underline{C} + \underline{A} = \underline{D}.$$

In the case illustrated in Fig. B.10, vector $\underline{B}$ is added to vector $\underline{A}$ which is added to vector $\underline{C}$ to obtain the resultant vector $\underline{D}$. That is:

$$\underline{C} + \underline{A} + \underline{B} = \underline{D}.$$

Note that regardless of the order in which vectors $\underline{A}$, $\underline{B}$, and $\underline{C}$ are added, the resultant vector $\underline{D}$ has the same magnitude and the same direction.

The parallelogram method of adding the same vectors is illustrated in Fig. B.11. In this case, two ($\underline{B}$ and $\underline{C}$) of the three vectors are added together first. The resultant (vector $\underline{E}$) of this operation is then added to vector $\underline{A}$ and the overall resultant (vector $\underline{D}$) is determined. The sequence of additions illustrated in Fig. B.11 can be mathematically expressed as:

$$\underline{A} + (\underline{B} + \underline{C}) = \underline{A} + \underline{E} = \underline{D}.$$

Similar to the tail-to-tip method used to add vectors $\underline{A}$, $\underline{B}$, and $\underline{C}$, there are also different ways to add these vectors using the parallelogram method and by taking into consideration the sequence of their addition. Again, regardless of the sequence chosen, the magnitude and the direction of the resultant vector $\underline{D}$ are going to be the same.

B.8 Projection of Vectors

As illustrated in Fig. B.12, let s represent a line and $\underline{A}$ be a vector whose line of action makes an angle θ with s. Assume that line s and vector $\underline{A}$ lie on a plane surface. To determine the projection or the component of vector $\underline{A}$ on s, drop a straight line from the tip (point Q) of the vector that cuts the line defined by s at right angles. If R denotes the point of intersection of these two lines and A_s is the length of the line segment between points P and R, then A_s is the *projection* or *scalar component* of vector $\underline{A}$ on s. Note that points P, Q, and R define a right-triangle, and therefore:

$$A_s = A \cos \theta. \tag{B.3}$$

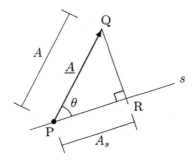

Fig. B.12 Projection of a vector on a given direction

B.9 Resolution of Vectors

The resolution of a vector into its components is the reverse action of adding two vectors. In Fig. B.13, x and y indicate the horizontal and vertical axes, respectively. $\underline{A}$ is a vector acting on the x-plane. To determine the component of $\underline{A}$ along the x direction, drop a vertical line from the tip of the vector that cuts the horizontal line defined by x at right angles. Similarly, to determine the component of $\underline{A}$ along the y direction, draw a horizontal line passing through the tip of the vector that cuts the vertical line defined by y at right angles. These operations result in two similar right-triangles forming a rectangle (a parallelogram). The lengths of the sides of this rectangle represent the *scalar components* of vector $\underline{A}$ along the x and y directions, which

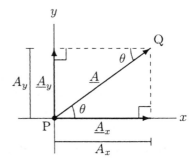

Fig. B.13 Components of a vector along two mutually perpendicular directions

can be determined by utilizing the properties of right-triangles:

$$A_x = A \cos \theta, \\ A_y = A \sin \theta. \tag{B.4}$$

In Eq. (B.4), θ is the angle that vector $\underline{A}$ makes with the horizontal. Vector $\underline{A}$ can also be represented as the sum of its *vector components* $\underline{A}_x$ and $\underline{A}_y$ along the x and y directions, respectively:

$$\underline{A} = \underline{A}_x + \underline{A}_y. \tag{B.5}$$

B.10 Unit Vectors

Usually, it is convenient to express a vector $\underline{A}$ as the product of its magnitude A and a vector $\underline{a}$ of unit magnitude that has the same direction as vector $\underline{A}$ (Fig. B.14). $\underline{a}$ is called the *unit vector*. For a given vector, its unit vector can be obtained by dividing that vector with its magnitude:

$$\underline{a} = \frac{\underline{A}}{A}. \tag{B.6}$$

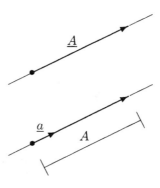

Fig. B.14 $\underline{A} = A\underline{a}$

The original vector $\underline{A}$ can now be expressed as:

$$\underline{A} = A\underline{a}. \qquad (B.7)$$

Since the magnitude of $\underline{a}$ is equal to the magnitude of $\underline{A}$ (which is A) divided by A, the magnitude of vector $\underline{a}$ is always equal to one. Unit vectors are used as a convenient way of describing directions.

B.11 Rectangular Coordinates

To be able to define the position of an object in space and to be able to analyze changes in position over time, measurements must be made relative to a reference frame or a coordinate system. There are a number of widely used coordinate systems. Among these, the *Cartesian* or *rectangular* coordinate system is the one most commonly used.

The two-dimensional Cartesian coordinate system consists of two mutually perpendicular axes x and y dividing the plane surface into four so-called quadrants, as illustrated in Fig. B.15. The point of intersection of the x and y axes is called the origin of the coordinate system, and it is usually denoted as O.

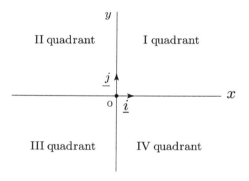

Fig. B.15 Two-dimensional Cartesian coordinate system

In the first quadrant, both x and y coordinates are positive. In the second quadrant, the y coordinates are positive and the x coordinates are negative. In the third quadrant, both the x and y coordinates are negative, and in the fourth quadrant, the x coordinates are positive and the y coordinates are negative. Two coordinates (x and y) are usually required to uniquely specify or define the position of a point in the two-dimensional space. For example, the two coordinates $x = -3$ and $y = 5$ define the position of point P in the second quadrant of the two-dimensional coordinate system. As illustrated in Fig. B.16, to determine the position of point P, a vertical line is drawn through point (-3) on the horizontal x axis and a horizontal line is then drawn through point (5) on the vertical y axis. The point of intersection of these two lines uniquely defines the point P.

As illustrated in Fig. B.17, the three-dimensional Cartesian coordinate system consists of three mutually perpendicular

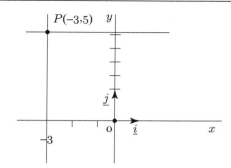

Fig. B.16 The position of point P in a two-dimensional coordinate system

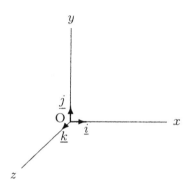

Fig. B.17 Three-dimensional Cartesian or rectangular coordinate system

axes, x, y, and z. Three coordinates (x, y, z) are usually required to uniquely define the position of a point in three-dimensional space. As illustrated in Fig. B.18, the position of point P is defined by three coordinates: $x = 3$, $y = 4$, and $z = 6$.

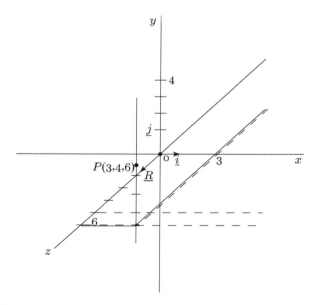

Fig. B.18 The position of point P in a three-dimensional coordinate system

To determine the position of point P in the three-dimensional space, two lines are drawn first through points with coordinates $x = 3$ and $z = 6$ parallel to the x and z axes, respectively. Then the third line is drawn in the positive y direction, through the point of intersection of these two lines. Point P is located four units above the point of intersection.

The concept of the unit vector has an important application in the construction of coordinate systems. The unit vectors along the Cartesian coordinate axes are so frequently used that they have widely accepted symbols. Symbols $\underline{i}$, $\underline{j}$, and $\underline{k}$ are commonly used to refer to *unit coordinate vectors* indicating positive x, y, and z directions, respectively.

B.12 Addition of Vectors: Trigonometric Method

We have seen the addition and subtraction of vectors, and the resolution of vectors into their components by means of graphical methods that may be time-consuming and not accurate. Faster and more precise results can be obtained through the use of trigonometric identities.

To apply the *trigonometric method* of addition and subtraction of vectors, one must first resolve each vector into its components. For example, consider the two vectors shown in Fig. B.19. Vectors $\underline{A}$ and $\underline{B}$ have magnitudes equal to A and B, and they make angles α and β with the horizontal (x axis). The scalar components of $\underline{A}$ and $\underline{B}$ along the x and y directions can be determined by utilizing the properties of right-triangles. For vector $\underline{A}$:

$$A_x = A \cos \alpha,$$
$$A_y = A \sin \alpha. \tag{B.8}$$

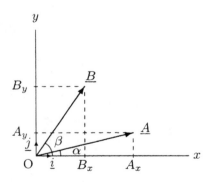

Fig. B.19 Vectors $\underline{A}$ and $\underline{B}$

A_x and A_y are the scalar components of $\underline{A}$ along the x and y directions, respectively. Making use of the unit vectors

$\underline{i}$ and $\underline{j}$ that identify positive x and y directions, the vector components of $\underline{A}$ in rectangular coordinates can be determined easily from its scalar components:

$$\underline{A}_x = A_x \underline{i},$$
$$\underline{A}_y = A_y \underline{j}. \tag{B.9}$$

Now, vector $\underline{A}$ can alternately be expressed as:

$$\underline{A} = \underline{A}_x + \underline{A}_y,$$
$$= A_x \underline{i} + A_y \underline{j},$$
$$= A \cos \alpha \, \underline{i} + A \sin \alpha \, \underline{j}. \tag{B.10}$$

Note that A_x, A_y, and A correspond to the sides of a right-triangle with A being the hypotenuse. Therefore:

$$A = \sqrt{(A_x)^2 + (A_y)^2}. \tag{B.11}$$

Similarly, vector $\underline{B}$ can be expressed as:

$$\underline{B} = \underline{B}_x + \underline{B}_y,$$
$$= B_x \underline{i} + B_y \underline{j},$$
$$= B \cos \beta \, \underline{i} + B \sin \beta \, \underline{j}. \tag{B.12}$$

Once the vectors are expressed in terms of their components, the next step is to add (or subtract) the x components of all vectors together. This will yield the x component of the resultant vector. Similarly, adding the y components of all vectors will give the y component of the resultant vector. For example, the addition of vectors $\underline{A}$ and $\underline{B}$ can be performed as:

$$\underline{A} + \underline{B} = (A_x \underline{i} + A_y \underline{j}) + (B_x \underline{i} + B_y \underline{j}),$$
$$= (A_x + B_x) \underline{i} + (A_y + B_y) \underline{j}, \tag{B.13}$$
$$= (A \cos \alpha + B \cos \beta) \, \underline{i} + (A \sin \alpha + B \sin \beta) \, \underline{j}.$$

If $\underline{C}$ refers to the vector sum of $\underline{A}$ and $\underline{B}$, then:

$$\underline{C} = \underline{A} + \underline{B},$$
$$= \underline{C}_x + \underline{C}_y,$$
$$= C_x \underline{i} + C_y \underline{j}. \tag{B.14}$$

Comparing Eqs. (B.13) and (B.14), one can conclude that:

$$C_x = A_x + B_x = A \cos \alpha + B \cos \beta,$$
$$C_y = A_y + B_y = A \sin \alpha + B \sin \beta. \tag{B.15}$$

If A, B, α, and β are known, then C_x and C_y can be determined from (B.15). Note that C_x, C_y, and C also form a right-triangle, with C being the hypotenuse (Fig. B.20). Therefore:

$$C = \sqrt{(C_x)^2 + (C_y)^2}. \qquad (B.16)$$

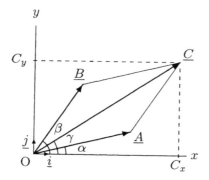

Fig. B.20 Vector sum of $\underline{A}$ and $\underline{B}$ is equal to $\underline{C}$

If γ represents an angle that vector $\underline{C}$ makes with the x axis, then:

$$\gamma = \tan^{-1}\left(\frac{C_y}{C_x}\right). \qquad (B.17)$$

Note that $\tan^{-1}$ is called the *inverse tangent* or *arctangent* (abbreviated as arctan).

Subtraction of one vector from another is as straightforward as adding the two. For example:

$$\begin{aligned}
\underline{A} - \underline{B} &= (A_x\underline{i} + A_y\underline{j}) - (B_x\underline{i} + B_y\underline{j}), \\
&= (A_x - B_x)\underline{i} + (A_y - B_y)\underline{j}, \qquad (B.18) \\
&= (A\cos\alpha - B\cos\beta)\underline{i} + (A\sin\alpha - B\sin\beta)\underline{j}.
\end{aligned}$$

Example B.1
Vector $\underline{A}$ in Fig. B.21 is such that its magnitude is $A = 5$ U and it makes an angle $\alpha = 36.87°$ with the horizontal.

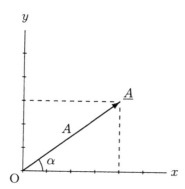

Fig. B.21 Example B.1

(a) Determine the scalar components of $\underline{A}$ along the horizontal and the vertical.
(b) Express $\underline{A}$ in terms of its components.

Solution
(a) Using Eq. (B.8):

$$\begin{aligned}
A_x &= A\cos\alpha = 5\cos(36.87°) = 4, \\
A_y &= A\sin\alpha = 5\sin(36.87°) = 3.
\end{aligned}$$

(b) From Eq. (B.10):

$$\underline{A} = A_x\underline{i} + A_y\underline{j} = 4\underline{i} + 3\underline{j}.$$

Example B.2
The component of vector $\underline{B}$ in Fig. B.22 along the positive x direction is measured as 8 U and its component in the negative y direction is 12 U.

(a) Express vector $\underline{B}$ in terms of its components.
(b) Determine the magnitude of vector $\underline{B}$ and angle β that it makes with the positive x direction.

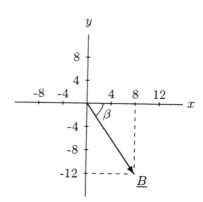

Fig. B.22 Example B.2

Solution
(a) Vector $\underline{B}$ can be expressed as:

$$\underline{B} = 8\underline{i} - 12\underline{j}$$

(b) Note that $B_x = 8$ and $B_y = 12$. Hence:

$$B = \sqrt{(B_x)^2 + (B_y)^2} = \sqrt{(8)^2 + (12)^2} = \sqrt{208} = 14.42.$$

$$\beta = \tan^{-1}\left(\frac{B_y}{B_x}\right) = \tan^{-1}\left(\frac{12}{8}\right) = 56.31°.$$

Example B.3
Consider vectors $\underline{A}$ and $\underline{B}$ shown in Fig. B.23. The scalar components of these vectors are measured as:

$$A_x = 15,\ B_x = 5,$$
$$A_y = 10,\ B_y = 10.$$

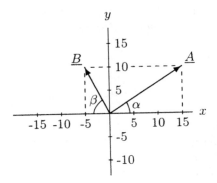

Fig. B.23 Example B.3

(a) Express $\underline{A}$ and $\underline{B}$ in terms of their components.
(b) Determine the magnitudes of $\underline{A}$ and $\underline{B}$, and angles α and β.
(c) Determine $\underline{A} + \underline{B}$.
(d) Determine $\underline{A} - \underline{B}$.
(e) Determine $\underline{B} - \underline{A}$.
(f) Determine $-\underline{A} - \underline{B}$.

Solution
(a)
$$\underline{A} = 15\underline{i} + 10\underline{j},$$
$$\underline{B} = -5\underline{j} + 10\underline{j}.$$

(b) Magnitudes of $\underline{A}$ and $\underline{B}$:

$$A = \sqrt{(A_x)^2 + (A_y)^2} = \sqrt{15^2 + 10^2} = 18.03,$$

$$B = \sqrt{(B_x)^2 + (B_y)^2} = \sqrt{5^2 + 10^2} = 11.18.$$

Angles α and β:

$$\alpha = \tan^{-1}\left(\frac{10}{15}\right) = 33.69°,$$

$$\beta = \tan^{-1}\left(\frac{10}{5}\right) = 63.43°.$$

(c) Let $\underline{C} = \underline{A} + \underline{B}$. Then:

$$\underline{C} = (15\underline{i} + 10\underline{j}) + (-5\underline{i} + 10\underline{j}),$$
$$= (15 - 5)\underline{i} + (10 + 10)\underline{j},$$
$$= 10\underline{i} + 20\underline{j}.$$

(d) Let $\underline{D} = \underline{A} - \underline{B}$. Then:

$$\underline{D} = (15\underline{i} + 10\underline{j}) - (-5\underline{i} + 10\underline{j}),$$
$$= (15 - 5)\underline{i} + (10 - 10)\underline{j},$$
$$= 20\underline{i}.$$

(e) Let $\underline{E} = \underline{B} - \underline{A}$. Then:

$$\underline{E} = (-5\underline{i} + 10\underline{j}) - (-15\underline{i} + 10\underline{j}),$$
$$= (-5 - 15)\underline{i} + (10 - 10)\underline{j},$$
$$= -20\underline{i}.$$

(f) Let $\underline{F} = -\underline{A} - \underline{B}$. Then:

$$\underline{F} = (-15\underline{i} + 10\underline{j}) - (-5\underline{i} + 10\underline{j}),$$
$$= (-15 + 5)\underline{i} + (-10 - 10)\underline{j},$$
$$= -10\underline{i} - 20\underline{j}.$$

The resultant vectors $\underline{C}$, $\underline{D}$, $\underline{E}$, and $\underline{F}$ are illustrated in Fig. B.24. A careful examination of these results shows that vectors $\underline{C}$ and $\underline{F}$ form a pair of negative vectors with equal magnitude and opposite directions, and so are vectors $\underline{D}$ and $\underline{E}$. Also note that vectors $\underline{D}$ and $\underline{E}$ have no components along the y direction.

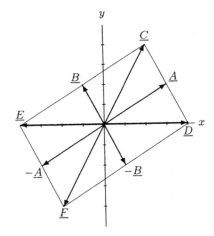

Fig. B.24 Addition and subtraction of vectors

B.13 Three-Dimensional Components of Vectors

The trigonometric method of adding and subtracting vectors with components in the x, y, and z directions is an extension of the principles introduced for two-dimensional systems. With respect to the Cartesian coordinate system, to define a vector quantity such as $\underline{A}$ uniquely, either all three components A_x, A_y, and A_z of vector $\underline{A}$, or the magnitude A

of the vector along with the angles the vector makes with the x, y, and z directions, must be provided (Fig. B.25). We can alternatively express $\underline{A}$ as:

$$
\begin{aligned}
\underline{A} &= \underline{A}_x + \underline{A}_y + \underline{A}_z, \\
&= A_x \underline{i} + A_y \underline{j} + A_z \underline{k}.
\end{aligned}
\tag{B.19}
$$

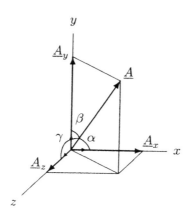

Fig. B.25 Components of a vector in three directions

If α, β, and γ refer to the angles that vector $\underline{A}$ makes with the x, y, and z directions, respectively, then:

$$
\begin{aligned}
A_x &= A \cos \alpha, \\
A_y &= A \cos \beta, \\
A_z &= A \cos \gamma.
\end{aligned}
\tag{B.20}
$$

Assume that there is a second vector, $\underline{B}$, that is to be added to vector $\underline{A}$. If $\underline{C}$ represents the resultant vector, then:

$$
\begin{aligned}
\underline{C} &= \underline{A} + \underline{B}, \\
&= (A_x \underline{i} + A_y \underline{j} + A_z \underline{k}) + (B_x \underline{i} + B_y \underline{j} + B_z \underline{k}), \\
&= (A_x + B_x) \underline{i} + (A_y + B_y) \underline{j} + (A_z + B_z) \underline{k}.
\end{aligned}
\tag{B.21}
$$

Note that the components of the resultant vector $\underline{C}$ in the x, y, and z directions are:

$$
\begin{aligned}
C_x &= A_x + B_x, \\
C_y &= A_y + B_y, \\
C_z &= A_z + B_z,
\end{aligned}
\tag{B.22}
$$

As is true for coplanar systems, the addition of two or more vector quantities having components in all three directions requires addition of the x components of all vectors together. The same procedure must be repeated for the y and z components. Subtraction of two or more three-dimensional vector quantities follows a similar approach.

B.14 Dot (Scalar) Product of Vectors

Two of the most commonly encountered vector quantities in mechanics are the force and displacement vectors. In mechanics, *work done* by a force is defined as the product of the force component in the direction of displacement and the magnitude of displacement. Although force and displacement are vector quantities, their product—work—is a scalar quantity. An operation that represents such situations concisely is called the *dot* or *scalar product*.

The dot product of any two vectors is defined as a scalar quantity equal to the product of magnitudes of the two vectors multiplied by the cosine of the smaller angle between the two. For example, consider vectors $\underline{A}$ and $\underline{B}$ in Fig. B.26. The dot product of these vectors is:

$$
\underline{A} \cdot \underline{B} = AB \cos \theta.
\tag{B.23}
$$

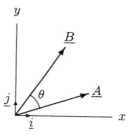

Fig. B.26 Dot (scalar) product of two vectors

In Eq. (B.23), A and B are the magnitudes of the two vectors, and θ is the smaller angle between them. The operation given by Eq. (B.23) is representative of first projecting vector $\underline{A}$ onto the line of action of vector $\underline{B}$ (or vice versa) and then multiplying the magnitudes of the projected component and the other vector.

Note that the dot product may result in positive or negative quantities depending on whether the smaller angle between the vectors is less or greater than $90°$. Since $\cos 90° = 0$, the dot product of two vectors is equal to zero if the vectors are perpendicular to one another. Unit vectors $\underline{i}$ and $\underline{j}$, which define the positive x and y directions, respectively, are also vector quantities with magnitudes equal to unity. The concept of the dot product is applicable to unit vectors as well:

$$
\begin{aligned}
\underline{i} \cdot \underline{i} &= \underline{j} \cdot \underline{j} = \underline{k} \cdot \underline{k} = 1, \quad 1 \cos 0° = 1, \\
\underline{i} \cdot \underline{j} &= \underline{j} \cdot \underline{k} = \underline{k} \cdot \underline{i} = 1, \quad 1 \cos 90° = 0.
\end{aligned}
\tag{B.24}
$$

To take the dot product of the vectors shown in Fig. B.26, we can first represent each vector in terms of its components along the x and y directions, and then apply the dot product to the unit vectors:

$$\underline{A} \cdot \underline{B} = (A_x\underline{i}+A_y\underline{j}) \cdot (B_x\underline{i}+B_y\underline{j}),$$
$$= A_xB_x(\underline{i} \cdot \underline{i}) + A_yB_y(\underline{i} \cdot \underline{j}) + A_yB_x(\underline{j} \cdot \underline{i}) + A_yB_y(\underline{j} \cdot \underline{j}),$$
$$= A_xB_x + A_yB_y.$$

(B.25)

The following are some of the properties of the dot product.

• The dot product of two vectors is a commutative operation:

$$\underline{A} \cdot \underline{B} = \underline{B} \cdot \underline{A}.$$

• The dot product is a distributive operation:

$$\underline{A} \cdot (\underline{B} + \underline{C}) = \underline{A} \cdot \underline{B} + \underline{A} \cdot \underline{C}.$$

• A vector multiplied by itself as a dot product is equal to the square of the magnitude of the vector:

$$\underline{A} \cdot \underline{A} = A^2$$

• The dot product of 2 three-dimensional vector quantities $\underline{A}$ and $\underline{B}$ expressed in terms of their Cartesian components is:

$$\underline{A} \cdot \underline{B} = A_xB_x + A_yB_y + A_zB_z.$$

• The scalar component of a vector along a given direction is equal to the dot product of the vector times the unit vector along that direction. For example, the x and y components of a vector $\underline{A}$ are:

$$A_x = \underline{A} \cdot \underline{i}, \quad A_y = \underline{A} \cdot \underline{j}.$$

B.15 Cross (Vector) Product of Vectors

There are other interactions between vector quantities that result in other vector quantities. An example of such interactions is the moment or torque generated by an applied force. The mathematical tool developed to define such interactions is called the *cross* or *vector product*.

Consider the vectors $\underline{A}$ and $\underline{B}$ in Fig. B.27. The cross product of these vectors is equal to a third vector, say $\underline{C}$. The commonly used mathematical notation for the cross product is:

$$\underline{A} \times \underline{B} = \underline{C}.$$

(B.26)

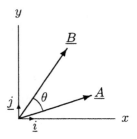

Fig. B.27 Two vectors $\underline{A}$ and $\underline{B}$

The vector $\underline{C}$ has a magnitude equal to the product of the magnitudes of vectors $\underline{C}$ and $\underline{B}$ times the sine of the smaller angle (angle θ in Fig. B.27) between the two:

$$C = AB \sin \theta.$$

(B.27)

Vector $\underline{C}$ has a direction perpendicular to the plane defined by vectors $\underline{A}$ and $\underline{B}$. For example, if both $\underline{A}$ and $\underline{B}$ are in the xy-plane, then $\underline{C}$ acts in the z direction.

The sense of vector $\underline{C}$ can be determined by the *right-hand-rule*. To apply this rule, first the fingers of the right hand are pointed in the direction of vector $\underline{A}$ (the first vector), and then they are curled toward vector $\underline{B}$ (the second vector) in such a way as to cover the smaller angle between $\underline{A}$ and $\underline{B}$. The extended thumb points in the direction of the product vector $\underline{C}$.

Note again that the concept of the cross product is applicable to the Cartesian unit vectors $\underline{i}, \underline{j}$, and $\underline{k}$ that are mutually perpendicular. Since $\sin 0° = 0$ and $\sin 90° = 1$, one can write:

$$\underline{i} \times \underline{i} = \underline{j} \times \underline{j} = \underline{k} \times \underline{k} = 0,$$
$$\underline{i} \times \underline{j} = \underline{k}, \quad \underline{j} \times \underline{i} = -\underline{k},$$
$$\underline{j} \times \underline{k} = \underline{i}, \quad \underline{k} \times \underline{j} = -\underline{i},$$
$$\underline{k} \times \underline{i} = \underline{j}, \quad \underline{i} \times \underline{k} = -\underline{j}.$$

(B.28)

Using the above relations between the unit vectors, one can determine mathematical relations to evaluate the vector products that would include both the magnitude and the direction of the product vector. For example, if $\underline{A}$ and $\underline{B}$ are two vectors in the xy-plane (Fig. B.28), then:

$$\underline{C} = \underline{A} \times \underline{B},$$
$$= (A_x\underline{i}+A_y\underline{j}) \times (B_x\underline{i}+B_y\underline{j}),$$
$$= A_xB_x(\underline{i} \times \underline{i}) + A_xB_y(\underline{i} \times \underline{j}) + A_yB_x(\underline{j} \times \underline{i}) + A_yB_y(\underline{j} \times \underline{j}),$$
$$= A_xB_x(0) + A_xB_y(\underline{k}) + A_yB_x(-\underline{k}) + A_yB_y(0),$$
$$= (A_xB_y - A_yB_x)\underline{k}.$$

(B.29)

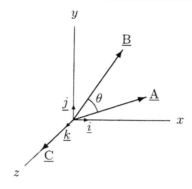

Fig. B.28 $\underline{C} = \underline{A} \times \underline{B}$

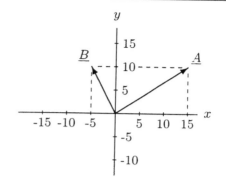

Fig. B.29 Example B.4

The following are some of the properties of the cross product.

- The cross product is not a commutative operation:

$$\underline{A} \times \underline{B} \neq \underline{B} \times \underline{A}.$$

But

$$\underline{A} \times \underline{B} = -\underline{B} \times \underline{A}.$$

- The cross product is a distributive operation:

$$\underline{A} \times (\underline{B} + \underline{C}) = \underline{A} \times \underline{B} + \underline{A} \times \underline{C}.$$

- The vector product of 2 three-dimensional vector quantities $\underline{A}$ and $\underline{B}$ expressed in terms of their Cartesian components is:

$$\underline{A} \times \underline{B} = (A_y B_z - A_z B_y)\underline{i} + (A_z B_x - A_x B_z)\underline{j} + (A_x B_y - A_y B_x)\underline{k}.$$

Example B.4

Vectors $\underline{A}$ and $\underline{B}$ shown in Fig. B.29 are given in terms of their Cartesian components:

$$\underline{A} = 15\underline{i} + 10\underline{j},$$
$$\underline{B} = -5\underline{i} + 10\underline{j}.$$

(a) Evaluate the scalar product $c = \underline{A} \cdot \underline{B}$.
(b) Evaluate the vector product $\underline{D} = \underline{A} \times \underline{B}$.
(c) Evaluate the vector product $\underline{E} = \underline{B} \times \underline{A}$.

Solution

(a)

$$c = \underline{A} \cdot \underline{B},$$
$$= A_x B_x + A_y B_y,$$
$$= (15)(-5) + (10)(10),$$
$$= 25.$$

(b)

$$\underline{D} = \underline{A} \times \underline{B},$$
$$= (A_x B_y - A_y B_x)\underline{k},$$
$$= [(15)(10) - (10)(-5)]\underline{k},$$
$$= 200\underline{k}.$$

(c)

$$\underline{E} = \underline{B} \times \underline{A} = -\underline{D}$$
$$= (B_x A_y - B_y A_x)\underline{k},$$
$$= [(-5)(10) - (10)(15)]\underline{k},$$
$$= -200\underline{k}.$$

Note that the product vectors $\underline{D}$ and $\underline{E}$ have an equal magnitude of 200 U. Vector $\underline{D}$ has a counterclockwise direction, whereas vector $\underline{E}$ is clockwise. Both vectors are acting in the direction perpendicular to the surface of the page, $\underline{D}$ out of the page and $\underline{E}$ into the page.

B.16 Exercise Problems

Problem B.1

There are four points A, B, C, and D on the two-dimensional coordinate system shown in Fig. B.30. If coordinates of these points are A (2.5, 3.0), B (−3.5, −1.0), C (1.5, −5.0), and D (−4.0, 3.5), locate them on the coordinate system.

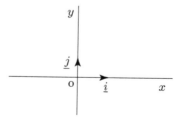

Fig. B.30 Two-dimensional coordinate system

Problem B.2

There are three points A, B, and C on the three-dimensional coordinate system shown in Fig. B.31. If the coordinates of these points are A (2.0, 2.5, −3.0), B (−1.5, 2.0, 3.5), and C (−3.0, −2.0, −3.0), locate them on the coordinate system.

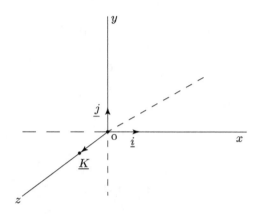

Fig. B.31 Three-dimensional coordinate system

Problem B.3

Two vectors are given such that their magnitudes are $A = 9$ and $B = 6.5$. Vector $\underline{A}$ makes an angle $\alpha = 66°$ and vector $\underline{B}$ makes an angle $\beta = 52°$ with the horizontal, as shown in Fig. B.32.

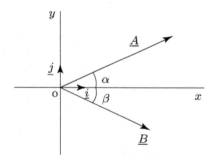

Fig. B.32 Problem B.3

(a) Determine the components of vectors $\underline{A}$ and $\underline{B}$ in the horizontal and vertical direction.
(b) Represent vectors $\underline{A}$ and $\underline{B}$ in terms of their components.
Answers:
(a)

$$A_x = 3.66; \quad A_y = 8.22; \quad B_x = 4.0; \quad B_y = 5.12.$$

(b)

$$\underline{A} = 3.66\underline{i} + 8.22\underline{j}; \quad \underline{B} = 4\underline{i} - 5.12\underline{j}.$$

Problem B.4

If the horizontal and vertical components of vector $\underline{A}$ are equal to 2 and 3 U, respectively, determine an angle α that the vector $\underline{A}$ makes with the horizontal.
Answer: $\alpha = 56.3°$

Problem B.5

A component of vector $\underline{A}$ in the horizontal direction is given such that its magnitude is $A_y = 4.75$. The vector $\underline{A}$ makes an angle α with the horizontal as shown in Fig. B.33, such that $\tan \alpha = 0.7$.

(a) Determine the component of vector $\underline{A}$ in the horizontal direction.

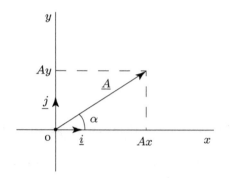

Fig. B.33 Problem B.5

(b) Determine the magnitude of vector $\underline{A}$.
(c) Represent vector $\underline{A}$ in terms of its components.
(d) Determine an angle α.
Answers:
(a) $A_x = 6.8$; (b) $A = 8.3$; (c) $\underline{A} = 6.8\underline{i} + 4.8\underline{j}$; (d) $\alpha = 35°$.

Problem B.6

Consider two vectors A and B as shown in Fig. B.34. If components of these vectors in the horizontal and vertical directions are given as:

$$A_x = 5.4; \quad A_y = 3.3,$$
$$B_x = 3.6; \quad B_y = 6.0.$$

(a) Represent vectors $\underline{A}$ and $\underline{B}$ in terms of their components.

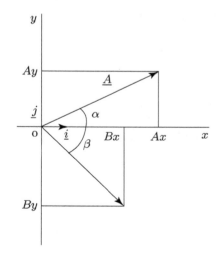

Fig. B.34 Problem B.6

(b) Determine angles α and β that vectors $\underline{A}$ and $\underline{B}$ make with the horizontal.

(c) Determine the magnitude of vector $\underline{C} = \underline{A} + \underline{B}$.

(d) Determine an angle γ that vector $\underline{C}$ makes with the horizontal.

Answers:

(a)
$$\underline{A} = 5.4\underline{i} + 3.3\underline{j}; \quad \underline{B} = 3.6\underline{i} - 6\underline{j}.$$

(b)
$$\alpha = 31.4°; \beta = 59°.$$

(c) $C = 9.4$

(d) $\gamma = 16.7°$

(d) Determine vector $\underline{F} = \underline{C} + \underline{B}$.

(e) Determine vector $\underline{G} = \underline{A} + \underline{C} - \underline{B}$.

(f) Determine vector $\underline{H} = \underline{A} + \underline{B} + \underline{C} + \underline{D}$.

(g) Calculate the magnitudes of $\underline{E}, \underline{F}, \underline{G}$, and $\underline{H}$.

Answers:

(a) $A_x = 4$, $A_y = 0$, $B_x = 6.06$, $B_y = 3.50$, $C_x = 0$, $C_y = 3$, $D_x = 3.54$, $D_y = 3.54$.

(b) $\underline{A} = 4\underline{i}$, $\underline{B} = 6.06\underline{i} + 3.50\underline{j}$, $\underline{C} = 3\underline{j}$, $\underline{D} = -3.54\underline{i} - 3.54\underline{j}$.

(c) $\underline{E} = 0.46\underline{i} - 3.54\underline{j}$.

(d) $\underline{F} = 6.06\underline{i} + 6.50\underline{j}$.

(e) $\underline{G} = -2.06\underline{i} - 0.50\underline{j}$.

(f) $\underline{H} = 6.52\underline{i} + 2.96\underline{j}$.

(g) $\underline{E} = 3.57$, $\underline{F} = 8.89$, $\underline{G} = 2.12$, $\underline{H} = 7.16$.

Problem B.7

Consider the system of four coplanar vectors $\underline{A}, \underline{B}, \underline{C}$, and $\underline{D}$ illustrated in Fig. B.35. The magnitudes of these vectors are such that $A = 4$, $B = 7$, $C = 3$, and $D = 5$. Vector $\underline{A}$ acts in the positive x direction, vector B makes an angle $30°$ with the positive x axis, vector $\underline{C}$ acts in the positive y direction, and vector $\underline{D}$ makes an angle $45°$ with the negative x axis.

(a) Determine the scalar components of vectors $\underline{A}, \underline{B}, \underline{C}$, and

Problem B.8

Consider vectors $\underline{A}$ and $\underline{B}$ illustrated in Fig. B.36. These vectors act on the xy-plane and have magnitudes $A = 10$ U and $B = 7$ U. Vector $\underline{A}$ makes an angle $\alpha = 15°$ with the positive x axis and vector $\underline{B}$ acts along the positive y direction.

(a) Express $\underline{A}$ and $\underline{B}$ in terms of their components.

Fig. B.35 Problem B.7

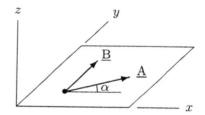

Fig. B.36 Problem B.8

$\underline{D}$ along the x and y directions.

(b) Express vectors $\underline{A}, \underline{B}, \underline{C}$, and $\underline{D}$ in terms of their components along the x and y directions.

(c) Determine vector $\underline{E} = \underline{A} + \underline{D}$.

(b) Evaluate the scalar product $m = \underline{A} \cdot \underline{B}$.

(c) Evaluate the vector product $\underline{C} = \underline{A} \times \underline{B}$.

Answers:

(a) $\underline{A} = 9.66\underline{i} + 2.59\underline{j}$, $\underline{B} = 7\underline{j}$.

(b) $m = 18.13$.

(c) $\underline{C} = 67.62\underline{k}$.

Appendix C: Calculus

C.1 Functions

Function is one of the most fundamental concepts in mathematics. The term "function" is used to denote the dependence of one quantity on another. In general, as a result of a series of experiments, a relationship between two quantities may be established. This relationship may be represented in the form of a table, a graph, or a mathematical equation called *function*. Once a function relating the two quantities is established, changes in one as a result of changes in the other quantity may be predicted without resorting to additional experiments.

Most physical laws are expressed in the form of mathematical equations. For example, velocity is defined as the time rate of change of position. If we can measure the change of position of an object over time and express that change in terms of a function, then we can also determine the velocity of that object simply by subjecting the function to certain mathematical operations without actually measuring the velocity of the object.

Consider two quantities X and Y. Assume that these quantities are related, such that a change in quantity X causes quantity Y to vary. Assume that a series of ten experiments are conducted, where by varying quantity X, corresponding Y values are measured. The variations in Y with respect to variations in X can be represented by various schemes. For example, the data collected can be presented in tabular form (Table C.1). However, information can be extracted more easily from a diagram than from a set of numbers. A diagram can be constructed by assigning the horizontal coordinate (abscissa) to the input X and the vertical coordinate (ordinate) to the output Y (Fig. C.1). Each pair of X and Y values recorded in Table C.1 will then correspond to a point on this diagram. A curve can be obtained by connecting these points, which will represent the graph of the data obtained.

Another way of representing the relationship between X and Y may be by means of a function. Functions are usually denoted by single letters, such as f or g. For example, $Y = f(X)$ implies that Y is a function of X. X in $Y = f(X)$ is the "input"

Table C.1 Experimental data

Experiment	X	Y
1	0.5	2.0
2	1.0	2.9
3	1.5	4.1
4	2.0	5.2
5	2.5	6.0
6	3.0	6.9
7	3.5	8.0
8	4.0	8.9
9	4.5	10.0
10	5.0	11.0

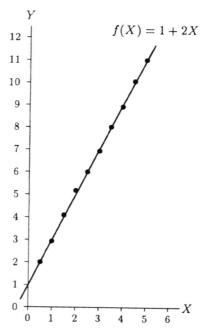

Fig. C.1 Y is a function of X, such that $Y = f(X) = 1 + 2X$

N. Özkaya et al., *Fundamentals of Biomechanics: Equilibrium, Motion, and Deformation*, DOI 10.1007/978-1-4614-1150-5, © Springer Science+Business Media, LLC 2012 or "cause" of an operation or process, while Y is the "output" or "effect." Usually, the input of a function is the *independent variable* and the output is the *dependent variable*. There are various ways of determining functions relating two or more quantities. One method is to compare the graph obtained as a result of an experiment or observation (such as the one in Fig. C.1) with the graphs of known functions. For example, the straight line in Fig. C.1 indicates that quantity Y is a "linear" function of X, and that the relationship between X and Y can be represented by the equation $Y = 1 + 2X$. Therefore, the function relating X and Y is $f(X) = 1 + 2X$. It is clear from this discussion that to be able to establish the most suitable function, one has to be familiar with the characteristics of commonly encountered functions.

There are a few basic functions in calculus that may be sufficient to describe a large number of physical phenomena. These basic functions include constant, power, trigonometric, logarithmic, and exponential functions. Basic functions can be combined together in various ways to construct more complex functions.

C.1.1 Constant Functions

Constant functions can be represented in the following general form:

$$f(X) = c. \tag{C.1}$$

Here, c is a symbol with a constant numerical value. For example, $Y = 2$ and $Y = 4.5$ are constant functions. Graphs of constant function are horizontal lines, as illustrated in Fig. C.2.

C.1.2 Power Functions

Power functions can be represented in the following general form:

$$f(X) = X^r. \tag{C.2}$$

In Eq. (C.2), r can be any real or integer number including zero and the ratio of numbers. Note that a constant function can be considered a power function for which $r = 0$, because $X^0 = 1$. The following are examples of power functions:

$$f(X) = X^1 = X,$$
$$f(X) = X^3 = X \cdot X \cdot X,$$
$$f(X) = X^{-1} = \frac{1}{X},$$
$$f(X) = X^{-2} = \frac{1}{X^2},$$
$$f(X) = X^{0.5} = X^{\frac{1}{2}} = \sqrt{X}.$$

Graphs of some power functions are illustrated in Fig. C.3. For a given function, its graph can be obtained by assigning a value to X, substituting that value to the function, solving the function for the corresponding Y value, repeating this for a number of X values, plotting each pair of X and Y values on the graph paper, and connecting them with a curve.

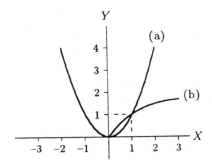

Fig. C.3 Example of power functions. (a) $f(X) = X^2$ and (b) $f(X) = \sqrt{X}$

C.1.3 Linear Functions

Linear functions can be represented in the following general form:

$$f(X) = a + bX. \tag{C.3}$$

In Eq. (C.3), a and b are some constant coefficients. The graph of a linear function is a straight line with coefficient a representing the point at which the straight line intersects the

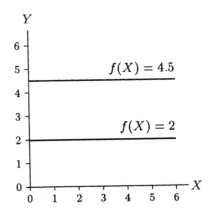

Fig. C.2 Constant functions

Y axis and coefficient b is the slope of the line. Examples of linear functions include:

$$f(X) = 1 + 2X,$$
$$f(X) = 0.5 - 5X,$$
$$f(X) = X.$$

The graphs of some linear functions are illustrated in Fig. C.4. Note that you need only two points to draw the graph of a linear function.

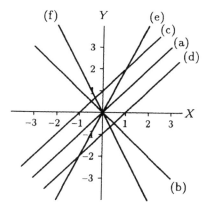

Fig. C.4 Linear functions. (a) $Y = X$, (b) $Y = -X$, (c) $Y = 1 + X$, (d) $Y = -1 + X$, (e) $Y = 2X$, and (f) $Y = -2X$

C.1.4 Quadratic Functions

Quadratic functions can be represented in the following general form:

$$f(X) = a + bX + cX^2. \qquad (C.4)$$

In Eq. (C.4), a, b, and c are real or integer, positive or negative numbers. Coefficients a and b can be zero. Examples of quadratic functions include:

$$f(X) = 2 + X - 0.5X^2,$$
$$f(X) = 5 + X^2,$$
$$f(X) = -3X + 4X^2,$$
$$f(X) = X^2.$$

The distinguishing characteristic of these functions is that the highest power of X appearing in these equations is two. However, for example, $f(X) = 1 + X^{0.5} - 3X^2$ is not a quadratic function, because of the term that carries $X^{0.5}$. Since $X^0 = 1$ and $X^1 = X$, quadratic functions can also be represented as:

$$f(X) = aX^0 + bX^1 + cX^2.$$

Graphs of quadratic functions are parabolas. The graphs of selected quadratic functions are illustrated in Fig. C.5.

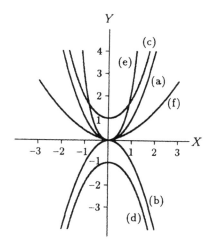

Fig. C.5 Quadratic functions. (a) $Y = X^2$, (b) $Y = -X^2$, (c) $Y = 1 + X^2$, (d) $Y = -1 - X^2$, (e) $Y = 3X^2$, and (f) $Y = X^2/3$

C.1.5 Polynomial Functions

A *polynomial function* is one for which

$$f(X) = A_0 + A_1 X + A_2 X^2 + A_3 X^3 + \cdots + A_n X^n \qquad (C.5)$$

Coefficients $A_0, A_1, \ldots A_n$ in Eq. (C.5) are real or integer, positive or negative, zero or nonzero numbers, and n is a positive integer number corresponding to the highest power of X. Power n defines the "order" of the polynomial. For example,

$$f(X) = 1 - X - 2X^2 + 5X^3$$

is a polynomial of order 3 with coefficients $A_0 = 1$, $A_1 = -1$, $A_2 = -2$, and $A_3 = 5$. The function

$$f(X) = 2 + 3X^2$$

is also a polynomial of order 2 with coefficients $A_0 = 2$, $A_1 = 0$, and $A_2 = 3$. It is also a quadratic function with coefficients $a = 2$, $b = 0$, and $c = 3$.

Note that constant, linear, and quadratic functions are special forms of polynomial functions. A constant function is also a zero-order polynomial, a linear function is a first-order polynomial, and a quadratic function is a second-order polynomial.

C.1.6 Trigonometric Functions

If a quantity Y depends on another quantity X through a trigonometric relationship, such as $Y = \sin(X)$ or $Y = \cos(X)$, then

Y is said to be a *trigonometric function* of X. $f(X) = \sin(X)$ and $f(X) = \cos(X)$ are called the *sine* and *cosine functions*, respectively. The graphs of these functions are illustrated in Figs. C.6 and C.7.

Fig. C.6 $Y = f(X) = \cos X$

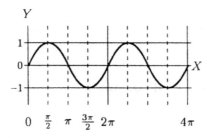

Fig. C.7 $Y = f(X) = \sin X$

X in $Y = \sin(X)$ and $Y = \cos(X)$ can be measured either in *degrees* or in *radians*. An angle of $180°$ is called π radians with $\pi = 3.1416$. Degrees can be converted into radians using:

$$\text{radian} = \frac{\pi}{180} \times \text{degree}.$$

For example, $0° = 0$ rad, $45° = \pi/4$ *rad*, $90° = \pi/2$ rad, $180° = \pi$ *rad*, $270° = 3\pi/2$ rad, $360° = 2\pi$ rad, and $720° = 4\pi$ rad.

Trigonometric functions are cyclic or periodic in the sense that their graphs repeat a pattern. The graphs of $Y = \sin(X)$ and $Y = \cos(X)$ in Figs. C.6 and C.7 repeat after every 2π radians or $360°$. This means that the *period* of $Y = \sin(X)$ and $Y = \cos(X)$ is 2π radians. Furthermore, Y in $Y = \sin(X)$ and $Y = \cos(X)$ assume values between -1 and $+1$. Therefore, the *amplitude* of $Y = \sin(X)$ and $Y = \cos(X)$ is 1.

$Y = \sin(X)$ and $Y = \cos(X)$ are the simplest forms of trigonometric functions. Sine functions can be expressed in a more general form as:

$$f(X) = a \sin(bX). \tag{C.6}$$

Here, a and b are some constants. The sine function defined by Eq. (C.6) has an amplitude a and a period $2\pi/b$. For example, as illustrated in Fig. C.8, the amplitude and period of the function $f(X) = 3 \sin(x/2)$ are 3 and 4π, respectively. The sine function in Eq. (C.6) can further be generalized as:

$$f(X) = a \sin(bX + c). \tag{C.7}$$

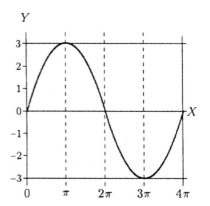

Fig. C.8 $Y = 3 \sin(x/2)$

Here, a is the amplitude, $2\pi/b$ is the period, and the graph of this function is shifted by c to the right or left as compared to the graph of the function $f(X) = a \sin(bX)$. For example, the graph of $f(X) = 3 \sin(x/2 - \pi/2)$ is the one shown in Fig. C.9.

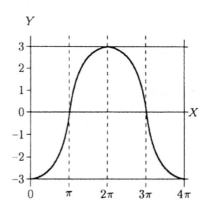

Fig. C.9 $Y = 3\sin(x/2 - \pi/2)$

Note that the graph of a sine function that is shifted by $\pi/2$ is essentially the graph of a negative cosine function. In other words, $\sin(x/2 - \pi/2) = -\cos(X/2)$. There are a number of other trigonometric identities and formulas that are useful in handling trigonometric functions. Some of these formulas are provided in Sect. C.4.

C.1.7 Exponential and Logarithmic Functions

Functions such as 3^X and $(1/2)^2$ are called *exponential functions*. The general form of exponential functions is b^X, where b is called the *base*. The most popular base in calculus is e, an irrational number between 2.71 and 2.72, and the function e^X or $\exp X$ is often referred to as the exponential function.

The exponential function $\exp X$ has an inverse, called the *natural logarithmic function* denoted by $\ln X$. It can also be written as $\log_e X$ and called the logarithm with base e. The properties of $\exp X$ and $\ln X$ are such that:

If $\ln X = Y$ then $\exp Y = X$

$\ln(\exp X) = X$ and $\exp(\ln X) = X$

$\ln 1 = 0$ and $\exp 0 = 1$

Graphs of $\exp X$ and $\ln X$ are shown in Fig. C.10. Additional properties of exponential and logarithmic functions include:

$$(\exp X)(\exp Y) = \exp(X + Y),$$

$$\frac{\exp X}{\exp Y} = \exp(X - Y),$$

$$\exp(-X) = \frac{1}{\exp X},$$

$$(\exp X)^Y = \exp(XY),$$

$$\ln(XY) = \ln X + \ln Y,$$

$$\ln\left(\frac{X}{Y}\right) = \ln X - \ln Y,$$

$$\ln\left(\frac{1}{X}\right) = -\ln X,$$

$$\ln X^Y = Y \ln X.$$

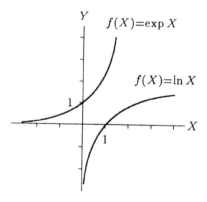

Fig. C.10 Exponential and logarithmic functions

Note that it is not allowed to take the "ln" of a negative number, but it is possible for $\ln X$ to come out negative. In other words, $\ln X$ is defined for X greater than zero. On the other hand, $\exp X$ is defined for all X. However, $\exp X$ is always positive.

C.2 The Derivative

The derivative is one of the fundamental mathematical operations used extensively to determine slopes of curves, maximums, and minimums, and it has many other applications. The process of finding the derivative of a function is called *differentiation*. The branch of calculus dealing with the derivative is called *differential calculus*.

The derivative of a function represents the slope or slopes of the graph of that function. For example, consider the linear function $Y = 1 + 2X$ whose graph is shown in Fig. C.11. The slope of the line representing $Y = 1 + 2X$ can be determined by considering any two points 1 and 2 along the line, such as those points with coordinates $X_1 = 3$ and $Y_1 = 7$, and $X_2 = 5$ and $Y_2 = 11$. The slope of a line is defined by the tangent of the angle that the line makes with the horizontal. For the line shown in Fig. C.11:

$$\text{slope} = \tan \theta = \frac{Y_2 - Y_1}{X_2 - X_1} = \frac{11 - 7}{5 - 3} = 2.$$

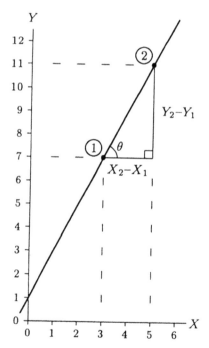

Fig. C.11 The derivative represents slope

Therefore, the derivative of the function $Y = 1 + 2X$ must be equal to 2. We shall demonstrate that this is the same 2 in front of the X in $Y = 1 + 2X$.

The slope of a straight line is constant. The curve that is not a straight line has many slopes, and it is usually difficult to predict the varying slopes of curves. Differentiation makes it easier to determine the slopes of curves.

In general, the derivative of a function is another function. For a function $Y = f(X)$, there are many symbols used to denote the derivative. For example:

$$f', \quad f'(X), \quad \frac{df}{dX}, \quad Y', \quad \frac{dY}{dX}.$$

C.2.1 Derivatives of Basic Functions

The graph of a constant function $f(X) = c$ is a horizontal line whose slope is zero. Therefore, the derivative of constant functions is always zero:

$$\frac{d}{dX}(c) = 0. \qquad (C.8)$$

Graph of the linear function $f(X) = X$ is a straight line with a slope equal to 1. Therefore:

$$\frac{d}{dX}(X) = 1. \qquad (C.9)$$

It is easy to find the derivatives of functions such as $f(X) = c$ and $f(X) = X$ using slopes. However, the graphs of power functions such as X^3, $\sqrt{X}$, and X^{-2} have varying slopes, and it is not easy to predict their derivatives. To differentiate a power function X^r, multiply the function by its power and reduce the power by 1:

$$\frac{d}{dX}(X^r) = rX^{r-1}. \qquad (C.10)$$

Equation C.10 is known as the Power Rule. The following examples illustrate the use of the Power Rule.

Function, $f(X)$	Derivative, f'
X^4	$4X^3$
X^{-2}	$-2X^{-3}$
$X^{2.3}$	$2.3X^{1.3}$
$\sqrt{X} = X^{\frac{1}{2}}$	$\frac{1}{2}X^{-\frac{1}{2}} = \frac{1}{2\sqrt{X}}$

If the power r is zero, then $X^0 = 1$ regardless of what X is. Therefore, 1 is X^0 or a power function with $r = 0$, and the Power Rule as given in Eq. (C.10) can be applied to differentiate it:

$$\frac{d}{dX}(1) = \frac{d}{dX}(X^0) = 0X^{-1} = 0.$$

Note that anything multiplied by zero is equal to zero. The function $f(X) = X$ is also a power function with $r = 1$. Therefore:

$$\frac{d}{dX}(X) = \frac{d}{dX}(X^1) = 1X^0 = 1.$$

The definition of the derivative can be utilized for the differentiation of other basic functions. We shall adopt the following definitions without presenting any proofs:

$$\frac{d}{dX}(\sin X) = \cos X, \qquad (C.11)$$

$$\frac{d}{dX}(\cos X) = -\sin X, \qquad (C.12)$$

$$\frac{d}{dX}(\exp X) = \exp X, \qquad (C.13)$$

$$\frac{d}{dX}(\ln X) = \frac{1}{X}. \qquad (C.14)$$

C.2.2 The Constant Multiple Rule

To differentiate a function in the form of a product of a constant c and another function $f(X)$, take the derivative of the function and then multiply it with the constant:

$$\frac{d}{dX}[cf(X)] = c\frac{df}{dX} = cf'. \qquad (C.15)$$

The following examples illustrate the use of Eq. (C.15).

Function, $f(X)$	Derivative, f'
$2X^2$	$4X$
$-1.25X^{-3}$	$3.75X^{-4}$
$0.5\cos X$	$-0.5\sin X$
$-3\exp X$	$-3\exp X$

C.2.3 The Sum Rule

The derivative of a function in the form of the sum of two functions is equal to the sum of the derivatives of the functions. If $f_1(X)$ and $f_2(X)$ represent two functions, then the derivative of the function $f(X) = f_1(X) + f_2(X)$ is:

$$\frac{d}{dX}[f_1(X) + f_2(X)] = f_1' + f_2'. \qquad (C.16)$$

For example:

$$\frac{d}{dX}(3X - 2\sin X) = \frac{d}{dX}(3X) + \frac{d}{dX}(-2\sin X)$$
$$= 3 - 2\cos X.$$

The sum rule can also be applied to take the derivative of linear and quadratic functions. If a, b, and c are constants, then:

$$\frac{\mathrm{d}}{\mathrm{d}X}(a + bX) = b, \qquad (C.17)$$

$$\frac{\mathrm{d}}{\mathrm{d}X}(a + bX + cX^2) = b + 2cX. \qquad (C.18)$$

Recall that linear and quadratic functions are special forms of polynomial functions. If $A_0, A_1, A_2, \ldots,$ and A_n are some constants, then the derivative of a polynomial is:

$$\frac{\mathrm{d}}{\mathrm{d}X}(A_0 + A_1X + A_2X^2 + \cdots + A_nX^n)$$

$$= A_1 + 2A_2X + \cdots + nA_nX^{n-1}. \qquad (C.19)$$

Note that differentiation reduces the order of the polynomial by one. For example, the derivative of a quadratic function (second-order polynomial) is a linear function (first-order polynomial). Similarly, the derivative of a third-order polynomial is a quadratic function. For example:

$$\frac{\mathrm{d}}{\mathrm{d}X}(3 - 2X + 5X^2 - X^3) = -2 + 10X - 3X^2.$$

C.2.4 The Product Rule

The derivative of a function in the form of a product of two functions is equal to the first function multiplied by the derivative of the second function plus the second function multiplied by the derivative of the first function. If $f_1(X)$ and $f_2(X)$ are two functions, then the derivative of the function $f(X) = f_1(X)f_2(X)$ is:

$$\frac{\mathrm{d}}{\mathrm{d}X}[f_1(X)f_2(X)] = f_1f_2' + f_1'f_2. \qquad (C.20)$$

For example:

$$\frac{\mathrm{d}}{\mathrm{d}X}(X^2 \cos X) = (X^2)\frac{\mathrm{d}}{\mathrm{d}X}(\cos X) + (\cos X)\frac{\mathrm{d}}{\mathrm{d}X}(X^2),$$

$$= (X^2)(-\sin X) + (\cos X)(2X),$$

$$= 2X \cos X - X^2 \sin X.$$

Note that the application of the product rule can be expanded to include functions in the form of the product of more than two functions. For example:

$$\frac{\mathrm{d}}{\mathrm{d}X}[f_1(X)f_2(X)f_3(X)] = f_1f_2f_3' + f_1f_2'f_3 + f_1'f_2f_3.$$

C.2.5 The Quotient Rule

The derivative of a function in the form of a ratio of two functions is equal to the derivative of the function in the numerator times the function in the denominator, minus the function in the numerator times the derivative of the function in the denominator, all divided by the square of the function in the denominator. If $f_1(X)$ and $f_2(X)$ represent two functions, then the derivative of the function $f(X) = f_1(X)/f_2(X)$ is:

$$\frac{\mathrm{d}}{\mathrm{d}X}\left[\frac{f_1(X)}{f_2(X)}\right] = \frac{f_1'f_2 - f_1f_2'}{f_2^2}. \qquad (C.21)$$

For example:

$$\frac{\mathrm{d}}{\mathrm{d}X}\left(\frac{2 + 3X^2}{X^2}\right) = \frac{(X^2)\frac{\mathrm{d}}{\mathrm{d}X}(2 + 3X^2) - (2 + 3X^2)\frac{\mathrm{d}}{\mathrm{d}X}(X^2)}{(X^2)^2},$$

$$= \frac{(X^2)(6X) - (2 + 3X^2)(2X)}{X^4},$$

$$= \frac{6X^3 - 4X - 6X^3}{X^4},$$

$$= -\frac{4}{X^3}.$$

Alternatively:

$$\frac{\mathrm{d}}{\mathrm{d}X}\left(\frac{2 + 3X^2}{X^2}\right) = \frac{\mathrm{d}}{\mathrm{d}X}\left(\frac{2}{X^2} + 3\right),$$

$$= \frac{(X^2)\frac{\mathrm{d}}{\mathrm{d}X}(2) - (2)\frac{\mathrm{d}}{\mathrm{d}X}(X^2)}{(X^2)^2} + \frac{\mathrm{d}}{\mathrm{d}X}(3),$$

$$= \frac{(X^2)(0) - (2)(2X)}{X^4} + 0,$$

$$= -\frac{4}{X^3}.$$

Note that trigonometric functions $\tan X$, $\cot X$, $\sec X$, and $\csc X$ are various quotients of $\sin X$ and $\cos X$. Therefore, the quotient rule can be applied to determine their derivatives. For example:

$$\frac{\mathrm{d}}{\mathrm{d}X}(\tan X) = \frac{\mathrm{d}}{\mathrm{d}X}\left(\frac{\sin X}{\cos X}\right)$$

$$= \frac{\cos X \frac{\mathrm{d}}{\mathrm{d}X}(\sin X) - (\sin X)\frac{\mathrm{d}}{\mathrm{d}X}(\cos X)}{(\cos X)^2},$$

$$= \frac{(\cos X)(\cos X) - (\sin X)(-\sin X)}{\cos^2 X},$$

$$= \frac{\cos^2 X + \sin^2 X}{\cos^2 X}.$$

Since $\cos^2 X + \sin^2 X = 1$:

$$\frac{d}{dX}(\tan X) = \frac{1}{\cos^2 X} = \sec^2 X. \qquad (C.22)$$

Similarly:

$$\frac{d}{dX}(\cot X) = -\frac{1}{\sin^2 X} = -\csc^2 X. \qquad (C.23)$$

$$\frac{d}{dX}(\sec X) = \frac{\sin X}{\cos^2 X} = \sec X \tan X. \qquad (C.24)$$

$$\frac{d}{dX}(\csc X) = -\frac{\cos X}{\sin^2 X} = -\csc X \cot X. \qquad (C.25)$$

C.2.6 The Chain Rule

Sometimes functions appear in forms other than those analyzed in previous sections. None of the above rules can be applied to differentiate functions such as $\cos(X^3)$, $\exp(2 - 5X)$, and $\ln(4X)$. For example, consider the function $f(X) = \exp(2 - 5X)$. We have already seen the derivatives of $\exp X$ and $2 - 5X$, but the derivative of $\exp(2 - 5X)$ follows a different rule. To take the derivative of $f(X) = \exp(2 - 5X)$, define the terms in the parentheses as $Z = 2 - 5X$, so that the original function can be reduced to a form $f(X) = \exp Z$. The *chain rule* states that:

$$\frac{df}{dX} = \frac{df}{dZ}\frac{dZ}{dX}. \qquad (C.26)$$

Now, we can take the derivative of $f(X) = \exp(2 - 5X)$.

$$\begin{aligned}
\frac{d}{dX}[\exp(2 - 5X)] &= \frac{d}{dZ}(\exp Z)\frac{d}{dX}(Z) \\
&= \frac{d}{dX}(\exp Z)\frac{d}{dX}(2 - 5X), \\
&= (\exp Z)(-5), \\
&= -5\exp(2 - 5X).
\end{aligned}$$

The chain rule has a vast number of applications. For example, to take the derivative of $Y = \sin(X^2)$, let $Z = X^2$ and $Y = \sin Z$. The derivative of Y with respect to Z is $\cos X$, and the derivative of Z with respect to X is $12X$. Therefore:

$$\begin{aligned}
\frac{d}{dX}[\sin(X^2)] &= \frac{d}{dZ}(\sin Z)\frac{d}{dX}(Z), \\
&= (\cos Z)(2X), \\
&= 2X\cos(X^2).
\end{aligned}$$

Consider the function $Y = (3 + X^2)^2$. There are two ways to take the derivative of this function. One way is by expanding the parentheses, writing the function as $Y = 9 + 6X^2 + X^4$, and then taking the derivative:

$$\begin{aligned}
\frac{d}{dX}[(3 + X^2)^2] &= \frac{d}{dX}(9 + 6X^2 + X^4), \\
&= 0 + 12X + 4X^3, \\
&= 12X + 4X^3.
\end{aligned}$$

The second way is to let $Z = 3 + X^2$ so that $Y = Z^2$, taking individual derivatives, and then applying the chain rule:

$$\begin{aligned}
\frac{d}{dX}[(3 + X^2)^2] &= \frac{d}{dZ}(Z^2)\frac{d}{dX}(3 + X^2), \\
&= (2Z)(2X) = 2(3 + X^2)(2X), \\
&= 12X + 4X^3.
\end{aligned}$$

For any Z that is a function of X, applications of the chain rule can be summarized in the following form:

$$\frac{d}{dX}(Z^n) = nZ^{n-1}\frac{dZ}{dX}, \qquad (C.27)$$

$$\frac{d}{dX}(\sin Z) = \cos Z\frac{dZ}{dX}, \qquad (C.28)$$

$$\frac{d}{dX}(\cos Z) = -\sin Z\frac{dZ}{dX}, \qquad (C.29)$$

$$\frac{d}{dX}(\exp Z) = \exp Z\frac{dZ}{dX}, \qquad (C.30)$$

$$\frac{d}{dX}(\ln Z) = \frac{1}{Z}\frac{dZ}{dX}. \qquad (C.31)$$

C.2.7 Implicit Differentiation

Sometimes a quantity may be an *implicit function* of another quantity. For example, in

$$Y^2 - X^2 = 4.$$

Y is an implicit function of X. The same equation can be rewritten in an *explicit* form as:

$$Y = (4 + X^2)^{1/2}.$$

The derivative of Y with respect to X can now be determined by applying the chain rule:

$$\frac{dY}{dX} = \frac{1}{2}(4 + X^2)^{-1/2}(2X) = \frac{X}{(4 + X^2)^{1/2}}.$$

Note that the derivative of Y with respect to X could also be determined directly from the implicit expression, by taking the derivative of both sides of the equation with respect to X:

$$2Y\frac{dY}{dX} - 2X = 0,$$

$$\frac{dY}{dX} = \frac{X}{Y} = \frac{X}{(4 + X^2)^{1/2}}.$$

C.2.8 Higher Derivatives

The derivative $f'(X)$ of a function $f(X)$ is also a function. The derivative of $f'(X)$ is yet another function, called the *second derivative* of $f(X)$ and is denoted by $f''(X)$. Some of the notations used for the second derivative are:

$$f'', \quad f''(X), \quad \frac{d^2f}{dX^2}, \quad Y'', \quad \frac{d^2Y}{dX^2}.$$

Similarly, $f'''(X)$ refers to the first derivative of $f''(X)$, the second derivative of $f'(X)$, and the third derivative of $f(X)$.

Consider the following examples:

$$Y = 1 + 3X - X^2 + 2X^3,$$
$$Y' = 3 - 2X + 6X^2,$$
$$Y'' = -2 + 12X,$$
$$Y''' = 12.$$

$$Y = X\sin(2X),$$
$$Y' = \sin(2X) + 2X\cos(2X),$$
$$Y'' = 2\cos(2X) + 2\cos(2X) - 4X\sin(2X),$$
$$= 4\cos(2X) - 4X\sin(2X),$$
$$Y''' = -8\sin(2X) - 4\sin(2X) - 8X\cos(2X),$$
$$= -12\sin(2X) - 8X\cos(2X).$$

C.3 The Integral

In Sect. C.2, we concentrated on finding the derivative $f'(X)$ of a given function $f(X)$. We have seen that the differentiation of basic functions is relatively simple and straightforward, and

the differentiation of relatively complex functions is possible using the derivatives of basic functions along with the rules for the derivatives of combinations (sums, products, and quotients) of functions.

Next we want to determine the functions whose derivatives are known. The reversed operation of differentiation is called *integration*. As compared to differentiation, integration is more difficult. There are no standard product, quotient, or chain rules for integration. In the absence of sufficient rules to integrate combinations of functions, it is a common practice to use integral tables.

Integration has many applications, such as area and volume calculations, work and energy computations.

The integral of a function $Y = f(X)$ with respect to X can be expressed in two ways:

$$\int f(X)dX, \tag{C.32}$$

$$\int_a^b f(X)dX. \tag{C.33}$$

Equation (C.32) is called the *indefinite* integral and Eq. (C.33) is called the *definite* integral. The integral symbol $\int$ is an elongated "S" which stands for summation, the function $f(X)$ to be integrated is called the *integrand*, dX is an increment in X, and a and b in Eq. (C.33) are called the *lower* and *upper limits* of integration, respectively.

Consider the function $Y = f(X) = 2X$ whose graph is shown in Fig. C.12. Also consider $F_1(X) = X^2$. The derivative of $F_1(X)$ is essentially equal to $f(X)$, and therefore,

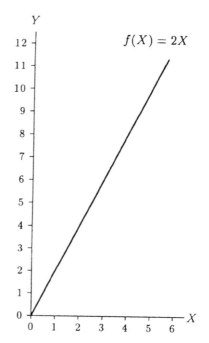

Fig. C.12 Function $Y = 2X$

$F_1(X)$ can be the integral of $f(X) = 2X$. Now consider another function $F_2(X) = X^2 + c_0$ where c_0 is a constant. The derivative of $F_2(X)$ is again equal to $f(X) = 2X$ because the derivative of a constant is zero. Note that F_1 is in fact a special form of F_2 for which $c_0 = 0$. Therefore, the indefinite integral of $f(X) = 2X$ is:

$$\int (2X)\mathrm{d}X = X^2 + c_0.$$

Here, c_0 is called the *constant of integration*. Note that the indefinite integral of a function is another function that is not unique. There are different solutions for different values of c_0. These different solutions have parallel graphs (Fig. C.13), all of which have slopes expressed by the function $2X$.

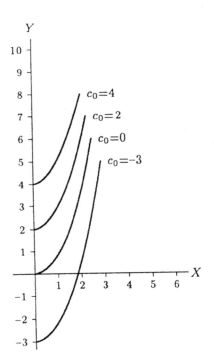

Fig. C.13 Slope of functions $Y = X^2 + c_0$ for different c_0 values is equal to $2X$

The definite integral of a given function is unique. To evaluate the definite integral of a function $f(X)$ between a and b, first take the integral of the given function. For definite integrals, the constant of integration cancels out during the course of integration. Therefore, it can be assumed that the constant of integration is zero. If $F(X)$ represents the integral of $f(X)$, then evaluate the values of $F(X)$ at $X = a$ and $X = b$ by subsequently substituting the numerical values of a and b wherever you have X in $F(X)$. In other words, evaluate $F(a)$ and $F(b)$. The definite integral of $f(X)$ between a and b is equal to $F(b)$ minus $F(a)$:

$$\int_a^b f(X)\mathrm{d}X = F(b) - F(a).$$

For example, consider the function $Y = f(X) = 2X$. Assume that the integral of $f(X)$ between $X = 0$ and $X = 3$ is to be evaluated. That is, $a = 0$ and $b = 3$. Recall that the integral of $f(X) = 2X$ is $F(X) = X^2$ (with $c_0 = 0$). The steps to be followed while evaluating the definite integral are as follows:

Step 1: $\int_0^3 (2X)\mathrm{d}X = [X^2]_0^3$.

Step 2: $= [(3^2) - (0^2)]$.

Step 3: $= (9 - 0) = 9$

That is, determine $F(X)$ by taking the integral of the given function (Step 1), evaluate $F(b)$ and $F(a)$ by substituting the upper and lower limits of integration (Step 2), and evaluate the difference $F(b) - F(a)$ (Step 3).

The physical meaning of the definite integral $\int_a^b f(X)\mathrm{d}X$ is such that it represents the area bounded by the given function $f(X)$, the X axis, and the vertical lines passing through $X = a$ and $X = b$. For example, the integral of function $f(X) = 2X$ between $X = 1$ and $X = 3$ is equal to the shaded area in Fig. C.14, which is 8.

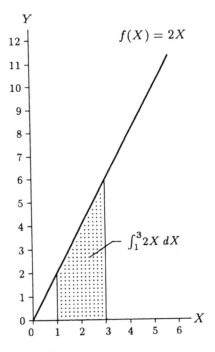

Fig. C.14 The integral represents area

C.3.1 Properties of Indefinite Integrals

Integrals of basic functions:

$$\int dX = X + c_0,$$

$$\int c\,dX = cX + c_0 \quad \text{(constant } c\text{)}, \qquad \text{(C.34)}$$

$$\int X\,dX = \frac{X^2}{2} + c_0,$$

$$\int X^2\,dX = \frac{X^3}{3} + c_0,$$

$$\int X^r\,dX = \frac{X^{r+1}}{r+1} + c_0 \quad (r \neq -1), \qquad \text{(C.35)}$$

$$\int \sin X\,dX = -\cos X + c_0, \qquad \text{(C.36)}$$

$$\int \cos X\,dX = \sin X + c_0, \qquad \text{(C.37)}$$

$$\int \exp X\,dX = \exp X + c_0, \qquad \text{(C.38)}$$

$$\int \frac{1}{X}\,dX = \ln X + c_0 \quad (X > 0). \qquad \text{(C.39)}$$

Constant multiple and sum rules:

$$\int [cf(X)]\,dX = c\int f(X)\,dX, \qquad \text{(C.40)}$$

$$\int [f_1(X) + f_2(X)]\,dX = \int f_1(X)\,dX + \int f_2(X)\,dX. \qquad \text{(C.41)}$$

Examples:

$$\int 2\cos X\,dX = 2\int \cos X\,dX = 2\sin X + c_0,$$

$$\int \left(5 - \frac{1}{2}X^3\right)\,dX = 5\int dX - \frac{1}{2}\int X^3\,dX,$$

$$= 5(X) - \frac{1}{2}\left(\frac{X^4}{4}\right) + c_0,$$

$$= 5X - \frac{1}{8}X^4 + c_0.$$

$$\int \left(\frac{1}{X^2} + \sin X\right)\,dX = \int X^{-2}\,dX + \int \sin X\,dX,$$

$$= \left(\frac{X^{-1}}{-1}\right) + (-\cos X) + c_0,$$

$$= -\frac{1}{X} - \cos X + c_0.$$

C.3.2 Properties of Definite Integrals

Let $F(X)$ represent the integral of a function $f(X)$ with respect to X. That is:

$$F(X) = \int f(X)\,dX.$$

Also let $f_1(X)$ and $f_2(X)$ be two other functions. Then:

$$\int_a^b f(X)\,dX = F(b) - F(a), \qquad \text{(C.42)}$$

$$\int_a^b [cf(X)]\,dX = c\int_a^b f(X)\,dX = c[F(b) - F(a)], \qquad \text{(C.43)}$$

$$\int_a^b [f_1(X) + f_2(X)]\,dX = \int_a^b f_1(X)\,dX + \int_a^b f_2(X)\,dX, \qquad \text{(C.44)}$$

$$\int_a^b f(X)\,dX + \int_b^c f(X)\,dX = \int_a^c f(X)\,dX, \qquad \text{(C.45)}$$

$$\int_a^b f(X)\,dX = -\int_b^a f(X)\,dX. \qquad \text{(C.46)}$$

Definite integrals of basic functions:

$$\int_a^b dX = [X]_a^b = b - a,$$

$$\int_a^b X\,dX = \left[\frac{X^2}{2}\right]_a^b = \left[\frac{b^2}{2} - \frac{a^2}{2}\right] = \frac{1}{2}(b^2 - a^2),$$

$$\int_a^b X^2\,dX = \left[\frac{X^3}{3}\right]_a^b = \left[\frac{b^3}{3} - \frac{a^3}{3}\right] = \frac{1}{3}(b^3 - a^3),$$

$$\int_a^b \sin X\,dX = [-\cos X]_a^b = -\cos b + \cos a,$$

$$\int_a^b \cos X\,dX = [\sin X]_a^b = \sin b - \sin a.$$

$$\int_a^b \exp X \mathrm{d}X = [\exp X]_a^b = \exp b - \exp a,$$

$$\int_a^b \frac{1}{X} X \, \mathrm{d}X = [\ln X]_a^b = \ln b - \ln a.$$

Examples:

$$\int_{45°}^{90°} \cos X \mathrm{d}X = [\sin X]_{45°}^{90°} = \sin 90° - \sin 45° = 0.3,$$

$$\int_1^2 (4X + 9X^2) \, \mathrm{d}X = \int_1^2 (4X) \, \mathrm{d}X + \int_1^2 (9X^2) \, \mathrm{d}X,$$
$$= 4\left[\frac{X^2}{2}\right]_1^2 + 9\left[\frac{X^3}{3}\right]_1^2,$$
$$= 2[X^2]_1^2 + 3[X^3]_1^2,$$
$$= 2(2^2 - 1^2) + 3(2^3 + 1^3),$$
$$= 2(4 - 1) + 3(8 - 1),$$
$$= 2(3) + 3(7) = 27.$$

Caution

Consider the linear function $f(X) = X$. The graph of this function is shown in Fig. C.15. Assume that the definite integral of this function between $X = -2$ and $X = 2$ will be evaluated. Since the definite integral represents the area enclosed by the function, the X axis, and vertical lines passing through the lower and upper limits, the integration of $f(X) = X$ between $X = -2$ and $X = 2$ should yield the shaded area $(A_1 + A_2)$ in Fig. C.15. From the geometry, we can readily determine that $(A_1 = A_2) = 2$, or the total shaded area is equal to 4 U. Now, we evaluate the integral:

$$\int_{-2}^2 X \mathrm{d}X = \left[\frac{X^2}{2}\right]_{-2}^2 = \frac{1}{2}[(2)^2 - (-2)^2] = \frac{1}{2}[4 - 4] = 0.$$

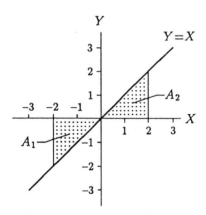

Fig. C.15 Caution while evaluating definite integrals

This unexpected result occurred due to the fact that in the interval between the lower and upper limits of integration, the function crosses the X axis, and also that a negative value is calculated for the area designated as A_1 in Fig. C.15 during the process of integration. To avoid such a mistake, the proper evaluation of the definite integral should follow the steps provided below:

$$\int_{-2}^2 X \mathrm{d}X = \left|\int_{-2}^0 X \mathrm{d}X\right| + \left|\int_0^2 X \mathrm{d}X\right|,$$
$$= \left|\left[\frac{X^2}{2}\right]_{-2}^0\right| + \left|\left[\frac{X^2}{2}\right]_0^2\right|,$$
$$= \frac{1}{2}|(0)^2 - (-2)^2| + \frac{1}{2}|(2)^2 - (0)^2|,$$
$$= \frac{1}{2}|-4| + \frac{1}{2}|4| = \frac{4}{2} + \frac{4}{2} = 2 + 2 = 4.$$

Therefore, before evaluating the definite integral of a function $f(X)$ with respect to X, one must first draw a simple graph of the function to be integrated to check whether the graph of the function crosses the X axis in the interval between the lower and upper limits of integration. For example, if a function $f(X)$ will be integrated with respect to X between a and b, and the curve of the function crosses the X axis at c and d such that $a < c < d < b$, then the integral should first be expressed as:

$$\int_a^b f(X) \, \mathrm{d}X = \left|\int_a^c f(X) \mathrm{d}X\right| + \left|\int_c^d f(X) \mathrm{d}X\right| + \left|\int_d^b f(X) \, \mathrm{d}X\right|.$$

C.3.3 Methods of Integration

There is no standard rule for integrating functions in the form of the product and quotient of other functions whose integrals are known. For such functions, there are integral tables and a number of methods of integration, including:

- Integration by substitution
- Integration by parts
- Integration by trigonometric substitution
- Integration by partial fraction decomposition
- Numerical integration

For example, assume that the following integral will be evaluated:

$$\int 2X \sin(X^2) \, \mathrm{d}X.$$

The function $2X\sin(X^2)$ can be integrated by observing that $2X$ is the derivative of X^2. If we let $Z = X^2$, then $dZ = 2XdX$. By substituting Z and dZ into the given integral, we can write:

Substitution: $\int 2X \sin(X^2)\,dX = \int \sin Z\,dZ$,

Integration: $= -\cos Z + c_0$,

Back substitution: $= -\cos(X^2) + c_0$.

The method used here is called "integration by substitution." Detailed descriptions of these methods are beyond the scope of this text. For further information, the interested reader should refer to calculus textbooks.

C.4 Trigonometric Identities

Negative angle formulas:

$$\sin(-X) = -\sin(X),$$
$$\cos(-X) = \cos(X).$$

Addition formulas:

$$\sin(X + Y) = \sin(X)\cos(Y) + \cos(X)\sin(Y),$$
$$\sin(X - Y) = \sin(X)\cos(Y) - \cos(X)\sin(Y),$$
$$\cos(X + Y) = \cos(X)\cos(Y) - \sin(X)\sin(Y),$$
$$\cos(X - Y) = \cos(X)\cos(Y) + \sin(X)\sin(Y).$$

Product formulas:

$$2\sin(X)\cos(Y) = \sin(X + Y) + \sin(X - Y),$$
$$2\cos(X)\sin(Y) = \sin(X + Y) - \sin(X - Y),$$
$$2\cos(X)\cos(Y) = \cos(X + Y) + \cos(X - Y),$$
$$2\sin(X)\sin(Y) = \cos(X - Y) - \cos(X + Y).$$

Factoring formulas:

$$\sin(X) + \sin(Y) = 2\cos\left(\frac{X - Y}{2}\right)\sin\left(\frac{X + Y}{2}\right),$$

$$\sin(X) - \sin(Y) = 2\cos\left(\frac{X + Y}{2}\right)\sin\left(\frac{X - Y}{2}\right),$$

$$\cos(X) + \cos(Y) = 2\cos\left(\frac{X + Y}{2}\right)\cos\left(\frac{X - Y}{2}\right),$$

$$\cos(X) - \cos(Y) = 2\cos\left(\frac{X + Y}{2}\right)\sin\left(\frac{X - Y}{2}\right).$$

Double-angle formulas:

$$\sin(2X) = 2\sin(X)\cos(X),$$
$$\cos(2X) = \cos^2(X) - \sin^2(X).$$

Half-angle formulas:

$$2\sin^2\left(\frac{X}{2}\right) = 1 - \cos(X),$$

$$2\cos^2\left(\frac{X}{2}\right) = 1 + \cos(X).$$

Pythagorean identity:

$$\sin^2(X) + \cos^2(X) = 1.$$

Reduction formulas:

$$\cos\left(\frac{\pi}{2} - X\right) = \sin(X), \quad \cos(\pi - X) = -\cos(X),$$

$$\sin\left(\frac{\pi}{2} - X\right) = \cos(X), \quad \sin(\pi - X) = \sin(X).$$

Sine and cosine are the basic trigonometric functions. Other trigonometric functions, namely the tangent (tan), cotangent (cot), secant (sec), and cosecant (csc), can be derived from sine and cosine using the following definitions:

$$\tan(X) = \frac{\sin(X)}{\cos(X)}, \quad \cot(X) = \frac{\cos(X)}{\sin(X)} = \frac{1}{\tan(X)},$$

$$\sec(X) = \frac{1}{\cos(X)}, \quad \csc(X) = \frac{1}{\sin(X)}.$$

C.5 The Quadratic Formula

An algebraic equation such as $3x = 7$ is a *linear equation* because the unknown x appears to the first power. This equation can be solved to determine that $x = 7/3$. An equation such as $x^2 + 3x = 5$, on the other hand, is a *quadratic equation*. In this case, the highest power of x is two. Note that $x^2 + 3x^{0.5} = 5$ is not a quadratic equation, because of the term that carries $x^{0.5}$. Quadratic equations can be written in the following general form:

$$ax^2 + bx + c = 0. \tag{C.47}$$

In Eq. (C.47), a, b, and c are some known numbers, and x is the unknown parameter. The general solution of Eq. (C.47) for x is:

$$x = \frac{-b \pm \sqrt{b^2 - 4ac}}{2a}. \tag{C.48}$$

The $\pm$ sign indicates that there are two solutions for x.

Caution

Note that in Eq. (C.48), b^2 must be greater than or equal to $4ac$, so that $(b^2 - 4ac)$ has a positive value. Otherwise $(b^2 - 4ac)$ will have a negative value. The square root of a negative number is an "imaginary" number as opposed to a "real" number.

For example, assume that the following quadratic equation will be solved for x:

$$x^2 + 2x = 8.$$

First, rewrite this equation in the form given in Eq. (C.47) by subtracting 8 from both sides:

$$x^2 + 2x - 8 = 0.$$

Compare this with Eq. (C.47) to observe that:

$$a = 1, b = 2, c = -8.$$

Substitute a, b, and c into Eq. (C.48):

$$x = \frac{-(2) \pm \sqrt{(2^2) - 4(1)(-8)}}{2(1)} = \frac{-2 \pm \sqrt{36}}{2} = \frac{-2 \pm 6}{2}.$$

Consider the plus sign:

$$x = \frac{-2 + 6}{2} = \frac{4}{2} = 2.$$

Consider the minus sign:

$$x = \frac{-2 - 6}{2} = \frac{-8}{2} = -4.$$

Note that $x = 2$ and $x = -4$ are the "roots" of the quadratic equation. The original equation can now be expressed as:

$$(x - 2)(x + 4) = 0.$$

C.6 Exercise Problems

Problem C.1

Show that slopes of the graph of the function $Y = 1 - X^2$ at $X = -2$, $X = 0$, and $X = 2$ are 4, 0, and -4, respectively.

Problem C.2

Evaluate the derivatives of the following functions with respect to X.

Functions	Answers
$Y = 5x^2$	$10x$
$Y = -4.5x^{-2}$	$9/x^3$
$Y = 3\sin x$	$3\cos x$
$Y = (2 + 8\sin x)$	$8\cos x$
$Y = X^2 \sin X$	$2X\sin X + X^2 \cos X$
$Y = \dfrac{X}{\cos X}$	$\dfrac{\cos X + X\sin X}{\cos^2 X}$
$Y = \dfrac{\cos X}{X}$	$\dfrac{-X\sin X - \cos X}{X^2}$
$Y = \sqrt{X - X^3}$	$\dfrac{1 - 3X^2}{2\sqrt{X - X^3}}$
$Y = \sqrt{2x - 5x^2}$	$\dfrac{2 - 10x}{2\sqrt{2x - 5x^2}}$

Problem C.3

Show that the integral of the function $Y = \cos X$ with respect to X between $X = 0°$ and $X = 180°$ is 2. (Examine the graph of the function before evaluating the integral.)

Problem C.4

Evaluate the following integrals.

Integrals	Answers
$\int 5\,dx$	$5x + c_0$
$\int 8x\,dx$	$4x^2 + c_0$
$\int 6\sin x\,dx$	$-6\cos x + c_0$
$\int (X - \sin X)\,dX$	$\dfrac{X^2}{2} + \cos X + c_0$
$\int \left(1 - \frac{1}{X^2}\right)\,dX$	$X + \frac{1}{X} + c_0$
$\int_1^2 \left(1 - \frac{1}{X^2}\right)\,dX$	0.5
$\int_{0°}^{45°} (\cos X + \sin X)\,dX$	1
$\int_2^4 6x^2\,dx$	112
$\int_0^2 2x^2\,dx + \int_2^6 2x^2\,dx$	144

Problem C.5

Show that the roots of the quadratic equation $x^2 - 7.5x = 4$ are $x = 8$ and $x = -0.5$.

Index

Note: Page numbers followed by f indicate figures; those followed by t indicate table.

N. Özkaya et al., *Fundamentals of Biomechanics: Equilibrium, Motion, and Deformation*,
DOI 10.1007/978-1-4614-1150-5, © Springer Science+Business Media, LLC 2012

CPSIA information can be obtained
at www.ICGtesting.com
Printed in the USA
LVHW101835211218
601381LV00002B/4/P